IP Routing Protocols

IP Routing Protocols

Link-State and Path-Vector Routing Protocols

James Aweya

CRC Press
Taylor & Francis Group
Boca Raton London New York

CRC Press is an imprint of the
Taylor & Francis Group, an **Informa** business

First edition published 2021
by CRC Press
6000 Broken Sound Parkway NW, Suite 300, Boca Raton, FL 33487-2742

and by CRC Press
2 Park Square, Milton Park, Abingdon, Oxon, OX14 4RN

CRC Press is an imprint of Taylor & Francis Group, LLC

Library of Congress Cataloging-in-Publication Data
Names: Aweya, James, author.
Title: IP routing protocols / James Aweya.
Description: First edition. l Boca Raton : CRC Press, 2021. l Includes bibliographical references and index.
l Contents: v. 1. Fundamentals -- v. 2. Link-state and patth-vector routing protocols. l Summary:
"This two-volume book describes the most common IP routing protocols used today (RIPv2, EIGRP, OSPFv2, IS-IS, and BGPv4), explaining the underlying concepts of each protocol and how the protocol components and processes fit within the typical router"-- Provided by publisher.
Identifiers: LCCN 2020052474 (print) l LCCN 2020052475 (ebook) l ISBN 9780367709624 (v. 1 ; paperback) l ISBN 9780367710415 (v. 1 ; hardback) l ISBN 9780367709631 (v. 2 ; paperback) l ISBN 9780367710361 (v. 2 ; hardback) l ISBN 9781003149040 (v. 1 ; ebook) l ISBN 9781003149019 (v. 2 ; ebook)
Subjects: LCSH: TCP/IP (Computer network protocol)
Classification: LCC TK5105.585 .A94 2021 (print) l LCC TK5105.585 (ebook) l DDC 004.6/65--dc23
LC record available at https://lccn.loc.gov/2020052474
LC ebook record available at https://lccn.loc.gov/2020052475

ISBN: 978-0-367-71036-1 (hbk)
ISBN: 978-0-367-70963-1 (pbk)
ISBN: 978-1-003-14901-9 (ebk)

Typeset in Times
by SPi Global, India

Contents

Preface

An Interior Gateway Protocol (IGP) is a type of routing protocol used by routers within an Autonomous System for exchanging routing information between themselves. Based on the method by which IGP routers within the Autonomous System exchange routing information and compute best paths to network destinations, IGPs are divided into two categories: distance-vector routing protocols and link-state routing protocols.

Open Shortest Path First (OSPF) and Intermediate System-to-Intermediate System (IS-IS) are the two main link-state routing protocols in use today. These protocols were developed to meet the inability of the Routing Information Protocol (RIP) to scale beyond a maximum hop count of 15 routers. OSPF and IS-IS have fast convergence, support Variable-Length Subnet Mask (VLSM), make efficient use of network bandwidth (by propagating routing changes using triggered routing updates instead of periodic updates), and support large network sizes. Link-state routing protocols send periodic link-state refresh, but only at longer intervals (such as every 30 minutes for OSPF LSA Refresh interval) to conserve network bandwidth. RIP and Enhanced Interior Gateway Routing Protocol (EIGRP) are classified as distance-vector routing protocols while Border Gateway Protocol (BGP) is a path-vector routing protocol as discussed in Volume 1 of this two-part book.

Each router in a link-state routing protocol network discovers and establishes an adjacency or neighbor relationship with each of its link-state neighbor routers. The adjacency information for each neighbor is maintained in an adjacency database (also called a neighbor table). This database on any given router keeps a record of the parameters and attributes of the router's immediate neighbors. Each router then begins sending link-state messages (called Link-State Advertisements [LSAs] in OSPF, and Link-State Packets [LSPs] in IS-IS) to its neighbors. Each neighbor in turn, receives link-state messages and forwards them to every neighbor except the one that sent the message (a process referred to as flooding). When a network topology change occurs, the link-state router experiencing the change creates a link-state message advertising that event. Routers receiving the link-state message will immediately forward it to their neighbors.

Each neighbor router receives copies of these link-state messages and stores them in a topological table called the Link-State Database (LSDB). The LSDB contains a list of all network destinations and the costs to reach them. If the state of the overall network converges and is stable, the LSDB in all routers should be identical. The LSDB describes a complete view or graph of the entire network allowing all routers in the link-state network to have the same complete picture of the network. This allows each link-state router in the network to independently make a best path routing decision based on the same accurate picture of the network topology. The link-state routers use link-state refresh (LSAs) to confirm the network topology information before the information in the LSDB ages out. Using the Dijkstra algorithm and the information in the LSDB, each link-state router then computes the shortest path to each known network destination, and enters these best paths in its IP Routing Table (also referred to the IP Routing Information Base [RIB]).

The information in the IP Routing Table is the final routing information used for forwarding IP packets. Typically, the information in the IP Routing Table is distilled into a more compact database called the IP Forwarding Table (also referred to the IP Forwarding Information Base [FIB]). This is because the information in the Routing Table is very extensive and contains information that is not directly relevant for IP packet forwarding. The IP Forwarding Table weeds out non-essential packet forwarding information and allows for faster routing information lookups and packet forwarding.

Link-state routers must know the detailed topology of their network in order to be able to compute best routes to network destinations. The link-state message (LSA or LSP) is the basic messaging unit for communicating routing information among routers in a link-state network. LSAs/LSPs are used to describe the topology details of a network and also form the building blocks of the LSDB a link-state router maintains. Each LSA/LSP serves as a database record that contains pieces of network topology information (i.e., link-state information) about a portion of the network that is described in the LSDB. The LSAs/LSPs in the LSDB, collectively, describe the entire topology of the link-state network.

Implementing a link-state routing protocol in a large network exposes some serious issues particularly regarding the processing and memory requirements in the routers. Therefore, partitioning a large network into areas is a way to address the three concerns commonly expressed about link-state routing protocols, which are the memory requirements for storing the LSDBs, CPU cycles (time) required to process the relatively more complex link-state routing algorithms, and the effects of LSA flooding on the available bandwidth in the network, particularly in unstable networks.

An area is simply a partition of the network that contains a subset of the routers that make up the entire network. The characteristics of an area include localizing the impact of a network topology change within an area, preventing link-state messages from being flooded beyond the boundary of an area, minimizing Routing Table entries, and providing means for hierarchical network design. Both OSPF and IS-IS support the grouping of routers in an Autonomous System into areas and hierarchical routing. Both protocols use a two-tier hierarchical area structure.

BGP and its extension (Multiprotocol Extensions for BGP or simply Multiprotocol BGP [MBGP]), is the only path-vector routing protocol in use today. BGP is used for exchanging routing information between Autonomous Systems. BGP carries as part of its routing information, the Autonomous System Numbers (ASNs) the routing information has passed through (for a set of network destination address prefixes sharing the same path attributes). This routing information (listing the Autonomous System traversed) is used by the various routers along the path to the network destination, for preventing routing loops.

An Autonomous System is a group of interconnected routers and network address prefixes owned or under the control of one or more network operators (e.g., organizations), but are managed by a single administrative entity (e.g., Internet Service Provider [ISP]). Furthermore, these interconnections of network prefixes are presented by the administrative entity/domain to the Internet under a common, clearly defined routing plan or policy.

A routing policy here refers to how routing decisions are made within the administrative entity/domain. Using its routing policy, the Autonomous System presents a consistent and coherent view of the network destinations that can be reached through it to other Autonomous Systems. A routing or network prefix represents, here, a group of IP addresses that can be reached through the network of the administrative entity, for example, the ISP's network. A network address prefix is a contiguous set of the most significant bits in the IP address that collectively represent a set of systems within a network.

Although BGP is an Exterior Gateway Protocol (EGP), the networks within many organizations have become so complex that BGP is used to simplify the design and administration of these networks. It is not uncommon nowadays for an organization's network to be structured as an internetwork of smaller BGP Autonomous Systems. BGP routers that exchange routing messages with each other are referred to as BGP peers. Using internal BGP (iBGP) peering sessions, BGP peers within the same organization are able to exchange routing information. BGP peers that are in different Autonomous Systems are referred to as external BGP (eBGP) peers, while BGP peers that are in the same Autonomous System are iBGP peers.

BGP peers exchange network reachability information using BGP routing updates. The routing update information contains the network IP address prefixes, BGP path-specific attributes, and a list of ASNs that a route has traversed to reach a destination network (path-vector routing). A BGP router sets and communicates BGP routing policy to other routers using BGP path attributes. BGP path attributes contain the characteristics of a BGP route that a BGP router advertises to its BGP peers. BGP routers also manipulate/modify BGP path attributes when implementing path control tools; that is, customize BGP's behavior and routing policy, filter routing information in and out of BGP, prefer certain paths over others, and so on.

BGP also supports mechanisms for guaranteeing loop-free exchange of routing information between Autonomous Systems. To prevent routing loops from occurring, a BGP router rejects any routing update that contains its local ASN. A received routing update that contains the local ASN, indicates that the routing update has already passed through that Autonomous System, and a routing loop would therefore be created if the update is accepted.

The companion Volume 1 of this two-part book focuses on the fundamental concepts of IP routing and distance-vector routing protocols (RIPv2 and EIGRP). This volume focuses on OSPFv2, IS-IS, and BGPv4. OSPFv2 was developed for IPv4 routing, and OSPFv3 for IPv6 routing.

The goal of this book is to provide an easily readable and understandable discussion of link-state and path-vector routing protocols using detailed illustrations and real-world examples. This is to enable readers to readily draw on these examples. The discussion is presented in a simple style to make it comprehensible and appealing to undergraduate and graduate level students, research and practicing engineers, scientists, IT personnel, and network engineers. It is geared toward readers who want to understand the concepts and theory of IP routing protocols, yet want these to be tied to clearly illustrated and close-to-real-world example systems and networks. The discussion aims to explain in simple language how routing protocols are designed and work, and is not meant to replace the material presented in the various international and proprietary standards accompanying each routing protocol.

Chapters 1–3 discuss the most common routing protocols OSPFv2, IS-IS, and BGPv4, respectively. The discussion covers their most identifying characteristics, operations, and the databases they maintain. Each routing protocol maintains a number of databases which hold information about the local router's neighbor routers, and the routing information it has learned from other routers in the network. Each database type is used for specific operations as defined by the particular routing protocol. For each routing protocol, we discuss the main components which include data structures, routing protocol messages, and best path computation algorithm. Each chapter also covers a high-level router architecture, processes, and databases for the particular routing protocol being discussed.

Chapter 1 begins with a review of the main concepts of link-state routing protocols and a discussion on the main features of OSPFv2. The first part of the chapter provides a detail description of the OSPF area types and hierarchical routing, and the different types of routers used in an OSPF Autonomous System. The chapter then describes in detail the different OSPF packet types, and inbound and outbound processing of OSPF LSAs. The chapter also provides a detailed description of the different OSPF network types, OSPF neighbor discovery and maintenance, LSDB synchronization, OSPF routing metrics, OSPF route summarization, and OSPF authentication. The discussion includes a high-level view of the OSPF router and components, plus an overview of OSPF inbound and outbound message processing.

Chapter 2 starts by providing an overview of the main features of IS-IS. The discussion includes a review of the basic concepts underlying the design of IS-IS. The chapter then continues to provide a detail description of IS-IS area types and hierarchical routing, and the different types of routers used by IS-IS. The chapter discusses the two-level routing hierarchy for controlling the distribution of intradomain (Level 1) and interdomain (Level 2) routing information within an IS-IS routing domain. The chapter then describes in detail IS-IS network address formats, IS-IS routing metrics, IS-IS packet types, IS-IS network types and adjacency formation, IS-IS LSDB and synchronization, and IS-IS authentication. The chapter includes a discussion on the high-level architecture of the IS-IS router and components, and inbound and outbound processing of messages.

Chapter 3 starts by reviewing the main concepts of path-vector routing protocols, and a discussion on the main features of BGP; interior versus exterior routing, BGP Autonomous System types, and BGP peering. The chapter then describes in detail BGP packet types, BGP session states and Finite State Machine, BGP path attributes types, and BGP ASNs. Other topics discussed include BGP route summarization, alternative solutions to the iBGP full mesh requirement, BGP best path selection process, and BGP authentication. The discussion in this chapter also includes a high-level view of the typical BGP router and its components, and inbound and outbound message processing.

Author

James Aweya, PhD, is a chief research scientist at the Etisalat British Telecom Innovation Center (EBTIC), Khalifa University, Abu Dhabi, UAE. He was a technical lead and senior systems architect with Nortel, Ottawa, Canada, from 1996 to 2009. He was awarded the 2007 Nortel Technology Award of Excellence (TAE) for his pioneering and innovative research on Timing and Synchronization across Packet and TDM Networks. He has been granted 68 US patents and has published over 54 journal papers, 40 conference papers, and 43 technical reports. He has authored four books including this book and is a Senior Member of the Institute of Electrical and Electronics Engineers (IEEE).

1 Open Shortest Path Routing (OSPF) Protocol

1.1 INTRODUCTION

A router running a link-state routing protocol develops a complete map of the network topology by exchanging link-state information with all other routers in the network through a processed called *flooding*. All the routers then use the link-state information they have gathered (in databases called Link-State Databases [LSDBs] or Topology Tables) to construct identical topology maps of the entire network, and from which they calculate the best paths to all known network destinations. Using the same topology map, each router is able to independently calculate the shortest path to every destination in the network.

OSPF **[RFC2328]** and IS-IS **[ISO/IEC10589]** both maintain routing information in LSDBs. These protocols determine which adjacent neighbor routers are operational, and exchange routing information in order to construct the network topology map. Neighbor routers are any two routers that have interfaces connected to a common network segment (which could be a simple point-to-point (P2P) link, a broadcast segment, or non-broadcast segment). In OSPF, for example, neighbors are discovered and maintained, usually dynamically, using OSPF's Hello protocol. Both OSPF and IS-IS use the Dijkstra algorithm to determine the best routes to network destinations, and install these in the router's IP Routing Table.

This chapter and the next, describe the most common link-state routing protocols (OSPFv2 and IS-IS), their characteristics, operations, and the databases they support. Both OSPF and IS-IS are relatively complex routing protocols compared to RIP, and use more protocol packet types, perform protocol handshakes not seen in RIP, and have relatively more routing information advertisement types. The characteristics of OSPFv2 are discussed in greater detail in this chapter.

1.2 OVERVIEW OF OSPF

OSPFv2 **[MOYJT1998] [RFC2328]** is routing protocol for IPv4 networks, while OSPFv3 **[RFC5340]** provides enhancements for IPv6 routing. Henceforth, OSPF in this chapter refers to OSPFv2 unless otherwise specified. OSPF is a link-state routing protocol and an Interior Gateway Protocol (IGP) for distributing routing information within an autonomous system. It is important to note that autonomous system as used in the OSPF context here, means a collection of routers in a routing domain that exchange OSPF routing information and under a common administrative control. In addition, the autonomous system may be divided into smaller subrouting domains called *areas*. The correct meaning of an autonomous system is the one used in Chapter 1 of Volume 1 of this two-part book.

As a link-state routing protocol, routers exchange topology information with their neighbors by flooding their directly connected networks and the topology information advertised by other routers, throughout the network (specifically, an area). This allows every router within the network to have complete knowledge or map of the network topology. Flooding in OSPF (and similarly in IS-IS) means the process by which OSPF routers distribute or disseminate routing information among themselves with the goal of synchronizing their LSDBs, that is, creating identical LSDBs. Using a variant of the Dijkstra algorithm and the LSDB (i.e., the same topology map), each OSPF router then calculates the best paths to all known destinations in the network.

One important advantage of link-state routing protocols like OSPF and IS-IS (over RIP) is that, by allowing all routers to have a complete map of the network topology, they can calculate best routes to satisfy specific network criteria, for example, determine routes that satisfy a number of network traffic engineering constraints. Routers will be able to calculate best routes with constraints that meet some defined quality-of-service (QoS) or routing policy requirements.

However, the main disadvantage of link-state routing protocols is that, they do not scale well as the network grows and more routers are added to it. Adding more routers tend to increase the rate and size of the link-state routing updates, and also the computational resources and time required to calculate best routes. The lack of scalability is one of the reasons why link-state routing protocols (like OSPF and IS-IS) are not suitable for communicating routing information across the public Internet, but rather used as IGPs within a single autonomous system.

Using Link-State Advertisement (LSA) messages, each OSPF router distributes routing information to other routers that consists of its local link-state (i.e., OSPF-enabled interfaces and their reachable neighbors, and the cost associated with each interface). An LSA is a unit of data information that describes the local state of a router link, and the type of network it is connected to. Each router will send LSAs that include the state of its interfaces and the nature of its adjacencies. Each LSA is propagated throughout the routing domain (or area). Each receiving router uses the received link-state routing messages plus its own local link-state, to construct a local LSDB that is identical to those maintained by other routers (in the same routing domain or area). The LSDB, when synchronized with others, describes the complete topology of the entire routing domain (or area).

From the LSDB, each router calculates best routes to network destinations using a Shortest Path First (SPF) algorithm (a variant of the Dijkstra algorithm), and then installs these best routes in its local IP Routing Table. The IP Routing Table contains all the destinations along with their corresponding next hop IP addresses, outgoing interfaces, and other useful information required for routing packets to their destinations.

1.3   OSPF CONCEPTS

OSPF operates directly over IP (similar to EIGRP, IGMP, ICMP, TCP, UDP, etc.) and uses Protocol Number 89 in the IP header. Unlike RIP and BGP, OSPF does not use the services of Transport Layer protocols such as TCP or UDP. Also, OSPF messages are sent as multicast and unicast transmissions (when required) rather than as broadcasts. OSPF Hello messages, for example, are sent via multicast transmissions on a network segment to the "All OSPF Routers" IPv4 multicast address 224.0.0.5, while messages to Designated Routers (DRs) and Backup Designated Routers (BDRs) in a network segment, are sent to the IPv4 multicast addresses 224.0.0.6 (i.e., the All OSPF Designated Routers address).

The equivalent IPv6 multicast addresses are FF02::5 for the "All OSPFv3 Router" multicast address, and FF02::6 for transmissions to a DR and BDR on a network segment. If the underlying network does not support broadcast capabilities naturally (e.g., using Ethernet technology), the OSPF neighbor relationship has to be established using an IPv4 unicast address. For an IPv6 network, this configuration will require a Link-Local IPv6 address. OSPF supports Variable-Length Subnet Masks (VLSMs), and Classless Inter-Domain Routing (CIDR) where the routing updates carry network mask information.

The first time an OSPF router discovers a new neighbor router, it will transmit full routing updates carrying all the link-state information it has learned. The ultimate goal of the routing information exchange is to allow all routers in the OSPF area to have identical and synchronized link-state information in their LSDBs. When the OSPF network has converged and all routers have a consistent view of the network topology, a router will send only a partial routing update to all its neighbors when a new neighbor/link comes up or becomes unavailable. The routing update reflecting the new change will then be flooded to all OSPF routers within the area.

The operation of OSPF is characterized by five main fundamental components: neighbor router discovery and maintenance, exchange of router link-state information, building of LSDBs, best route computations to network destinations, and building and maintenance of the IP Routing Table. These components, which are discussed in detail in this chapter, are summarized as follows:

- **Establishment and Maintenance of OSPF Neighbor Adjacencies**: An adjacency is a relationship established between two neighbor routers for the purpose of exchanging routing information. OSPF routers must discover and establish adjacencies with their neighbors before they can exchange routing information with those neighbors. An OSPF router transmits Hello packets out its interfaces to determine which neighbors are reachable and available on those interfaces. If a neighbor is reachable, the router attempts to establish an OSPF neighbor relationship with that neighbor. However, not every pair of neighbor routers can form an adjacency – this depends on the OSPF network type.
- **Exchange of LSAs**: After two routers have established an adjacency, they start exchanging LSAs that contain the state and cost of each of their directly connected interfaces. Each router stores and then floods its LSAs to directly connected and reachable neighbors, and each adjacent neighbor receiving the LSA, in turn, stores and immediately floods the LSA to other directly reachable neighbor. The flooding continues until all OSPF routers in the area have received all advertised LSAs. The routers propagate the LSAs to all neighbors using a special multicast address.

- **Construction and Maintenance of the LSDB (or Topology Table)**: After the OSPF routers have received the LSAs, they each construct their LSDB (or Topology Table) based on the link-state information contained in them. Each piece of information in a router's LSDB also includes its local state (e.g., the router's active interfaces, link costs, and the neighbors that are reachable over these interfaces). In addition to LSAs flooded by other routers, the router also propagates its local state to other routers in its area via LSAs. The LSDB of each router, upon network convergence, eventually contains all the information about the entire network topology. Most important, all routers in a particular OSPF area converge to have the same view of the network topology (i.e., have identical or synchronized LSDBs) – the same routing information in their LSDBs.
- **Execution of the SPF algorithm**: After an OSPF router has built its LSDB, it will then execute the SPF algorithm (a variant of the Dijkstra algorithm) on the LSDB, with the result being a Shortest Path Tree (SPT) (with the router itself as root) linking the router to all currently known network destinations.
- **Construction and Maintenance of the IP Routing Table**: Each router then inserts the best paths to each network destination (derived from the SPF) into its IP Routing Table. The router then makes forwarding decisions for each packet it receives using the routes in the IP Routing Table. The flooding of LSAs ensures that all routers can update their LSDBs, and Routing Tables to reflect new network topology changes.

A link in an OSPF network is viewed as an interface on a router, and the state of the link is a description of the properties and parameters associated with that interface and its relationship with neighbor OSPF routers. Generally, interface and link are used interchangeably and mean the same thing. Either of these terms means a connection between a router and at least one of its neighbors over an attached network segment.

A router interface (or link) has associated with it some state information (referred to as link-states) which the router derives using the routing protocol itself, and the underlying lower level protocols the interfaces uses (e.g., Ethernet, ATM, and PPP). The link-states on an OSPF router are descriptions/indications of the state (including parameters) of each interface on the router (belonging to a specific OSPF area) and its associated adjacencies. Each router will generate a link-state for each of its interfaces belonging to a particular area, and propagate these via various types of LSAs to other routers.

A description of a router interface/link includes the IPv4 address assigned to the interface, network mask, Area Identifier (ID), type of network connected to the interface (e.g., P2P, broadcast multiaccess, and Virtual Link), the OSPF neighbor routers connected to that network segment (or link), etc. An OSPF router gathers all the link-states (carried via LSAs) from all routers in an OSPF area (including itself) to form its LSDB. The OSPF router then determines the best paths to all known network destinations based on the states of the links that connect the router and those destinations.

After all the OSPF routers in an area have established adjacencies between themselves, each router will periodically send new instances of existing LSAs to refresh its adjacencies, or LSA updates to report any changes in its link-state. When a router detects a change in a link-state (such as interface failure), it will create an LSA describing that change. By comparing the state of established adjacencies in its LSDB with received LSAs, a router can determine changes in the area's topology, and then update its LSDB accordingly.

1.4 OSPF AREAS AND HIERARCHICAL ROUTING

When OSPF is run in a small or simple network with a small number of routers and links, the memory and processing requirements for best paths computations are generally manageable. However, these requirements can be relatively high and complex in larger networks with many routers and links. The SPF computations (which require examining the entire network topology [maintained in the LSDB]) to determine best paths in this case can easily become a complex and time-consuming exercise for an OSPF router.

One method that is generally used to reduce the memory and processing requirements on OSPF routers, as well as, the size of LSAs, LSDB, and Routing Tables, is to carve out or partition an OSPF network into smaller routing domains (or subdomains) called *areas* (Figure 1.1). It is important to note that, dividing an OSPF autonomous system into areas is simply a logical grouping of routers, links, and hosts into smaller routing domains within the autonomous system. An area includes the internal OSPF routers as well as the interfaces that connect to other devices within the area.

By structuring or partitioning a large OSPF network into smaller routing domains (or areas), the administration of the network is simplified, while at the same time allowing network traffic and resource utilization to be optimized. The use of areas

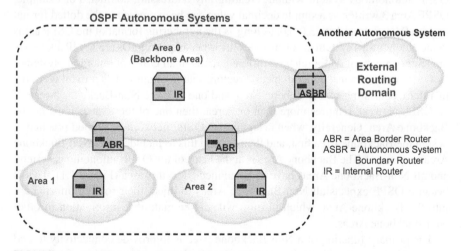

FIGURE 1.1 OSPF Areas and Hierarchical Routing

also reduces the frequency and processing time the SPF algorithm takes to complete best path computations. Also, routing within an OSPF area (referred to as intra-area routing) is based solely on the area's own topology, thus, protecting the area from bad out-of-area routing information.

All OSPF routers within an area exchange routing information with the goal of maintaining identical LSDBs. Each router within the area maintains a separate but identical LSDB whose routes may be summarized and advertised to the rest of the OSPF autonomous system by the router that connects the area to the other parts of the autonomous system. This structuring makes the routers outside the area unaware of the topological details of the area.

The concepts of OSPF areas makes the topology details of an OSPF area not visible to the rest of the autonomous system – such details are visible only to routers within that OSPF area. This results in a significant reduction in the routing traffic exchanged between the various parts of the OSPF autonomous system. Routers in a different area will populate their Routing Tables with summary routes rather than with exact routes from the originating areas, which results in not only smaller Routing Tables but also lower destination address lookup times during packet forwarding.

Thus, routers within an area exchange detailed link-state information in order to build identical LSDBs for that area. However, the routing information that one area exchanges with another contains only summary information (or summary LSAs) of the originating area's LSDB and not details of the network topology of that area. Routers in the receiving area receive the summary LSAs, and inject the routing information directly into their Routing Tables without having to rerun the SPF algorithm.

Each area within an OSPF autonomous system must be uniquely identified by a 32-bit number. This number or Area ID is commonly written in the dot-decimal notation used in IPv4 addressing. Area Identifiers (IDs) are not IP addresses, even if written in the dot-decimal format, and may duplicate any existing IPv4 address in the OSPF autonomous system without creating any addressing conflict. For example, OSPF Area 3 written as a simple decimal value can be expressed in the dotted format as Area 0.0.0.3 or as Area 3.0.0.0, as long as the numbering format of the OSPF areas in the autonomous system is consistent. This is because Area IDs are not IP addresses but simply a way of distinguishing different areas in an OSPF autonomous system.

OSPF networks use a hierarchical structure that is organized as a two-layer area hierarchy consisting of a Backbone Area, and one or more Non-Backbone Areas. If an OSPF network requires more than one area, then one of these areas must be the Backbone Area. Generally, when designing an OSPF network, it is good practice to design the Backbone Area first, and then expand this to include other Non-Backbone Areas as needed. The Backbone Area is at the heart of the OSPF autonomous system, and all Non-Backbone Areas have to be connected to this area (Figure 1.1). This is because OSPF expects all Non-Backbone Areas to inject their routing information into the Backbone Area, which in turn, will disseminate that information to other Non-Backbone Areas.

The primary function of a Non-Backbone Area is to provide connectivity to end users and resources. Typically, Non-Backbone Areas are set up or constructed according to geographic or functional groupings of end users and resources. The key requirement here is that, traffic sent between different Non-Backbone Areas in the OSPF routing domain must always travel through the Backbone Area.

An OSPF autonomous system originates two types of routes, which are, intra-area routes and inter-area routes. Intra-area routes are routes that are originated by an OSPF area and advertised within the same area (i.e., destinations that belong to the same area). On the other hand, inter-area routes are routes that are originated by one OSPF area and advertised in other OSPF areas. Area Border Routers (ABRs) are responsible for injecting inter-area routes from one area into another area.

External routes are routes that are originated by other non-OSPF routing protocols and domains, and advertised or injected into the OSPF autonomous system via redistribution. External routing information is generally kept separate from the OSPF derived link-state information. The advertising router (called an Autonomous System Boundary Router [ASBR]) can also tag each external route, allowing additional information to be passed from the external routing domain to routers within the autonomous system. The different types of OSPF areas are described below.

1.4.1 BACKBONE AREA

The Backbone Area is also referred to as Area 0 (or Area 0.0.0.0 when expressed in the IPv4 dotted decimal address notation if OSPF Area ID's are formatted this way). The Backbone Area is uniquely identified as Area 0 or Area 0.0.0.0. One key requirement for a Backbone Area is that, it must have connectivity to all other Non-Backbone areas in the OSPF routing domain (sometimes referred to as OSPF autonomous system). In the rare case where, physical connectivity is not practical or possible, a Virtual Link can be used to establish a virtually/logically connection from a Non-Backbone area to the Backbone Area.

Another requirement is that, the Backbone Area must always be (logically) contiguous (not necessarily physically). This means the Backbone Area is not allowed to be split up even under any failure conditions (e.g., router or link failures). Virtual Links, where appropriate (mostly in rare situations), can also be used to create internal connections between disjoint parts of the backbone in order to make the Backbone Area appear as if it is a single physical or contiguous network segment.

In the (rare) case where a Virtual Link is used to connect any two disjoint Backbone Routers (that have physical connections [interfaces] to Non-Backbone Areas), the Virtual Link is considered part of or belongs to the same Backbone Area. Here, the Virtual Link is used to patch the discontinuity in the Backbone Area. OSPF treats the two Backbone Routers connected by the Virtual Link as if they were connected by an unnumbered P2P link in the Backbone Area. In this case, the OSPF traffic flowing over the Virtual Link represents intra-area routing traffic only (i.e., Backbone Area traffic).

As discussed above, the Backbone Area is responsible for providing routing connectivity between Non-Backbone Areas. Thus, all Non-Backbone Areas in the OSPF routing domain must be connected to the Backbone Area; and in the cases where a physical connection to the backbone is not possible, Virtual Links can be used. As shown in Figure 1.1, the Backbone Area always has interfaces to all ABRs. All Non-Backbone Areas must connect to the Backbone Area, either directly, or through other

OSPF areas. Inter-area routing in the OSPF routing domain takes place through the routers that connect both to the Backbone Area and the respective Non-Backbone Areas (i.e., the ABRs). All Non-Backbone Areas feed their routing information into the Backbone Area which then distributes that information to other Non-Backbone Areas. Generally, a Backbone Area does not connect directly to end users.

A Virtual Link can be configured to connect a Non-Backbone Area to the Backbone Area, for example, where a newly introduced Non-Backbone Area (from a new company) has no direct physical access to the Backbone Area. The Virtual Link provides a logical path to the Backbone Area for the disjoint Non-Backbone Area. Let us assume, for example, that Area 1 is physically connected to the Backbone Area, and Area 2 has no direct connection to it, but instead has a connection to Area 1. In this case, a Virtual Link can be used to connect Area 2 through Area 1 (serving as a Transit Area) to the Backbone Area. The Virtual Link has to be configured between the two ABRs that have Area 1 as their common area (i.e., the Transit Area), but with one of the ABRs directly connected to the Backbone Area. When the parameter *"TransitCapability"* in the Area Data Structure is set, an area becomes a Transit Area **[RFC2328]**.

1.4.2   STUB AREA, NOT-SO-STUBBY AREA, AND TOTALLY STUBBY AREA

In normal or standard OSPF routing, an ASBR located anywhere in the OSPF domain may generate Type 5 LSAs (AS-External-LSAs) that are flooded to normal OSPF areas to allow the routers in those areas to install external routes in their Routing Tables. However, in some cases, it is not necessary to provide routers in certain OSPF areas with external routing information. In such cases, it is sufficient only to provide those routers with a default route so that they can access internal OSPF routes and routes to destinations outside the OSPF autonomous system (i.e., external destinations). This is one of the main reasons why OSPF supports various types of Stub Areas.

Other than standard OSPF areas, other areas in an OSPF network may be defined as a Stub Area, Not-So-Stubby Area (NSSA), Totally Stubby Area, and Totally NSSA. An OSPF area can qualify as a Stub Area when it is essentially a dead-end OSPF area, and routing to destinations outside the area does not necessarily have to take an optimal route to those destinations. Also, a Stub Area may have multiple ABRs, that is, exit points, which will inject a default route into the area for communications with the outside world. This means routing to external destinations could take a suboptimal route via a more distant exit point to reach those destinations (which is not a problem in itself since the area is a dead-end area after all).

The main purpose of using any of these stub area types is to simply area design and LSA advertisement by introducing default routes into an OSPF area and, in some stub area types, do away with (all together) the need to inject/flood external and Summary-LSAs into the area. The use of a Stub Area in an OSPF network provides many benefits such as, a reduction in the amount of LSA flooding, and smaller LSDB and Routing Table sizes in the OSPF routers within the Stub Area.

For these reasons, it is recommended and at the same time beneficial to use stub area techniques when building OSPF networks, wherever possible. Stub area techniques provide significant OSPF performance improvements, as well as, memory savings in OSPF routers. The use of default routes reduces the size of the Routing Table in OSPF routers (because a single default route can replace multiple external routes), and consequently, reduces the router's memory and CPU utilization. The use of stub area techniques allows an OSPF network to scale to significantly larger sizes while providing the benefits discussed above. An OSPF router (in the Stub Area) may inject a default route unconditionally in its Routing Table, or use a default route present in the IP Routing Table **[TEAREDIA15]**.

We discuss in greater detail the characteristics of the different stub area types below and with Figure 1.2 presenting a feature summary.

1.4.2.1 Stub Area

This section describes the main features of an OSPF Stub Area:

- **Routes from Other OSPF Areas**: A Stub Area accepts summary routes advertised by other OSPF areas within the same autonomous system. This means Summary LSAs (Type 3 LSAs) from other areas can be injected into a Stub Area by the ABR connecting the area to the OSPF backbone.
- **External Routes from Outside the OSPF Autonomous System**: An OSPF Stub Area does not accept LSAs that advertise external routes to its OSPF autonomous system. Type 5 LSAs (AS-External-LSAs) which are advertised by an ASBR are not injected into a Stub Area by the ABR connecting the area to the OSPF backbone; the ABR stops all such LSAs from being flooded into the Stub Area. This has the added benefit that external route/link flaps will not be injected into the Stub Area.
- **Use of a Default Route**: A Stub Area relies entirely on a default route (with IP address 0.0.0.0 and network mask 0.0.0.0) for routing from within the area to destinations outside the autonomous system. Note that in normal IP routing, a default route is used when the Routing Table does not contain a matching entry for a network destination. Default routes are used in the Stub Area to prevent the routers within from learning specific routes to each external network destination.
  o If a router within the Stub Area needs to route traffic to destinations outside the autonomous system, it has to use a default route – a default route to an ABR replaces all external routes. If there are no external routes to be injected into an OSPF autonomous system, then there is no benefit or need to define any Stub Area. Stub Areas do not accept external routes but do receive routing information from other areas that belong to the same OSPF autonomous system.
- **Default Routes Are Injected into a Stub Area via Summary-LSA**: Routers within a Stub Area must rely on a default route (0.0.0.0 or *DefaultDestination* in OSPF) that is originated/advertised by an ABR attached to the area in order to reach destinations external to the OSPF autonomous system. The ABR will still advertise inter-area routing information into the Stub Area. Internal Routers will still have all intra-area routes, inter-area routes, and an inter-area default route for external destinations.

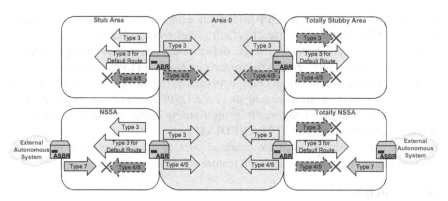

a) OSPF Area Types and Messages

Characteristics of OSPF Areas Types				
Area Type	Accepts Routes From Within Same OSPF Area	Accepts Routes from Other OSPF Areas	Accepts External Routes from Outside OSPF	Allows Attachment to an ASBR
Standard OSPF Area	Yes	Yes	Yes	Yes
Backbone Area	Yes	Yes	Yes	Yes
Stub Area	Yes	Yes	No (Uses Default Route)	No
Totally Stubby Area	Yes	No (Uses Default Route)	No (Uses Default Route)	No
NSSA	Yes	Yes	No (Uses Default Route)	Yes
Totally NSSA	Yes	No (Uses Default Route)	No (Uses Default Route)	Yes

b) Characteristics of OSPF Area Types

LSA Types Allowed in Area					
LSA Type	Standard OSPF Area	Stub Area	Totally Stubby Area	NSSA	Totally NSSA
1	Yes	Yes	Yes	Yes	Yes
2	Yes	Yes	Yes	Yes	Yes
3	Yes	Yes / For Default Route	X / For Default Route	Yes / For Default Route	X / For Default Route
4	Yes	X	X	X	X
5	Yes	X	X	X	X
7	X	X	X	Yes	Yes

c) LSAs Allowed in OSPF Area Types

FIGURE 1.2 Main Characteristics of OSPF Stubby Areas: (a) OSPF Area Types and Messages, (b) Characteristics of OSPF Area Types, (c) LSAs Allowed in OSPF Area Types

o The ABR automatically generates a default route via Summary LSAs (Type 3 LSAs) with the Link-State ID set to 0.0.0.0. The ABR will advertise this default route throughout the Stub Area so that routers in the area can communicate with external destinations, instead of (flooding) the external routes. External routes are not allowed to be injected into the Stub Area, and the Routing Tables of routers in the area do not contain any external routes. These routes are accessible via the injected intra-area default route.

- **Attachment to an ASBR**: A Stub Area can connect to an ABR (Figure 1.3) but *not* to an ASBR except in the case where the ABR it connects to also functions as an ASBR. The ABR of the Stub Area (even if functioning also as ASBR) does not inject external routes from other autonomous systems into the area.

o Because a Stub Area does not accept external routes, its ABR will delete ASBR-Summary-LSAs (Type 4 LSAs) and AS-External LSAs (Type 5 LSAs) trying to get to the Internal Routers of the Stub Area. The ABR will advertise to them, instead, a default route of 0.0.0.0 (Figure 1.4). The ABR plays the role of a default gateway to the Stub Area. AS-External LSAs (Type 5 LSAs) carry routing information from other routing protocols that are redistributed into OSPF, and are not allowed to be flooded into the Stub Area. Traffic to external destinations in another autonomous system will be first sent along the default route injected into the Stub Area by the ABR, and then along the default route injected into the Backbone Area by the ASBR leading to the destination.

In Figure 1.3, the routers in Area 3 do not need to know about all or any of the destinations that are external to the OSPF autonomous system – no advantage in learning

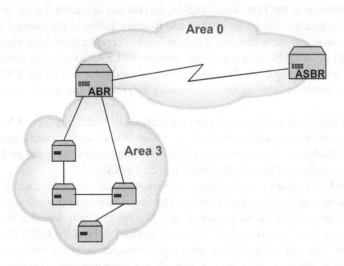

FIGURE 1.3 Defining a Stub Area

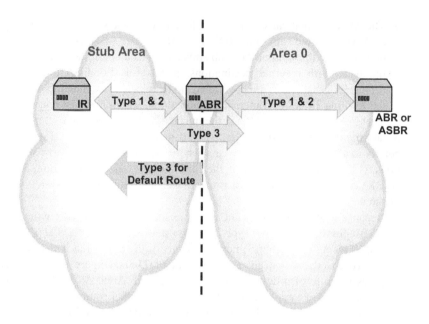

FIGURE 1.4   Characteristics of OSPF Stub Area

external routes. This is because the routers in Area 3 must still forward all packets to the ABR connected to it in order to reach the ASBR (or any other ASBR attached to the OSPF autonomous system), no matter where those external destinations are. Therefore, to simply routing and area design, Area 3 can simply be defined as a Stub Area in the OSPF autonomous system.

The deletion of the Type 4 and 5 LSAs, and the use of a default route (Figures 1.3 and 1.4) reduce the size of the LSDBs and Routing Tables in the Internal Routers of the Stub Area. For these reasons, it is beneficial to create a Stub Area when much of the LSDB of an OSPF area potentially consists of LSAs advertising routes external to the autonomous system. This reduces significantly the size of the LSDBs in the area and, therefore, the amount of memory required in the area's internal OSPF routers.

Some important restrictions on the applications of Stub Areas are: A Virtual Link cannot be created through a Stub Area, a Stub Area cannot contain internally an ASBR, a Backbone Area cannot be a Stub Area (and vice versa), and an OSPF area cannot be configured to be both a Stub Area and a NSSA.

For an OSPF area to be configured as Stub Area, all the OSPF routers inside the area must be configured as Stub Routers. The router interfaces that belong to the Stub Area exchange OSPF Hello packets with a flag bit that indicates that the interface belongs to a Stub Area (i.e., the E-bit in the Options field of the OSPF Hello packet is set to 0). The setting of this flag must match between neighbor routers for them to be able to establish an adjacency. If they do not have matching flag bits, then they will not be able form an adjacency and routing will not take effect.

1.4.2.2   Totally Stubby Area

The Totally Stubby Area is a Cisco proprietary extension to the Stub Area concept discussed above **[CISCID6208] [TEAREDIA15]**. The LSDBs of Internal Routers in a Stub Area can be further minimized by configuring ABRs to inject only Summary LSAs (Type 3 LSAs) describing a default route into the area. Figure 1.5 summarizes the type of LSAs accepted into the area:

- **Routes from Other OSPF Areas**: A Totally Stubby Area does not accept summary routes from other areas within the same OSPF autonomous system (i.e., inter-area routes); Summary LSAs (Type 3 LSAs) are not injected into the area.
- **External Routes from Outside the OSPF Autonomous System**: A Totally Stubby Area does not accept external routes from other autonomous systems. Type 4 LSAs (ASBR-Summary-LSAs) and Type 5 LSAs (AS-External-LSAs), which are advertised by an ASBR, are not injected into the area.
- **Use of a Default Route**: If an OSPF router within the Totally Stubby Area needs to send traffic to a destination external to the area, it does so via a default route injected into the area by the ABR attached to the area. The ABR automatically generates a default route via Summary LSAs (Type 3 LSAs) and with the Link-State ID set to 0.0.0.0. Routers within the Totally Stubby Area maintain in their LSDBs only information about routes within that area, in addition to, the default route. The Routing Tables of routers in the area do not contain inter-area and external routes; these routes are accessible via the injected intra-area default route.

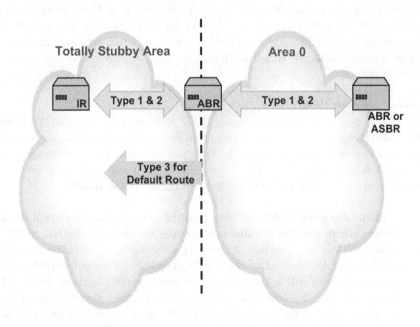

FIGURE 1.5   Characteristics of OSPF Totally Stubby Area

o Routers within the area must rely on the default route (0.0.0.0 or
DefaultDestination in OSPF) originated/advertised (via Summary-LSAs
(Type 3 LSAs) with the Link-State ID set to 0.0.0.0) by an ABR attached to
the area to reach destinations external to the area. The default route presents
a path to all external routes (to destinations outside the OSPF autonomous
system), as well as, all summarized and non-summarized routes from other
areas within the same OSPF autonomous system.

• **Attachment to an ASBR**: A Totally Stubby Area cannot connect to an ASBR
except where the NSSA ABR also functions as an ASBR.

o A Totally Stubby Area does not allow Summary-LSAs (Type 3 LSAs), and
ASBR-Summary-LSAs (Type 4 LSAs), which are both inter-area routes,
from being injected into the area. It also does not allow AS-External-LSAs
(Type 5 LSAs) from entering the area. By blocking these routes, a Totally
Stubby Area relies on only the default route and intra-area routes for communications with external destinations. Each ABR attached to the Totally
Stubby Area injects the default summary route into the area, and each router
in the area selects the closest ABR as its gateway to reach destinations outside the area.

The use of Totally Stubby Areas in an OSPF autonomous system further reduces
routing information than what can be achieved with Stub Areas, and also further
helps to increase/improve the scalability and stability of OSPF networks
[TEAREDIA15].

In Figure 1.3, routers in Area 3 do not need to know about all destinations that are
external to the area and advertised by the Backbone Area to all OSPF areas (via Type
5 LSAs or any Summary Type 3 and 4 LSAs). This is because the routers in Area 3
must always forward all packets to the ABR connected to the area in order to reach
any destination outside the area. Therefore, Area 3 can be conveniently configured as
a Totally Stub Area.

1.4.2.3 Not-So-Stubby Area (NSSA)

A NSSA is a modification of the basic OSPF Stub Area concept **[MOYJT1998]**
[RFC3101] (Figure 1.6). There are other proprietary variations that are supported by
a number of router vendors, such as the Totally Stubby Area (described above), and
the NSSA Totally Stubby Area. These two are both extensions of the basic Stub Area
concept and are implemented in Cisco Systems routers **[CISCID6208]**.

• **Attachment to an ASBR but with Limited Acceptance of External Routes
from Outside the OSPF Autonomous System**: Unlike a Stub Area, an NSSA
allows the (limited) injection of external routes into the area to some extent. An
NSSA provides the rest of the OSPF domain with some external routing
information, but will not accept external routes coming from the local OSPF
Backbone Area. An NSSA will accept external routes directly from another
(adjacent or directly attached) non-OSPF autonomous systems, and will
advertises these routes to other areas within its local OSPF autonomous system,

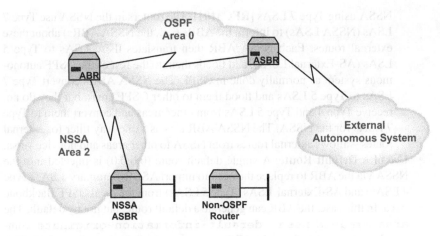

FIGURE 1.6 Defining an NSSA

but will not accept external routes advertised by other areas in its OSPF autonomous system via the local Backbone Area.

o This means OSPF routers within an NSSA will not accept external LSAs advertised from other areas in the local OSPF autonomous system, but the NSSA is allowed to propagate external routes (from an adjacent non-OSPF autonomous system) to those other areas for redistribution. In such a scenario, the NSSA is primarily used to connect a user network through an OSPF routing domain to an ISP (Internet Service Provider).

o For example, an ASBR attached to an NSSA (i.e., an NSSA ASBR) (see Figures 1.6 and 1.7) can import external routes and redistribute them in the

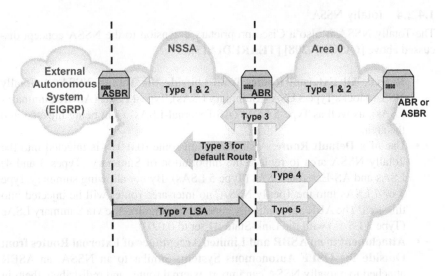

FIGURE 1.7 Characteristics of OSPF NSSA

NSSA using Type 7 LSAs [RFC3101]. The routers in the NSSA use Type 7 LSAs (NSSA LSAs) to inform the ABRs (i.e., the NSSA ABRs) about these external routes. Each NSSA ABR then translates these LSAs to Type 5 LSAs (AS-External LSAs), and floods them to the rest of the OSPF autonomous system as normally done in OSPF. The NSSA ABRs convert Type 7 LSAs to Type 5 LSAs and flood them to other OSPF areas, but they do not receive Type 4 and Type 5 LSAs from other areas and convert them to Type 7 LSAs for the NSSA. The NSSA ABR acts as a one-way filter for external routes – allows external routes from NSSA to other areas and not vice versa.

- **Use of a Default Route**: A single default route (0.0.0.0) is injected into the NSSA via the ABR to replace the need to import ASBR-Summary-LSAs (Type 4 LSAs) and AS-External-LSAs (Type 5 LSAs) from the local OSPF Backbone Area. In this case, the ABR can generate a default route, but not by default. The **area** *area-id* **nssa   default-information-originate** command can be used to force the ABR to generate the default route for the NSSA [CISCOSPFCOMD19].

In Figure 1.6, Area 2 can be defined as an NSSA to protect the area's internal OSPF routers from learning all the ASBR-Summary-LSAs (Type 4 LSAs) and AS-External (Type 5) LSAs advertised by the ASBR connected to OSPF Area 0, but still allow the connection of the non-OSPF router attached to the NSSA ASBR. The external routing information from the non-OSPF router is imported into the NSSA using Type-7 LSAs. NSSAs provide more flexibility in OSPF network design than what could have been achieved with Stub Areas only. This is because an NSSA can redistribute external routing information into the OSPF autonomous system and, thereby, provide transit service to non-OSPF routing domains that are adjacent to the OSPF autonomous system.

1.4.2.4   Totally NSSA

The Totally NSSA is also a Cisco proprietary extension to the NSSA concept discussed above [CISCID6208] [TEAREDIA15].

- **Rejects All External Routes from Outside the Totally NSSA**: A Totally NSSA blocks Type 3 LSAs (Summary LSAs), Type 4 LSAs (ASBR-Summary-LSAs), as well as Type 5 LSAs (AS-External-LSAs) from being injected into the area.
- **Use of a Default Route**: A single default route (0.0.0.0) is injected into the Totally NSSA area to replace the importation of Summary (Types 3 and 4) LSAs and AS-External-LSAs (Type 5 LSAs). By not allowing summary Type 3 or 4 LSAs into the Totally NSSA, no inter-area routes will be injected into this area. The ABR automatically generates a default route via Summary LSAs (Type 3 LSAs) with the Link-State ID set to 0.0.0.0.
- **Attachment to an ASBR and Limited Acceptance of External Routes from Outside the OSPF Autonomous System**: Similar to an NSSA, an ASBR attached to a totally NSSA can import external routes and redistribute them in the area using Type 7 LSAs. The Totally NSSA ABRs will convert Type 7 LSAs to Type 5 LSAs and flood them to other OSPF areas, but they will not

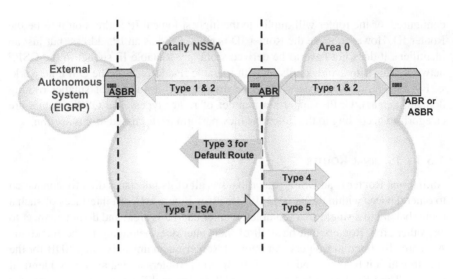

FIGURE 1.8 Characteristics of OSPF Totally NSSA

receive Type 5 LSAs from other areas and convert them to Type 7 LSAs for the Totally NSSA.

The ABR connected to the Totally NSSA (NSSA ABRs) must be configured to generate a default summary route to prevent the flooding of summary routes from other areas in the local autonomous system into the Totally NSSA. As illustrated in Figure 1.8, the NSSA ABR will advertise this default route (instead of the summary and external routes which are not allowed to be injected into the Totally NSSA) to allow routers in the area to communicate with external destinations. In Figure 1.6, Area 2 can be defined as a Totally NSSA to keep any Type 3 and 4 summary routes and AS-External (Type 5) routes from leaking into Area 2.

1.4.3 TRANSIT AREA

A Transit Area is an OSPF area used for transporting traffic from one adjacent area to another adjacent area. Any one of the adjacent areas can be of any OSPF area type including the Backbone Area. The traffic carried across the Transit Area does not originate from it nor is it destined to it. A Stub Area cannot be used as a Transit Area since it is essentially a dead-end area and cannot be used to interconnect two adjacent OSPF areas. This last point also means Virtual Links cannot be configured through a Stub Area (and turning it into a Transit Area) to interconnect two adjacent areas.

1.5 OSPF ROUTER TYPES

Each router in an OSPF network is assigned a unique identifier (a Router ID) which is customarily expressed in the dotted decimal address format used for IPv4 addresses. The Router ID must be established every time OSPF runs. The ID can be explicitly configured, or the router will duplicate the highest logical IP address on it to be the Router ID. However, since the Router ID is not viewed as an IP address (but just an identifier), it does not have to be derived from any routable IP subnet in the OSPF network. This is to avoid confusion with routable IP addresses in the OSPF network.

By allowing a bigger network to be divided into a number of smaller routing domains (areas), OSPF supports a number of router types. The OSPF routers are categorized according to the functions they perform as discussed in this section.

1.5.1 INTERNAL ROUTER

An Internal Router is an OSPF router that has all of its interfaces directly connected to other devices within the same OSPF area (see Figure 1.1). All interfaces of such a router belong to a single area (completely internal to this area) and do not connect to any other area. Routers that have all of their interfaces belonging to the Backbone Area are also Internal Routers. An Internal Router maintains only one LSDB for the area to which it belongs, and also all the Internal Routers in that area have identical LSDBs. The LSDB must be identical on all the Internal Routers within an area for them to be able to construct an accurate SPT to network destinations.

1.5.2 AREA BORDER ROUTER (ABR)

An ABR is an OSPF router that connects one or more OSPF areas to the Backbone Area (see Figures 1.1 and 1.9). An ABR must have either a direct or indirect (via a Virtual Link) connection to the Backbone Area. One interface of the ABR connects to the Backbone Area while each one of the other interfaces connects to a Non-Backbone Area. The ABR also maintains a separate LSDB for each OSPF area to which it is connected (including the Backbone Area). If an ABR is connected to four OSPF areas, it will maintain four different LSDBs, one for each area.

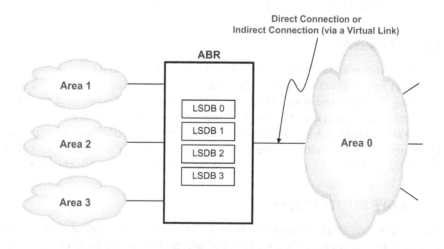

FIGURE 1.9 Structure of an ABR

Therefore, an ABR can have interfaces that belong to multiple OSPF areas (including the Backbone Area) and is a member of all the areas it is connected to. The ABR summarizes or aggregates the routing information it receives from each attached area and distributes this (via flooding) over the OSPF Backbone Area, which in turn, advertises this information to other OSPF areas. An ABR performs route computations for each area it connects to, and advertise this routing information to other areas. In particular, an ABR does not advertise routing information from an area to other areas until its LSDB is synchronized with all the other routers in that area.

1.5.3 BACKBONE ROUTER

A Backbone Router is an OSPF router that has at least one of its interfaces connected to the OSPF Backbone Area. A Backbone Router may also function as an ABR but does not have to. The design rules of OSPF require that all areas be connected to a single Backbone Area (Area 0).

1.5.4 AUTONOMOUS SYSTEM BOUNDARY ROUTER (ASBR)

An ASBR is a router that has at least one of its interfaces connected to an area within an OSPF autonomous system, and at least one of its other interfaces connected to a non-OSPF autonomous system. An ASBR acts as a gateway for redistributing routes between an OSPF autonomous system and routing domain running another routing protocol (e.g., EIGRP, and RIP). An ASBR exchanges routing information with non-OSPF routers belonging to another autonomous system, and advertises the external routing information (i.e., the external routes learned) throughout the OSPF autonomous system. External routes are routes learned by any other routing protocol and redistributed into OSPF.

Routers within the OSPF area where the ASBR connects must know the path to that ASBR. Any router outside this area only knows the path to the nearest ABR that connects to the area where the ASBR resides. Essentially, every router in the OSPF autonomous system has a path to each ASBR. An ASBR (depending on the location) may be an Internal Router (with the exception of routers in Stub Areas), an ABR, or a Backbone Router. A router within a Stub Area (internal Stub Routers) cannot be an ASBR because a Stub Area cannot contain any AS-External-LSAs (Type 5 LSAs). A router that connects to one or more OSPF areas, and also receives routes from a BGP router connected to another autonomous system, is both an ABR and an ASBR.

To internetwork with a non-OSPF routing domain, an ASBR may use static routes or run an Exterior Gateway Protocol (EGP) (like BGP) or both. The ASBR will receive and distribute routes from the external autonomous system throughout the OSPF autonomous system it is connected to. It will create AS-External-LSAs describing the external routes, and floods them to all OSPF areas (via their ABRs). Routers in other OSPF areas will use their ABRs as next hops to access these external routes. The ABRs will forward packets to the external destinations to the ASBR that advertised the external routes.

1.6   OSPF MESSAGE TYPES

All OSPF packets are directly encapsulated in IP (as illustrated in Figure 1.10) and use the IP protocol number 89. OSPF uses five distinct packet types which all share a common 24-byte OSPF packet header (Figure 1.11). The common header carries all the information necessary to handle an OSPF packet when it leaves the source and travels to a receiver. The OSPF packet header also contains all the information needed by a router to determine whether to accept a packet for further processing.

OSPF routers use the different message types for exchanging routing information, and for other types of communication with routers in an OSPF area or in the autonomous system. Specifically, OSPF routers use these message types to exchange routing information with the goal of obtaining identical (or synchronized) LSDBs,

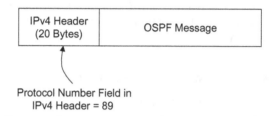

| IPv4 Header (20 Bytes) | OSPF Message |

Protocol Number Field in
IPv4 Header = 89

FIGURE 1.10   OSPF Message in an IPv4 Packet

0	7	15	23	31
Version	Type	Packet Length		
Source Router ID				
Area ID				
Checksum		Authentication Type		
Authentication				
Authentication				

Field	Meaning
Version (8 Bits)	Specifies the version of the OSPF protocol used in the packet (which is 2 as specified in RFC 2328)
Type (8 Bits)	Specifies the OSPF packet type (1 = Hello; 2 = Database Description; 3 = Link State Request; 4 = Link State Update; 5 = Link State Acknowledgment)
Packet Length (16 Bits)	Specifies the length of the OSPF packet (in bytes) including the standard OSPF header
Source Router ID (32 Bits)	Specifies the router ID that is the source of the OSPF packet
Area ID (32 Bits)	All OSPF packets are associated with a single area and this field identifies the area that the OSPF packet belongs to. OSPF packets sent over a virtual link are labelled with the backbone Area ID of 0.0.0.0.
Checksum (16 Bits)	Carries the standard IP checksum of the entire contents of the packet, starting with the OSPF packet header but excluding the 64-bit authentication field. This checksum is calculated as the 16-bit one's complement of the one's complement sum of all the 16-bit words in the packet, excluding the authentication field.
Authentication Type (16 Bits)	Identifies the authentication method to be used for the packet (0 = Null authentication; 1 = Simple password; 2 = Cryptographic authentication)
Authentication (64 Bits)	Carries data used by the authentication scheme specified

FIGURE 1.11   OSPF Packet Header

ultimately, helping to improve routing convergence, and the stability and scalability of the OSPF network.

OSPF does not define a fragmentation mechanism for its protocol packets, and instead depends on IPv4 fragmentation (if available) to fragment and transmit protocol packets larger than the interface Maximum Transmission Unit (MTU). Theoretically, the size of OSPF packets (including the IP header) can be up to 65,535 bytes. OSPF packet types such Database Description packets, Link-State Request packets, Link-State Update packets, and Link-State Acknowledgment packets are more likely to be large, and may be fragmented into several smaller protocol packets before transmission, without loss of OSPF functionality.

Fragmentation by the data source is more preferable and is recommended, while IP fragmentation at downstream routers should be avoided whenever possible. Generally, OSPF routers try to limit the sizes of OSPF packets transmitted over Virtual Links to 576 bytes unless the source router employs Path MTU Discovery **[RFC1191] [RFC8201]** to discover the MTU along a network path.

OSPF routers detect errors in OSPF packets by means of checksums carried in fields in the OSPF packet header and in the LSA header. If the OSPF router detects that the length of the OSPF packet is not an integral multiple of 16-bit words, it will pad the packet with one or more bytes consisting of all zeros before calculating the checksum. The checksum calculation and verification are considered to be part of the Null and Simple Password Authentication procedures for OSPF packets but not for the Cryptographic Authentication method where the checksum calculation is omitted.

1.6.1 WELL-KNOWN OSPF IPv4 ADDRESSES

Some OSPF messages are sent via multicast and use two distinct IPv4 multicast addresses:

- **"All OSPF Routers" IPv4 Multicast Address 224.0.0.5** (*AllSPFRouters* **Address**): All OSPF routers should be capable of receiving packets sent to this IP address. In broadcast multiaccess networks such as those based Ethernet, OSPF Hello packets (see discussion below) are always sent to this IP address. Also, during the LSA flooding process, some OSPF protocol packet types are sent to this IP address.
- **"OSPF All Designated Routers" IPv4 Multicast Address 224.0.0.6** (*AllDRouters* **Address**): Both the Designated Router and BDR in a network segment must be capable of receiving OSPF packets sent to this IP address. Some OSPF packet types are also sent to this IP address during the LSA flooding process.

OSPF packets sent to these IP multicast addresses are meant to travel over a single router hop only and should never be forwarded by any receiving router. To ensure that OSPF packets sent by a router will be limited to only a single-hop travel, their IP TTL (Time-to-Live) field values are set to 1.

1.6.2   OSPF HELLO PACKET

OSPF routers use Hello packets (Type 1 packets) to discover, establish, and maintain adjacencies with their OSPF neighbors (see Figure 1.12). Basically, OSPF routers use Hello packets as a form of "greeting" to allow them "to begin a conversation" and discover other neighbor routers on their local links and networks. The OSPF Hello Protocol forms a key part of the OSPF routing protocol, and is used to establish and maintain neighbor relationships including bidirectional communication between neighbors.

Each OSPF router periodically transmits Hello packets on all of its interfaces including Virtual Links to validate neighbor reachability, and to establish and maintain neighbor router relationships. On broadcast multiaccess and P2P networks, OSPF routers use the Hello Protocol to dynamically discover neighbor routers. On broadcast and non-broadcast multiple access (NMBA) network types, OSPF routers use Hello packets also to elect the Designated Router (DR) and BDR for the network segment.

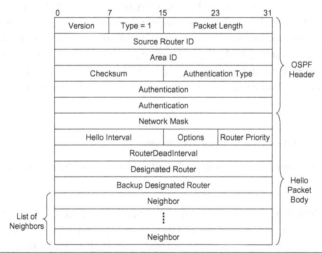

Field	Meaning
Network Mask (32 Bits)	Specifies the network mask associated with the router interface handling the Hello packet
Hello Interval (16 Bits)	Specifies the length of time, in seconds, the router has to wait before sending the next Hello packet out of the interface (how often Hello packets are sent out)
Options (8 Bits)	Specifies the optional capabilities supported by the router (see RFC 2328, Section A.2)
Router Priority (8 Bits)	Specifies the priority of the router, a parameter used in the election of the (Backup) Designated Router. If set to 0, the router will be ineligible to become a (Backup) Designated Router.
RouterDeadInterval (32 Bits)	Specifies the length of time (in seconds) the router has to wait for a Hello packet from a neighbour before declaring it as unavailable/unreachable
Designated Router (32 Bits)	Identifies the Designated Router for the network associated with the router sending the Hello packet. The sending router identifies the Designated Router by its IP interface address on the network. If set to 0.0.0.0, then there is no Designated Router.
Backup Designated Router (32 Bits)	Identifies the Backup Designated Router for the network associated with the router sending the Hello packet. The sending router identifies the Backup Designated Router by its IP interface address on the network. If set to 0.0.0.0, then there is no Backup Designated Router.
Neighbor (32 Bits)	Each of these fields specifies the Router ID from whom valid Hello packets have been received recently on the network, that is, within the last RouterDeadInterval seconds

FIGURE 1.12   OSPF Hello Packet Format

1.6.2.1   Communicating Key Parameters

The OSPF routers discover neighbors and communicate key parameters about how to establish adjacency relationships, as well as, how to communicate and exchange routing information with those discovered routers in the OSPF network. To establish an adjacency, two OSPF routers (or peers) at both ends of the link connecting them, or on the same network segment, must agree on a number of parameters contained in the Hello packets they exchange (e.g., Network Mask, Hello Interval, and *RouterDeadInterval*, as shown in Figure 1.12).

1.6.2.2   The Network Mask Field

The Network Mask field in a Hello packet contains the IP address mask of the network attached to a router interface. This field is set to 0.0.0.0 when the router interface connects to an unnumbered P2P link or a Virtual Link. Each router also maintains a number of timers that it uses to determine if neighbor routers are still reachable. When an OSPF router receives a Hello packet before a specific timer expires, it will reset it. But when no Hello packets are received before the timer expires, it will declare the neighbor router as unreachable.

1.6.2.3   The Neighbor Field

The Neighbor field in the Hello packet (Figure 1.12) contains the list of all neighbor routers on the network segment from which Hello packets have been received recently. A router learns the OSPF Router ID of a neighbor router (i.e., the Neighbor ID) when it receives Hello packets from that neighbor, or when Router ID is configured during the setup of a virtual adjacency.

Two routers become neighbors and establish bidirectional communication as soon as each one sees its own Router ID listed in the neighbor's Hello packet. This is a critical step toward guaranteeing a two-way communication between neighbors.

Only the primary IPv4 address of an interface can be used in neighbor negotiation. A router interface can be configured with multiple secondary IPv4 addresses but these have to belong to the same OSPF area as the configured primary IPv4 address.

Establishing an adjacency is the next step after the OSPF routers complete the neighbor negotiation process. Routers that have established an adjacency go beyond the simple process of exchanging Hello packets and proceed into the process of exchanging LSDBs. Routers that have established a full adjacency will have the same LSDBs (at the "Full State" of the OSPF neighbor formation process, see appropriate discussion below).

1.6.2.4   The Hello Interval and Router Dead Interval Fields

OSPF Hello packets carry two timer values (the Hello Interval and the *RouterDeadInterval* [or equivalently, Router Dead Interval]) that routers use to check the availability/reachability of neighbor routers (see Figure 1.12). Hello packets with these timer values serve as keepalive messages that an OSPF router sends with the purpose of announcing its presence to neighbors, or receives to acknowledge the presence of neighbors in the network.

The Hello Interval (and associated timer) specifies (in seconds) how often the OSPF router sends Hello packets on an interface. The Router Dead Interval (and timer) specifies how long the router has to wait (when it has not received a Hello packet from a neighbor) before declaring that neighbor router as down or unreachable. A smaller setting of the OSPF Hello Interval value results in faster detection of network topology changes, but has the downside of creating more routing information traffic on the link.

The setting of both the Hello Interval and *RouterDeadInterval* must be the same for all OSPF routers attached to a common network segment for them to be able to establish a full adjacency, elect the DR/BDR on the network segment, as well as, exchange link-state information and synchronize their LSDBs **[TEAREDIA15]**. The default value of the Hello Interval on Virtual Links, P2P links, and broadcast multiaccess networks is 10 seconds. The corresponding default setting is 30 seconds on all other OSPF network types (including NBMA and P2MP networks).

The default Router Dead Interval value on Virtual Links, P2P links, and broadcast multiaccess network is four times the Hello Interval, that is, equal to 40 seconds. The corresponding value is 120 seconds for all other OSPF network types (NBMA and P2MP networks). In Cisco System routers, when the Hello Interval is configured, the default setting of the Router Dead Interval is automatically adjusted accordingly to four times the Hello Interval **[TEAREDIA15]**.

The default OSPF timer values can be decreased on low-speed links in order to achieve faster network convergence. However, lowering the OSPF Hello Interval too much may result in increased network traffic overhead in the form of more frequent routing updates over the low-speed link. The higher routing update rate (although intended to achieve faster convergence) in turn, can cause more routing traffic to go over the link, in addition to, higher router CPU utilization.

1.6.2.5 The Options Field

The Options field in a Hello packet carries subfields (or bits) that describe the optional OSPF capabilities of a router **[RFC2328] [RFC5250]**. For example, the E-bit in the Options field is set if the OSPF area attached to the router interface is capable of processing AS-External-LSAs (i.e., if the OSPF area is not a Stub Area). The setting of the E-bit must match the "*ExternalRoutingCapability*" parameter of the OSPF area **[RFC2328]**.

The *ExternalRoutingCapability* parameter indicates whether an AS-External-LSA will be advertised into/throughout an OSPF area. Neighbor routers will not accept a Hello packet if the E-bit is set incorrectly (i.e., if there is a mismatch in the setting for the attached area). Also, bits in the Hello packet's Options field that are unrecognized by a router must be set to zero.

In addition to Hello packets, OSPF Database Description packets and all OSPF LSAs also contain the OSPF Options field. OSPF routers that encounter unrecognized Option bits in these packet types or in LSAs, must ignore the capability indicated, and process the packet or LSA normally.

1.6.2.6 The Router Priority, Designated Router, and Backup Designated Router Fields

The Designated Router field and BDR field in the Hello packet contain, respectively, a router's current choice for DR and BDR. If any of these fields contain a value of 0.0.0.0, then it means that no corresponding Designated Router has yet been selected.

The Router Priority field value (Figure 1.12) is one of the parameters used in the selection of the DR for a network segment. On the other hand, it must be noted that any two OSPF neighbor routers interconnected over a Virtual Link, P2P and Point-to-Multipoint (P2MP) network, automatically become adjacent whenever they can communicate directly. These network types do not require the election of a DR and BDR.

1.6.2.7 Sending Hello Packets on Broadcast and Non-Broadcast Network Segments

OSPF routers that share a common link or network segment (e.g., Ethernet LAN) can become neighbors. Neighbors are detected via the Hello protocol which also allows OSPF routers to send Hello packets periodically out each of their interfaces:

- On P2P, broadcast multiaccess, and P2MP (broadcast) networks, Hello packets are sent to the "All OSPF Routers" IPv4 multicast address 224.0.0.5.
- On Virtual Links, Hello packets are sent as unicasts and are addressed directly to the router (i.e., ABR) at the other end of the link.
- On P2MP (non-broadcast) networks, Hello packets are sent as unicast to each attached neighbor router.

On NMBA networks, static/manual configuration of connections between neighbors may be necessary in order for the Hello Protocol to function. Every router attached to the network that is eligible to be elected the DR must be aware of all of its neighbors on the network. Thus, each router is aware of the DR eligibility of its neighbor. The Interface State of a NBMA router interface must be at least in the "Waiting" State for it to send out any Hello packets [RFC2328]. Also, the Hello packet sending-behavior of the router varies and depends on whether the router itself is eligible to become the DR. A router sends Hello packets directly (as unicasts) to some subset of its neighbor routers on the NBMA network. Sometimes the router sends a Hello packet periodically at intervals determined by a timer, and at other times, it sends a Hello packet as a response to a received Hello Packet:

- If the router is eligible to be elected DR on the NBMA network, it must send out Hello packets periodically to all neighbor routers that are also eligible to be DR. Also, if the router itself is the current DR or BDR, it must also send out Hello packets periodically to all other neighbor routers. This means that any two routers on the NBMA network that are eligible to become DR always exchange Hello packets, a condition which is necessary for the correct operation of the DR election algorithm.
- If the router is not eligible to be elected DR on the NBMA network, it must send out Hello packets periodically to both the current DR and the BDR (if they exist). The (non-eligible) router must also send a Hello packet in reply to a Hello packet received from any neighbor that is eligible to become DR (except the current DR and DR). This additional requirement is needed to allow the router to establish an initial bidirectional communication/relationship with any potential DR.
- When a router on the NBMA network sends out Hello packets periodically to any neighbor router, the neighbor's state determines the interval between

sending Hello packets. If the neighbor is in the "Down" State, it will send Hello packets every *PollInterval* seconds. Otherwise, the router will send Hello packets every Hello Interval seconds. If a neighbor router has been declared unreachable or inactive (because Hello packets have not been received from it for *RouterDeadInterval* seconds), it may still be necessary for the other routers in the NBMA network to send Hello Packets to this inactive neighbor. Routers in this case send Hello packets at the reduced rate which is referred to as the *PollInterval*, which is set much larger than the Hello Interval.

1.6.3   OSPF DATABASE DESCRIPTION PACKET

After any two OSPF routers have established a neighbor adjacency, each will send Database Description packets (Type 2 packets) describing its LSDB, thus, allowing the other router to determine whether its local LSDB is in sync with the sender's own LSDB. Each Database Description packet sent describes a set of LSAs in the router's LSDB. Essentially, each carries pieces of the descriptions of the network topology of the OSPF area shared by the two neighbor routers as described in the sender's current LDSB (Figure 1.13).

OSPF routers use Database Description packets to convey to other routers the contents of their LSDBs for the area in which they belong; Backbone or Non-Backbone Areas. Each LSA listed in a router's LSDB is described by an LSA header. The LSA header is defined to have structure that contains all the information required for an OSPF router to uniquely identify a particular LSA and, also, its current instance. Therefore, as shown in Figure 1.13, each Database Description packet contains a number of individual LSA headers in the sender's LSDB. Specifically, Database Description packets contain summaries of the LSAs in the originating router's LSDB, and these are exchanged between neighbor routers to ensure that their LSDBs are synchronized. OSPF Database Description packets are similar to IS-IS Complete Sequence Number PDUs (CSNPs) as described in Chapter 2.

The neighbor routers exchange Database Description packets only during the initial or first LSDB synchronization they perform between themselves, that is, only during the ExStart and Exchange phase of the OSPF Neighbor State (see "OSPF Neighbor State" section below). After achieving this initial LSDB synchronization, the neighbors (as well as all other routers in the OSPF area that have synchronized LSDBs), will flood and acknowledge new LSAs, as well as, refresh and reflood LSAs without the need for exchanging more Database Description packets. Once the neighbors have transitioned past the ExStart and Exchange States, no more Database Description packets are used.

1.6.3.1   Rationale Behind the Use of Database Description Packets

After an OSPF router has established bidirectional communication with a neighbor, it will exchange Database Description packets with the neighbor that contain headers of individual LSAs in its LSDB. Basically, every record (LSA) in an OSPF LSDB is uniquely identified by its LSA header information. In theory and naively, if a router needs to compare its local LSDB with that of its neighbor, this would require the

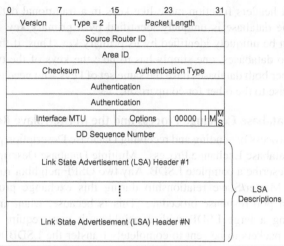

Field	Meaning
Interface MTU (16 Bits)	Specifies the size (in bytes) of the largest IP packet that can be sent out the associated interface, without fragmentation. The router must set the Interface MTU to 0 in Database Description packets sent over virtual links.
Options (8 Bits)	Specifies the optional capabilities supported by the router (see RFC 2328, Section A.2)
I-bit	This is the Init bit and when set to 1, indicates that this packet is the first in a sequence of Database Description Packets
M-bit	This is the More bit and when set to 1, indicates that more Database Description Packets are to follow
MS-bit	This is the Master/Slave bit and when set to 1, indicates that the router is the master during the Database Exchange process, otherwise, the router is the slave.
DD Sequence Number (32 Bits)	This is a value used to sequence a collection of Database Description (DD) packets. The router sets the initial value (indicated by the Init bit being set) to a unique value. The router then increments the DD sequence number until the complete database description has been sent.
LSA Header (20 Bytes) (Variable Number of LSA Headers)	The rest of the DD packet consists of a (possibly, partial) list of the LSA information pieces associated with each LSA in the database described by an LSA header. The LSA header contains all the information required to uniquely identify both the LSA and the LSA's current instance.

FIGURE 1.13   OSPF Database Description Packet

neighbor to transfer its entire LSDBs, something that is cumbersome, inefficient, and in many cases not feasible.

However, providing each router with just the neighbor's LSA headers (rather than the neighbor's entire LSDB) will allow each router to check whether both LSDBs have the same set of LSA headers. If a router detects that some LSA headers are missing, it will simply download those missing LSAs which are identified by the missing LSA headers. OSPF Database Description packets and IS-IS CSNPs were developed with this function in mind.

Thus, when any two routers want to synchronize their LSDBs, they will first exchange (via Database Description packets) the list of LSA headers that uniquely identify their individual LSAs and their instances. This does away with the need for any one of the routers to transfer entire LSAs (or equivalently, the entire LSDB) when they establish an adjacency. Database Description packets allows each router to request and transfer only those LSAs that are found to be either newer or missing during the Database Exchange Process.

The LSA headers function more like keys in a traditional database where each record in the database is uniquely identified by a primary key. Each record in a database can be uniquely identified by its primary key. Thus, if there is the need to compare two databases, one simply has to take the keys of the two databases and check whether both databases have the same set of keys – no need to transfer any one entire database to the other for comparison.

1.6.3.2 Database Exchange Process and the Master/Slave Relationship

The OSPF process of sending and receiving Database Description packets is referred to as the "Database Exchange Process". Multiple Database Description packets may be used to describe a complete LSDB. Any two OSPF neighbor routers also need to establish a Master/Slave relationship during this exchange process in order to implement a poll-response procedure. This is because, when an OSPF router is synchronizing a large LSDB with a neighbor, it may require several Database Description packets to be sent to completely transfer the LSDB information to the neighbor.

1.6.3.2.1 Understanding the Master/Slave Relationship

A key requirement of the Database Exchange Process is that, Database Description packets have to be transferred reliably and in the correct sequence to neighbors. However, OSPF does not have a separate packet type that can be used to acknowledge the successful receipt of Database Description packets. Note that OSPF uses Link-State Acknowledgment packets to acknowledge the receipt of Link-State Update packets but not for Database Description packets. As with all LSA flooding operations, OSPF routers flood LSAs (carried within Link-State Update packets), and these are acknowledged using Link-State Acknowledgment packets.

Therefore, in the absence of a special acknowledgement packet type, OSPF also uses Database Description packets to acknowledge Database Description packets themselves. Also, since Database Description packets have to be received in order to be acknowledged by the receiver, they are correctly sequenced. One of the two routers involved in the exchange process (seeking to synchronize its LSDB) will be designated Master, and the one allowed to set and increment the Database Description (DD) Sequence Number.

The router designated the Slave will be required to respond to each Database Description packet received from the Master by sending its own Database Description packet with the same DD Sequence Number as indicated in the Database Description packet received from the Master. If the Slave has no link-state information to send, it will send an empty Database Description packet (i.e., one with an empty body). In such cases, the Slave still has to respond to the Master's Database Description packets using these empty packets. The Master will always set and increment the DD Sequence Number in each subsequent Database Description packet while the Slave merely responds with the same DD Sequence Number, as a way of acknowledging the receipt of that particular Database Description packet from the Master.

1.6.3.2.2 Choosing the Master

OSPF has a simple mechanism for choosing the Master. At the beginning of the Database Exchange Process, each neighbor considers itself Master, and sends

Database Description packets to each other, indicating its own initial DD Sequence Number. However, because OSPF routers are assigned unique Router IDs, the router with the higher Router ID will become the Master while the one with the lower Router ID will revert to playing the role of Slave. This allows the Master to assert its own sequencing of Database Description packets.

The ExStart State is the phase of the OSPF Neighbor State (see "OSPF Neighbor State" section below) where the routers exchange Database Description packets and determine the Master/Slave roles. After the ExStart State, the routers transition to the Exchange State, where they send and receive Database Description packets and build a list of LSAs to be requested from the other neighbor.

During the Loading State, a router requests the missing LSAs from the neighbor, and when it receives all missing LSAs, its Neighbor State changes to the Full State. In the Full State, the routers do not exchange anymore Database Description packets because at this point, their LSDBs are identical, and routers can advertise incrementally all subsequent changes to the LSDBs by individually sending and acknowledging LSAs using the Link-State Update and Link-State Acknowledgment packets, respectively.

1.6.3.2.3   The I (Initialize), M (More), and MS (Master/Slave) Bits

The Master is the router to send the first/initial Database Description packet (which takes place in the ExStart State). The way Database Description packets are sent depends on the Neighbor State discussed in the "Neighbor Discovery and Maintenance" section below. This section discusses in detail the Database Exchange Process. In the ExStart State, the Master declares itself as Master by setting the MS (Master/Slave) bit to 1, and starts sending empty Database Description packets to the Slave, with the I (Initialize), M (More) and MS bits set. The Master retransmits these Database Description packets at intervals of *RxmtInterval* until the next Neighbor State (Exchange State) is entered (see "OSPF Neighbor State" section below).

In the Exchange State, the Database Description packets sent contain actual summaries of the link-state information contained in the sending router's LSDB. Each LSA in the LSDB of the sender's OSPF area (at the time the neighbor router transitions into the Exchange State) is listed in the Database Summary List associated with the neighbor. The Database Summary List contains the complete list of LSAs that are in the area LSDB of the sender at the time the sender enters the Exchange State with the neighbor. In the Exchange State, the determination of when a neighbor would send a Database Description packet depends on whether the neighbor is designated the Master or Slave.

1.6.3.2.4   Database Description (DD) Sequence Number Field

The Master sends Database Description packets (i.e., the polling process) to the Slave which then acknowledges them by sending its own Database Description packets to the Master (i.e., the response process). The response packets the Slave sends are linked to the poll packets sent by the Master via the DD Sequence Numbers carried in the Database Description packets. The sending router (designated as Master) will send Database Description packets (in the proper sequence each carrying a DD Sequence Number [Figure 1.13]) which must be explicitly acknowledged by the receiving router (designated as Slave).

Each time the Master sends a new Database Description Packet, it copies the packet's DD Sequence Number from the data structure associated with the Slave, and then describes the current top of the Database Summary List. The Master removes items from the Database Summary List when the Slave acknowledges the previous Database Description packet.

1.6.3.2.5 The M (More) Bit

As discussed above, after two routers have established an adjacency, they synchronize their LSDBs through the Database Exchange Process. Each router sends a set of Database Description packets to the other that describes its current LSDB. We have noted above that the sending router will list a set of LSA headers in its LSDB in the Database Description packets it sends. By going through this process, each router is able to determine whether the other router has missing or more recent LSA instances that it should request for via Link-State Request packets. A router considers the Database Exchange Process to be complete when it has received and sent a Database Description Packet with the M-bit set to 0.

1.6.3.3 The Options Field

A router can learn a neighbor's OSPF Options (also carried in Database Description packets) during the Database Exchange process (see Options field in Figure 1.13). The advertised Options described the optional OSPF capabilities supported by the neighbor router **[RFC2328] [RFC5250]**. For example, the E-bit in the Options field should be set (to 1) if the router interface attaches to an OSPF area that is not a Stub Area and as a result can send and receive AS external link-state advertisements on that interface.

A router can also learn a neighbor's optional OSPF capabilities by examining the Options field carried in Hello packets received from the neighbor (see Options field in Figure 1.12). This enables a router to reject received Hello Packets if it detects a mismatch in certain crucial OSPF capabilities (i.e., the router will not even start to form a neighbor relationship).

1.6.3.4 The Interface MTU Field

OSPF does not have any built-in packet fragmentation and reassembly mechanisms. An OSPF router specifies an IP MTU in Database Description packets (see Figure 1.13) to indicate to neighbor routers the maximum sized IPv4 packet that it can forward out the interface without fragmentation. In normal IPv4 fragmentation, if a packet with size larger than the specified IPv4 MTU arrives at the router interface, it will be discarded (if the "Don't Fragment (DF)" bit is set in the IPv4 packet header Flags field), otherwise it will be fragmented. On the other hand, IPv6 does not support the fragmentation of IPv6 packets as in IPv4, except by the source of the IPv6 packet.

OSPF for IPv4 (i.e., OSPFv2) completely relies on the fragmentation capabilities of IPv4 (if present) for the possible fragmentation of packets. RFC 2328 does not recommend the fragmentation of OSPF packets; however, there can be situations where this might be necessary, that is, when the size of the OSPF packet is larger than the IPv4 MTU of the router interface. The downside to this is that, if two OSPF neighbors have mismatched MTUs and fragmentation is allowed, this could cause some problems when they exchange link-state packets, most likely resulting in endless retransmissions of OSPF packets.

So, to prevent this from happening, OSPF best practice recommends that both sides of the link connecting the two neighbors be configured with the same IPv4 MTU. Otherwise, if the two OSPF neighbors are configured with mismatched IPv4 MTUs, they will not be able to establish a full adjacency (i.e., transition from the OSPF Down State to the Full State) and will be stuck in the *ExStart* State (see OSPF Neighbor States below). The MTU mismatch will result in the OSPF neighbor routers not being able to exchange full link-state information and synchronize their LSDBs. To form a full adjacency, the OSPF neighbors must have matched IPv4 MTU on both sides of the link.

1.6.4   OSPF LINK-STATE REQUEST PACKET

After exchanging Database Description packets with a neighbor router in order to carry out LSDB synchronization, a router might find out that it is still missing some link-state information (or list of LSAs) in its LSDB, or parts of the database is out-of-date. The router in this case will send Link-State Request packets (Type 3 packets) to inform its OSPF neighbors to send the most recent version of the missing or out-of-date information (see Figure 1.14). OSPF routers use Link-State Request packets

0	7	15	23	31
Version	Type = 3		Packet Length	
Source Router ID				
Area ID				
Checksum		Authentication Type		
Authentication				
Authentication				
LS Type				
Link State ID				
Advertising Router				
⋮				
LS Type				
Link State ID				
Advertising Router				

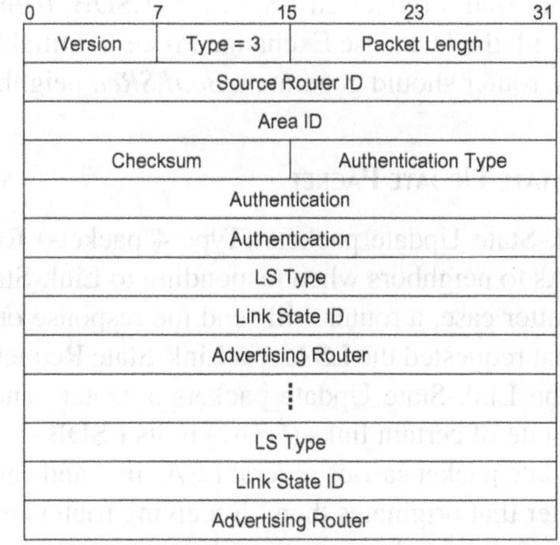

Field	Meaning
LS Type (32 Bits)	Specifies the type of the LSA. The LSA types defined in RFC 2328 are: 1 = Router-LSAs; 2 = Network-LSAs; 3 = Summary-LSAs (IP network); 4 = Summary-LSAs (ASBR); 5 = AS-external-LSAs. Each LSA type has a separate advertisement format.
Link State ID (32 Bits)	Identifies the portion of the network environment being advertised by the LSA. The actual contents of this field depend on the LSA's LS Type: **LS Type Link State ID** 1 The originating router's Router ID. 2 The IP address of interface of the network's Designated Router. 3 The destination network's IP address. 4 The Router ID of the described Autonomous System Boundary Router (ASBR). 5 The destination network's IP address.
Advertising Router (32 Bits)	Specifies the Router ID of the router that originated the LSA. In Router LSAs, this field is identical to the Link State ID field. In network-LSAs (LS Type 2 LSAs), this field is set to the Router ID of the network's Designated Router. Summary LSAs are originated by Area Border Routers (ABRs). AS external LSAs are originated by ASBRs.

FIGURE 1.14   OSPF Link-State Request Packet

to request updated information about a portion of the LSDB in which they have interest.

When an OSPF router detects that a portion of its LSDB is out of date, it will send a Link-State Request packet to a neighbor asking for a precise instance of the LSDB. The router may use multiple Link-State Request packets if necessary. The Link-State Request packets consist of an OSPF header plus additional fields (as shown in Figure 1.14) that uniquely identify the LSDB information that the sending router is seeking.

Each LSA requested is identified by its LS Type, Link-State ID, and Advertising Router (see LSA header format below). These parameters uniquely identify a particular LSA, but not its instance. Instead, each LSA instance is uniquely defined by its LS Age, LS Sequence Number, and LS Checksum. Although the Link-State Request packet itself does not specify these fields, the Link-State Request packets a router sends are understood to requests for the most recent LSA instance the receiver has. This means, the router sending the Link-State Request packet may end up receiving in response even more recent LSA instances.

The router that receives a Link-State Request packet will locate each LSA specified in the packets in its LSDB, and copy these into Link-State Update packets for transmission to the requesting neighbor. The responding router should not place these LSAs in its Link-State Retransmission List for the requesting neighbor. If the responding router cannot find a requested LSA in its LSDB, then something obviously has gone wrong with the Database Exchange Process (initial LSDB synchronization process), and the router should generate a *BadLSReq* neighbor event.

1.6.5   OSPF LINK-STATE UPDATE PACKET

OSPF routers use Link-State Update packets (Type 4 packets) for flooding LSAs, and also to convey LSAs to neighbors when responding to Link-State Request packets from them. In the latter case, a router will send the response only to the directly connected neighbors that requested the LSAs via Link-State Request packets (Figure 1.15). In both cases, the Link-State Update packets a router sends, carry updated information about the state of certain links (LSAs) in its LSDB.

Each Link-State Update packet carries several LSAs in it and can only be sent one hop away from the router that originates them. Receiving routers use the contents of the Link-State Update packets (where each packet can contain possibly multiple LSAs) to update the information in their LSDBs.

An OSPF router will multicast (flood) Link-State Update packets on network segments that support multicast or broadcast capabilities (e.g., Ethernet networks). A router that receives Link-State Update packets must acknowledge all of them. If retransmission is required, the sending router must retransmit the requested LSAs in unicast mode. Each Link-State Update packets consist of an OSPF header plus other fields that include the number of LSAs carried in the packet, the Link-State ID, and the LSAs themselves. The use of Link-State Request and Link-State Update packets can be explained using Figure 1.16.

When a neighbor router needs to flood LSAs that it has received from other routers, it is responsible for the re-encapsulation of those LSAs in new Link-State Update

0	7	15		31
Version	Type = 4		Packet Length	
Source Router ID				
Area ID				
Checksum			Authentication Type	
Authentication				
Authentication				
Number of LSAs				
Link State ID				
LSA Bodies ⋮				

FIGURE 1.15   OSPF Link-State Update Packet

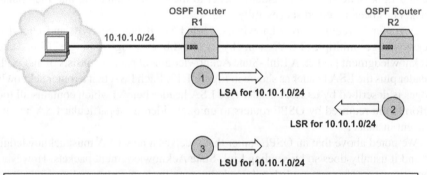

When two OSPF neighbour routers exchange routing information, they send to each other a list of all LSAs in their respective LSDB. Each router then checks its LSDB andsends a Link State Request (LSR) packet requesting all LSAs not found in its LSDB to the other router. The receiving router responds with a Link State Update (LSU) packet that contains all LSAs requested by the other neighbor:

1. Router R1 advertises an LSA for its directly connected network 10.10.1.0/24.
2. Router R2 check its LSDB and determines that it is missing routing information about that network. Router R2 then sends a LSR packet requesting further information about network 10.10.1.0/24.
3. Router R1 responds with LSU packet which contains information about network 10.10.1.0/24.

FIGURE 1.16   Using OSPF Link-State Update Packet

packets before flooding them. Thus, OSPF routers send Link-State Update packets as a way of communicating OSPF routing updates throughout the OSPF area.

1.6.6   OSPF LINK-STATE ACKNOWLEDGMENT PACKET

OSPF routers use Link-State Acknowledgment (LSAck) packets (Type 5 packets) to explicitly acknowledge the receipt of LSAs in order to make the flooding of LSAs

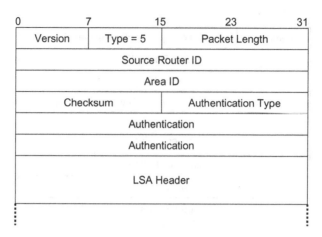

0   7   15   23   31

Version	Type = 5	Packet Length
Source Router ID		
Area ID		
Checksum	Authentication Type	
Authentication		
Authentication		
LSA Header		

FIGURE 1.17   OSPF Link-State Acknowledgment Packet

reliable (Figure 1.17). An OSPF router that receives a Link-State Update packet will send a Link-State Acknowledgment packet to the sender indicating that the update packet has been received successfully.

A router is required to acknowledge each LSA it has received, and can also acknowledge multiple LSAs sent by a neighbor using a single Link-State Acknowledgment packet. A Link-State Acknowledgment packet consists of an OSPF header plus the LSA header as shown in Figure 1.17. Each LSA that a router acknowledges is described by its LSA header (see LSA header below) which contains all the information required by OSPF routers to uniquely identify a particular LSA and its current instance.

We noted above that an OSPF router that receives a new LSA must acknowledge it, and it usually does so by sending Link-State Acknowledgment packets. However, the router can also accomplish acknowledgments by implicitly sending Link-State Update packets to the sending router. The router can also acknowledge LSAs by grouping multiple acknowledgments together into a single Link-State Acknowledgment packet and sending it back out the interface on which the LSA was received. The router can send the Link-State Acknowledgment packet in one of two ways: delay and send it on an interval timer, or send it directly to the particular neighbor. The precise strategies the router uses to acknowledge LSAs depend on the circumstances under which the LSA was received are described in detail in **[RFC2328]**.

When a router uses a delayed acknowledgments strategy, it accomplishes the following:

1. It allows the router to group multiple acknowledgments into a single Link-State Acknowledgment packet.
2. It enables the router to use a single Link-State Acknowledgment packet and multicast transmission to send acknowledgments to several neighbors at the same time.
3. It allows routers attached to a common network segment to randomize the transmission of Link-State Acknowledgment packets. The router must space

the delayed transmissions by a fixed interval that must be less than *RxmtInterval* (which specifies the number of seconds between the retransmissions of LSAs), otherwise unnecessary retransmissions may result.

The router sends delayed acknowledgments to all adjacent routers associated with the router interface on which the LSA was received. On broadcast multiaccess networks (such as Ethernet), the router accomplishes this by sending the delayed Link-State Acknowledgment packets as multicasts. The destination IP address carried in the multicast packets depends on the state of the router interface transmitting the packets.

If the router Interface State indicates that the router is a Designated Router or BDR, then the multicast packets will use as destination address the "All OSPF Routers" IPv4 multicast address 224.0.0.5 (i.e., *AllSPFRouters* address). In all other router Interface States, the multicast packets will use as destination address the "OSPF All Designated Routers" address 224.0.0.6 (i.e., *AllDRouters* address). On non-broadcast networks, the router will send delayed Link-State Acknowledgment packets as unicasts separately to each adjacent router (i.e., to a neighbor router whose Neighbor State is greater than or equal to the Exchange State).

A router will send acknowledgments directly to a particular neighbor router in response to the receipt of duplicate LSAs, and will send such acknowledgments immediately when the duplicate LSA is received. On multiaccess networks, the router will send the acknowledgments directly to the neighbor as a unicast using the neighbor's IP address.

When a router receives a Link-State Acknowledgment packet with one or more acknowledgments for LSAs that were sent, it has to make a number of consistency checks before handing over control to the local LSA flooding process to continue with its tasks. The router has to determine first which particular neighbor the packet is associated with. If the neighbor is in a Neighbor State that is less than the Exchange State, the Link-State Acknowledgment packet will be discarded. If the neighbor is in any other state, the router will perform the following steps for each acknowledgment in the Link-State Acknowledgment packet:

- The router will check if the LSA that is acknowledged has an instance in the neighbor's Link-State Retransmission List. If no LSA instance exists, the router will examine the next acknowledgment in the packet (if any), otherwise it will perform the following:
  o If the router determines that the acknowledgment is for the same LSA instance as contained in the Link-State Retransmission List, it will remove that instance from the list, and then examine the next acknowledgment in the packet (if any). Otherwise, the router will perform the following:
    - It will log this acknowledgment which it will deem as questionable, and then examine the next acknowledgment, if any.

1.6.7 LSA FORMATS

OSPF routers in an area must know the detailed topology of the OSPF area in order to be able to compute best routes to network destinations. The LSA is the basic messaging unit for communicating routing information among routers in an area and in the OSPF autonomous system. LSAs are used to describe the topology details of an OSPF network and also form the building blocks of the LSDB each OSPF router maintains. Basically, each LSA serves as database record in the LSDB that contains pieces of network topology information (i.e., link-state information about a portion of the network). The LSAs in the LSDB, collectively, describe the entire topology of an area in the OSPF autonomous system.

OSPF routers use different LSA types to exchange information about the topology of an OSPF area and autonomous system. Each router originates and also receives LSAs and then stores them in its local LSDB. After routers in an area have synchronized their LSDBs, each OSPF router then uses the SPF algorithm to calculate the best routes to each known network destination. Each OSPF router determines the routes with the total lowest cost to a destination and installs these as the best routes in the IP Routing Table.

The best routes within an OSPF area (i.e., intra-area routes) are computed individually by each router within that area (i.e., OSPF Internal Router). However, each Internal Router must rely on the best path information received from the ABRs connected to its area to determine/calculate the best inter-area routes (i.e., routes leading to destinations outside the local area).

The format of the common LSA header is described in Figure 1.18. Every LSA an OSPF router sends out begins with this common 20-byte header. The LS type, Link-State ID and Advertising Router fields in the LSA header, in combination, uniquely identifies an LSA. The Link-State ID in an LSA takes on the values shown in Figure

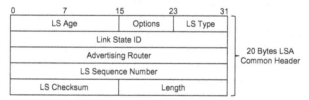

Field	Meaning
LS Age (16 Bits)	Specifies the age of the LSA (in seconds) since it was originated.
Options (8 Bits)	Specifies the optional capabilities supported by the router (see RFC 2328, Section A.2)
LS Type (8 Bits)	Specifies the type of LSA.
LS State ID (32 Bits)	Identifies the portion of the network environment being advertised by the LSA. The actual contents of this field depend on the LSA's LS Type
Advertising Router (32 Bits)	Specifies the Router ID of the router that originated the LSA. In router-LSAs (LS Type 1), this field is identical to the Link State ID field. In network-LSAs (LS Type 2), this field is set to the Router ID of the network's Designated Router. Summary LSAs (LS Type 3) and ASBR summary-LSAs (LS Type 4) are originated by Area Border Routers (ABRs). AS-external-LSAs (LS Type 5) are originated by AS Boundary Routers (ASBRs).
LS Sequence Number (32 Bits)	This is used to number and sequence successive instances of an LSA originated by a router. The successive LS sequence numbers (given to successive instances of an LSA) are used to detect old or duplicate LSAs
LS Checksum (16 Bits)	Carries the Fletcher checksum (see RFC 905, Annex B [RFC905]) of the complete contents of the LSA, including the LSA header but excluding the LS Age field.
Length (16 Bits)	Specifies the length (in bytes) of the LSA including the 20 byte common LSA header.

FIGURE 1.18   LSA Header

1.14 for the Link-State Request packet format depending on the LSA's LS Type. All OSPF packet types (except Hello packets) deal with important OSPF tasks that are related to the sending and/or receiving of LSAs.

When an LSA is originated, its LS Age is set to 0 and is incremented by *InfTransDelay* at every OSPF hop/router in the LSA flooding process. A router will also age LSAs as they are held in its LSDB. The LS Age is never incremented beyond the value *MaxAge*, and LSAs with age greater than *MaxAge* are not used in the calculation of best paths for the Routing Table. When the LS Age first reaches *MaxAge*, that LSA will be reflooded and will finally be flushed from the router's LSDB when it is no longer needed, to ensure correct LSDB synchronization.

The LS Age field is excluded from the LS Checksum (also referred to as the Fletcher Checksum) calculation so that the LS Age of an LSA can be incremented without having to update/recalculate the LS Checksum accordingly. The LSA header of an LSA also contains the Length field which indicates the total number of bytes in the LSA including the header. So, subtracting the size of the LS Age field (which is two bytes) from the Length field value gives the amount of data bytes over which the LS Checksum is calculated.

The LS Checksum is used to detect if any data in an LSA has been corrupted while the LSA is being flooded or held in a router's memory. The LS Checksum field value cannot be zero, a value that is considered a checksum failure. Also, the LS Checksum calculation for LSAs is not optional, and must be performed/verified when (a) a router receives a Link-State Update packet, and (b) during the time a router ages LSAs in the LSDB.

The scope of some of the LSA types are summarized as follows (Figure 1.19):

- Types 1 LSAs (Router-LSAs) are generated by each Internal Router within an OSPF area (generated per area link), and are flooded between routers within

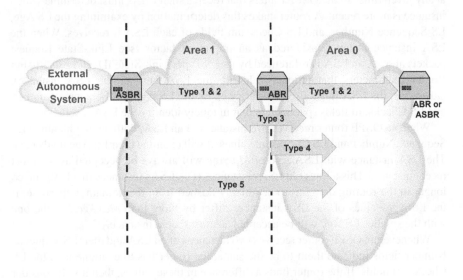

FIGURE 1.19   Scope of the Most Common OSPF LSA Types

that single OSPF area. Every router in an OSPF autonomous system originates a Router-LSA.

- Type 2 LSAs (Network-LSAs) are generated by a Designated Router on a multiaccess network segment, and the information they carry indicates all OSPF routers connected to that particular network segment. They are flooded only within the single OSPF area to which the Designated Router belongs.
- Type 3 LSAs (Summary-LSAs) are originated by ABRs to describe inter-area routes/destinations and are exchanged between OSPF areas via their ABRs.
- Type 4 LSAs (ASBR-Summary-LSAs) are flooded in an area by ABRs to inform all routers in their respective OSPF areas on how to reach a specific ASBR. These LSAs describes the path to a specific ASBR.
- Type 5 LSAs (AS-External-LSAs) are flooded in an entire OSPF autonomous system by an ASBR (except Stub Areas) to describe external routes (i.e., routes to destinations that are outside the OSPF autonomous system). Type 5 LSAs are the only type of LSAs that are flooded throughout an entire OSPF Autonomous System. All other LSA types are flooded only within a single OSPF area.

An OSPF router will add an LSA to its LSDB when the LSA is originated by the router itself or is received during the flooding process. The router deletes an LSA from its LSDB when any one of the following happens: an existing LSA has to be overwritten by a newer LSA instance during the flooding process; the router has originated a newer LSA instance of an LSA that the router itself has previously originated; an existing LSA has been aged out and has to be flushed from the routing domain. Whenever an LSA has been deleted from a router's LSDB, it must also be purged from the Link-State Retransmission List of all neighbor routers.

Several instances of an LSA may be travelling in an OSPF Autonomous System at any given time, so the OSPF routers that receive such LSAs must determine which instance is more recent. A router makes this determination by examining the LS Age, LS Sequence Number, and LS Checksum fields of each LSA it receives. When the LSA instance is not considered as an important factor (see Link-State Request packets above), an LSA is referenced by its LS Type, Link-State ID, and Advertising Router (parameters that uniquely identify the LSA but not the LSA instance). Otherwise, the LSA must also be referenced by the LS Age, LS Sequence Number, and LS Checksum fields (parameters that uniquely identify the LSA instance).

When an OSPF router receives two instances of an LSA, both having identical LS Sequence Number and LS Checksum values, it will examine their LS Age fields next. The LSA instance with LSA Age of *MaxAge* will always be accepted as the most recent instance. This allows old LSAs instances (i.e., LSAs instances that have stayed longer in the routing domain) to be quickly flushed from the domain. Otherwise, if the LS Age values of the LSA instances differ by more than *MaxAgeDiff*, the one with the smaller LS Age will be accepted as most recent instance.

Whenever an OSPF router receives two instances of an LSA and their LS Sequence Number fields indicate them to be the same, the router has to examine next the LS Checksum fields. If the router finds a difference in these values, then it will consider the LSA instance with the larger LS Checksum to be the most recent instance. We describe in greater detail next, the most common types of OSPF LSAs.

1.6.7.1 Type 1 LSA: Router-LSA

Every OSPF router generates Type 1 LSAs (Router-LSAs) to be flooded/exchanged only in the OSPF area in which the router belongs. The router sends Router-LSAs (Figure 1.20) to describe the state and cost of its links belonging to the area, and these LSAs are flooded only within that particular area. A single Router-LSA describes all of the router's links to the area. Each router in the area keeps copies of all Router-LSAs LSDB flooded, including those that it has generated locally. When a router has interfaces in multiple OSPF areas, that is, functioning as an ABR, its OSPF LSDB includes Router-LSAs from all these multiple areas.

One of the fields in the common 20-byte LSA header (Figure 1.18) is the Link-State ID field which is set to the originating Router ID (i.e., the Router ID of the router originating the LSA). So, an OSPF router identifies all the Router-LSAs it generates using a 32-bit Link-State ID. When a router generates a Router-LSA, it uses its own Router ID as the value for both the Link-State ID and the Advertising Router fields.

The router also sets appropriate bits (B and E bits as shown in Figure 1.20) in its Router-LSAs to indicate to other routers whether it is an ABR or ASBR. The setting of these bits enables any other OSPF router to save paths to ABRs and ASBRs in its Routing Table for the processing of Summary-LSAs and AS-External-LSAs that are sent to it later. The router sets the V bit in Router-LSA it originates for an OSPF area if the router itself is the endpoint of one or more Virtual Links and is using that OSPF area as a Transit Area. The router sets the V bit to enable other OSPF routers in the area through which the Virtual Link passes through (i.e., the Transit Area) to discover whether that area supports transit traffic.

The content of the Router-LSAs describe the kind of network a link is connected to (indicated according to the kind of network attached as specified in the Link Type field), and other configuration settings. Each router link is defined in the Link Type field as belonging to one of four types as shown in Figure 1.20. The Router-LSAs also describe the Link ID and Link Data fields that specify parameters that identify the object that the router link connects to. The Link ID identifies the entity that is attached to the other end of the link (Figure 1.20).

The Link Data field specifies extra information for each link. This field provides 32 bits of additional information for the link that in many cases is an IP address. For example, for a link connected to a transit network, numbered P2P link, or Virtual Link, the Link Data field specifies the IP address of the router interface associated with the link (which is information that is needed for the Routing Table calculation).

A link can be connected to another router (P2P), a transit network, a stub network, or it can be a Virtual Link. The transit network can be a multiaccess broadcast network (based on Ethernet), or a NMBA network segment which can consist of two or more OSPF routers. If the link is connected to a transit network (Link Type 2), then the Router-LSA also includes information about the IP address of the DR of that network segment (see Figure 1.20).

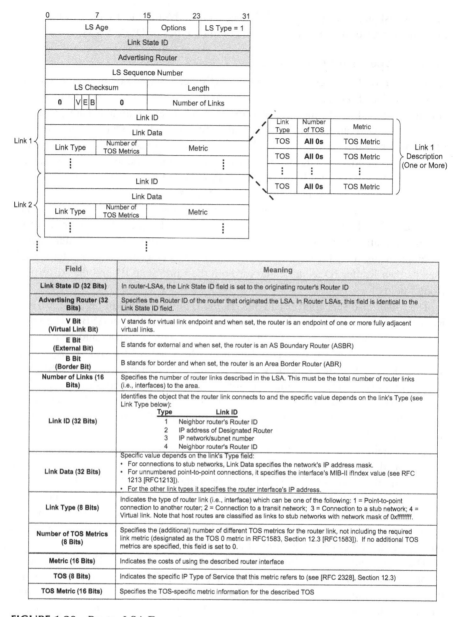

FIGURE 1.20 Router-LSA Format

Field	Meaning
Link State ID (32 Bits)	In router-LSAs, the Link State ID field is set to the originating router's Router ID
Advertising Router (32 Bits)	Specifies the Router ID of the router that originated the LSA. In Router LSAs, this field is identical to the Link State ID field.
V Bit (Virtual Link Bit)	V stands for virtual link endpoint and when set, the router is an endpoint of one or more fully adjacent virtual links.
E Bit (External Bit)	E stands for external and when set, the router is an AS Boundary Router (ASBR)
B Bit (Border Bit)	B stands for border and when set, the router is an Area Border Router (ABR)
Number of Links (16 Bits)	Specifies the number of router links described in the LSA. This must be the total number of router links (i.e., interfaces) to the area.
Link ID (32 Bits)	Identifies the object that the router link connects to and the specific value depends on the link's Type (see Link Type below): Type Link ID 1 Neighbor router's Router ID 2 IP address of Designated Router 3 IP network/subnet number 4 Neighbor router's Router ID
Link Data (32 Bits)	Specific value depends on the link's Type field: • For connections to stub networks, Link Data specifies the network's IP address mask. • For unnumbered point-to-point connections, it specifies the interface's MIB-II ifIndex value (see RFC 1213 [RFC1213]). • For the other link types it specifies the router interface's IP address.
Link Type (8 Bits)	Indicates the type of router link (i.e., interface) which can be one of the following: 1 = Point-to-point connection to another router; 2 = Connection to a transit network; 3 = Connection to a stub network; 4 = Virtual link. Note that host routes are classified as links to stub networks with network mask of 0xffffffff.
Number of TOS Metrics (8 Bits)	Specifies the (additional) number of different TOS metrics for the router link, not including the required link metric (designated as the TOS 0 metric in RFC1583, Section 12.3 [RFC1583]). If no additional TOS metrics are specified, this field is set to 0.
Metric (16 Bits)	Indicates the costs of using the described router interface
TOS (8 Bits)	Indicates the specific IP Type of Service that this metric refers to (see [RFC 2328], Section 12.3)
TOS Metric (16 Bits)	Specifies the TOS-specific metric information for the described TOS

1.6.7.2 Type 2 LSA: Network-LSA

In order to reduce the amount of routing information OSPF routers exchange on a multiaccess network segment, the routers elect one OSPF router to be a DR, and another to be the BDR. The BDR acts as a backup node sharing similar features with the DR and will take over when the DR goes down.

The election of the DR/BDR allows the routers on the network segment to have a central point of contact for exchanging routing information, instead of allowing each router to exchange routing updates with every other router on the segment. A broadcast multiaccess and an NBMA network segment that has at least two routers have to have a DR/BDR.

Every router on the network segment will exchange routing information with the DR/BDR, which will then relay that information to all other routers on the segment. Each router on the network segment is required to establish a full adjacency with only the DR/BDR. The BDR also needs to establish full adjacencies with all the other OSPF routers in the segment (including the DR).

The DR ensures that all the other routers on the segment have identical/ synchronized LSDBs. The idea of having a DR/BDR is to reduce the number of OSPF adjacencies required on a broadcast or NBMA network. This beneficially reduces the amount of routing updates exchanged over the network and the size of the LSDBs.

The other routers on the network segment also maintain a 2-Way State (which is partial-neighbor relationship) with the other non-DR or non-BDR routers on the segment. When an OSPF router joins a broadcast segment which already has an elected DR and BDR, it will proceed to the 2-Way State (of the neighbor relationship state) with all routers on the segment, including the DR and BDR. The newly joined router will proceed to establish full bidirectional adjacencies only with the elected DR and BDR.

DRs generate Type 2 LSAs (Network-LSAs) for the multiaccess network segments in which they have been elected (see appropriate discussion below on DRs). A DR will send Network-LSAs (Figure 1.21) to describe the set of OSPF routers that are attached to a particular multiaccess network segment and are fully adjacent to it (a list that includes the DR itself). Note that the DR floods the Network-LSAs only within the OSPF area in which the network segment belongs. The Link-State ID in the common 20-byte LSA header of the Network-LSA is set to the IP address of the DR's interface that originates/advertises the LSA.

Figure 1.22 shows the flow and scope of Network-LSAs in an OSPF area. The Network-LSAs are generated/originated by DRs in every transit broadcast multiaccess or NBMA network segment within the OSPF area. Recall that only the DR of the network segment is responsible for advertising Network-LSAs. A Network-LSA will list each of the OSPF routers that are connected to the network segment (including the DR itself) and the network mask of the connecting link interface. The Network-LSAs are then flooded to all routers within the OSPF area and never cross the boundary (i.e., the ABRs) of that OSPF area.

Specifically, the content of the Network-LSA describes the network segment and a list of all active nods: the IP address of the DR (in the Link-State ID and Advertising Router fields); the Router IDs of all other routers attached to the network segment and are fully adjacent (in the Attached Router fields); and the common network mask of the network segment (in the Network Mask field). This information is used by each router in the local OSPF area to construct a complete picture of the routers and links that make up that particular multiaccess network segment, a picture which cannot be fully described by exchanging just Router-LSAs.

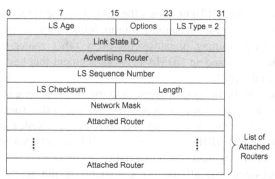

Field	Meaning
Link State ID (32 Bits)	In network-LSAs, the Link State ID is set to the IP interface address of the Designated Router
Advertising Router (32 Bits)	In network-LSAs (LS Type 2), this field is set to the Router ID of the network's Designated Router. Network-LSAs are originated by the network's Designated Router.
Network Mask (32 Bits)	Specifies the IP address mask for the network
Attached Router (32 Bits)	Each field specifies the Router ID of each of the routers attached to the network. Only routers that are fully adjacent to the Designated Router are listed. The Designated Router includes itself in this list. The number of routers carried in the can be determined from the LSA header's length field.

FIGURE 1.21   Network-LSA Format

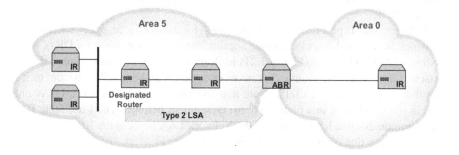

FIGURE 1.22   Flow and Scope of OSPF Type 2 LSAs (Network-LSAs)

1.6.7.3   Type 3 LSA: Summary-LSA

An ABR is a router that connects one or more areas within an OSPF autonomous system to the Backbone Area (Area 0) of that system. ABRs do not forward Router-LSAs and Network-LSAs between the OSPF areas they connect to because these LSAs are only flooded within their respective areas. However, routers in one area still need to know how to reach routers in other areas. Summary-LSAs describe inter-area routes, which are routes to destinations that are outside an OSPF area but are within the same OSPF autonomous system.

So, an ABR will learn/discover routing information in one OSPF area, and then summarizes this information for other areas in the same autonomous system (i.e., inter-area routes) using Type 3 LSAs (Summary-LSAs). An ABR will originate a single Summary-LSA for each known inter-area route it has discovered. The ABR will inject all information into the Backbone Area (Area 0) which, in turn, will pass that information to other OSPF areas.

Summary-LSAs are generated by ABRs and provide a way of disseminating network reachability information between OSPF areas:

- The summary information provided by the ABR helps to provide routing and network scalability by hiding detailed topology information of one OSPF area from other areas.
- Type 3 LSAs (Summary-LSAs) are sent to describe inter-area routes. The route is to a destination that is described as either an IP network or an IP prefix (i.e., a range of IP addresses). The route is external to the local area, but still belongs to the same OSPF Autonomous System.
- The routing information is summarized simply into a number of IPv4 address prefixes along with their routing metrics (Figure 1.23).
- Type 3 Summary-LSAs can also be used to advertise a (*per-area*) default route for OSPF Stub Areas to allow Internal Routers to reach external destinations/ routes **[RFC2328]**. Stub Areas use default summary routes to external destinations instead of requiring the flooding of a complete set of external routes into the areas.

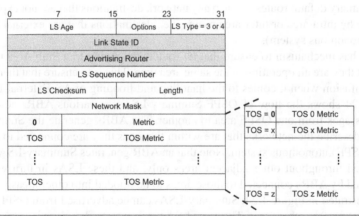

Field	Meaning
Link State ID (32 Bits)	• In summary-LSAs (LS Type 3), the Link State ID is set to the destination network's IP address • In ASBR summary-LSAs (LS Type 4), the Link State ID is set to the destination ASBR's Router ID. When the destination is an ASBR, an ASBR summary-LSA is used
Advertising Router (32 Bits)	This field specifies the Router ID of the LSA's originator: • Summary-LSAs are originated by ABRs. This field specifies the Router ID of the originating ABR. • ASBR summary-LSAs are originated by ABRs. This field specifies the Router ID of the originating ABR
Network Mask (32 Bits)	• In summary-LSAs, this specifies the destination network's IP address mask • In ASBR summary-LSAs, this field is not meaningful and must be set to zero
Metric (24 Bits)	This specifies the cost of using this route and is expressed in the same units as the interface costs in the router-LSAs
TOS (8 Bits)	Indicates the specific IP Type of Service that this metric refers to (see [RFC 2328], Section 12.3)
TOS Metric (24 Bits)	Specifies the TOS-specific metric information for the described TOS

FIGURE 1.23 Summary and ASBR-Summary-LSA Format

The Link-State ID in the common 20-byte LSA header of the Summary-LSAs (Figure 1.23) is set to the destination network's IP address, while the Network Mask field carries the associated destination network mask.

ABRs connected to a Stub Area can originate Summary-LSAs into the area, or may be configured to originate only a subset of the Summary-LSAs into the area. The benefits are, the fewer the number of LSAs originated, the smaller the size of the Stub Area's LSDB, which further reduces the demand placed on router computing and memory resources. In a Stub Area, instead of importing and advertising external (out-of-area) routes, an ABR can also advertise/inject Type 3 LSAs (Summary-LSAs) into a Stub Area to describe (per-area) default routes (i.e., via "default Summary-LSAs"). OSPF routers within the Stub Area would then use the default summary routes instead of a defined set of external routes.

To describe a default summary route, the ABR will always set the Link-State ID in the advertised Summary-LSAs to *DefaultDestination* (i.e., 0.0.0.0), the Network Mask to 0.0.0.0, and the metric to *StubDefaultCost* (which is a [per-area] configurable parameter). The *StubDefaultCost* parameter indicates the cost of the default summary route described in the Summary-LSA that the ABR advertises into the Stub Area. The *StubDefaultCost* parameter need not be configured to be identical on all of the Stub Area's ABRs.

The default summary routes are advertised only throughout that particular Stub Area, but not further. OSPF routers inside the Stub Area (Internal Routers) will use these summary default routes to reach any network destinations that are not explicitly reachable by intra-area or inter-area routes (i.e., destinations that are external to the OSPF autonomous system).

OSPF has mechanism to ensure that all routers belonging to a Stub Area understand that they are all operating in the same area type. This is to ensure that there will be no confusion when it comes to the handling and flooding of AS-External-LSAs. Figure 1.24 shows the flow of OSPF Summary-LSAs as various ABRs advertise these messages from one OSPF area to another. The ABRs generate the Summary-LSAs to describe the networks that are within the areas they are connected to in the overall OSPF autonomous system. Note that an ABR generates Summary-LSAs that are flooded throughout other adjacent areas only, and these LSAs in turn will be regenerated by ABRs connected to those areas to be flooded into other areas.

As illustrated in Figure 1.24, Summary-LSAs can be advertised from OSPF Area 3 through Area 0 to other areas. The same LSA types can also be advertised by ABRs

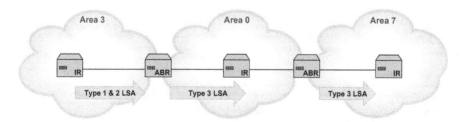

FIGURE 1.24   Flow and Scope of OSPF Type 3 LSAs (Summary-LSAs)

in the other direction, from Area 7 through Area 0 into Area 3. OSPF routers do not automatically summarize groups of contiguous IP addresses, and they also do not summarize a network to its classful IP address boundary as done in EIGRP.

An ABR may advertise a Summary-LSA into the Backbone Area for every network or subnet that is defined in the OSPF area to which it is attached, a situation which can cause excessive LSA flooding problems in larger networks. So, as OSPF best practice, network engineers generally use manual route summarization on ABRs to limit the amount of routing information that can be exchanged between OSPF areas.

1.6.7.4 Type 4 LSA: ASBR-Summary-LSA

ABRs flood Type 4 LSAs (ASBR-Summary-LSAs) to inform the rest of the routers in an OSPF autonomous system on how to get to a particular ASBR. ABRs have the task of propagating network reachability information to ASBRs connected to the OSPF autonomous system. This information indicates to routers within the OSPF autonomous system on how to reach external routes in other non-OSPF autonomous systems. Thus, an ABR will send ASBR-Summary-LSAs that carry information about the existence of a particular ASBR attached to the OSPF autonomous system. The Link-State ID in the LSA header of the ASBR-Summary-LSA (Figure 1.23) is set to the Router ID of the particular ASBR.

Figure 1.25 shows the flow and scope of ASBR-Summary-LSAs generated by an ABR when an ASBR is attached to its local OSPF autonomous system. An ASBR-Summary-LSA identifies the ASBR and indicates to routers in an area (connected to the ABR) the route to use to reach that ASBR. The ABR sets the Link-State ID is to the ASBR's Router ID. Routers sending traffic destined to an external autonomous system require knowledge of the ASBR that originated the routes to that external system.

In Figure 1.25, the ASBR will transmits a Router-LSA (Type 1 LSA) with the External (E) bit set to identify itself as an ASBR (see E bit in Figure 1.20). When the ABR (identified by the B bit in the Router-LSA) receives this LSA, it will construct an ASBR-Summary-LSA and flood/advertise it into the Backbone Area. Other ABRs in the OSPF autonomous system will then regenerate the ASBR-Summary-LSA and flood it into their respective areas to indicate routes to the ASBR. However, ASBR-Summary-LSAs are never injected into Stub Areas.

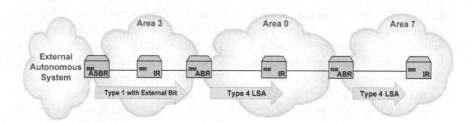

FIGURE 1.25   Flow and Scope of OSPF Type 4 LSA (ASBR-Summary-LSA)

1.6.7.5   Type 5 LSA: AS-External-LSA

For routers in an OSPF autonomous system to be able to utilize external routing information, they all have to know the paths to all the ASBRs advertising external routing information (except routers in Stub Areas). For this reason, the routes to all ASBRs connected to the OSPF autonomous system are summarized and advertised by the ABRs. An ASBR generates Type 5 LSAs (AS-External-LSAs) to inform routers within an OSPF autonomous system about routes to destinations that are external to that autonomous system.

An ASBR will originate a single AS-External-LSA for each known external route it has discovered. The ASBR can learn the external router either through another routing protocol such as BGP, or via system configuration information. As discussed above, an ASBR is a router that connects an OSPF autonomous system and an external routing domain. An ASBR sends AS-External-LSAs as an indication of routes to networks that are outside of the OSPF autonomous system. These external routes are injected into the OSPF autonomous system by the ASBR via route redistribution.

An AS-External-LSA with a Forwarding Address of 0.0.0.0 indicates to the routers in the OSPF autonomous system that the originating ASBR is itself the gateway that can reach the external networks. The Link-State ID in the LSA header of the AS-External-LSA (Figure 1.26) is set to the IPv4 address of the external network. An ASBR also uses AS-External-LSAs to describe a default route (to an external destination), particularly, for OSPF stubby types areas (Stub Areas, Totally Stubby Areas, NSSA, and Totally NSSA) **[RFC2328]**. In such cases, a default route is used when there is no specific route to the destination. When an AS-External-LSA (Type 5 LSA) is sent to describe a default route, the Link-State ID in the LSA header is set to 0.0.0.0 (*DefaultDestination*) and the Network Mask to 0.0.0.0.

AS-External-LSAs contain routing information that is imported into OSPF from other routing protocols. The routing metric that is advertised for an external route can be either a Type 1 or Type 2 External metric. For "Type 1 External" LSAs, routers in the OSPF system make routing decisions to an external destination using a cost which includes the external path cost (i.e., path cost to the external destination) and the sum of internal path costs to the ASBR that advertised the external route. The latter is the internal OSPF cost to the particular ASBR advertising the AS-External-LSA. For "Type 2 External" LSAs, the OSPF routers make routing decisions to an external destination using a cost that is based solely on the external path cost. OSPF metrics are discussed in detail in a section below.

The AS-External-LSAs are flooded everywhere within an OSPF autonomous system unchanged, except Stub Areas. Recall that routers in an NSSA do not receive external LSAs from ABRs via the local Backbone Area, but are allowed to advertise external routes learned by a directly attached ASBR (i.e., an NSSA ASBR) for redistribution by an ABR to the Backbone Area. The NSSA ASBR uses Type 7 LSAs to inform Internal Routers and the ABRs about these external routes, and each ABR will then translate these to Type 5 LSAs, and floods them as normal via the Backbone Area to the rest of the OSPF autonomous system (see Type 7 LSAs below).

Figure 1.27 shows a scenario where AS-External-LSAs are used to describe routes to networks that are outside the OSPF autonomous system. An ASBR originates AS-External-LSAs that are then flooded by other ABRs into their respective areas to

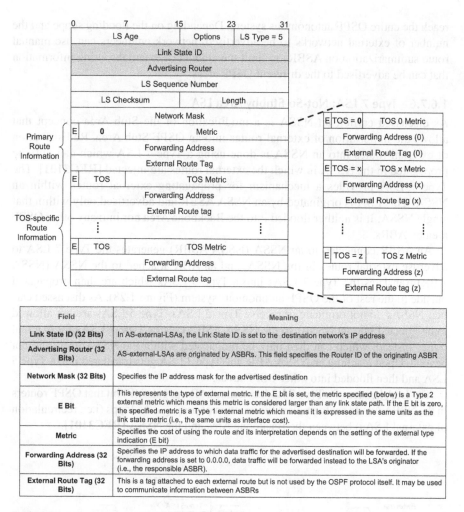

Field	Meaning
Link State ID (32 Bits)	In AS-external-LSAs, the Link State ID is set to the destination network's IP address
Advertising Router (32 Bits)	AS-external-LSAs are originated by ASBRs. This field specifies the Router ID of the originating ASBR
Network Mask (32 Bits)	Specifies the IP address mask for the advertised destination
E Bit	This represents the type of external metric. If the E bit is set, the metric specified (below) is a Type 2 external metric which means this metric is considered larger than any link state path. If the E bit is zero, the specified metric is a Type 1 external metric which means it is expressed in the same units as the link state metric (i.e., the same units as interface cost).
Metric	Specifies the cost of using the route and its interpretation depends on the setting of the external type indication (E bit)
Forwarding Address (32 Bits)	Specifies the IP address to which data traffic for the advertised destination will be forwarded. If the forwarding address is set to 0.0.0.0, data traffic will be forwarded instead to the LSA's originator (i.e., the responsible ASBR).
External Route Tag (32 Bits)	This is a tag attached to each external route but is not used by the OSPF protocol itself. It may be used to communicate information between ASBRs

FIGURE 1.26   AS-External-LSA Format

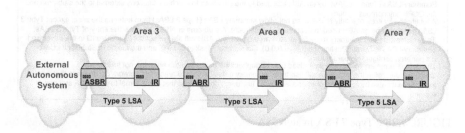

FIGURE 1.27   Flow and Scope of OSPF Type 5 LSAs (AS-External-LSAs)

reach the entire OSPF autonomous system. Depending on the flooding scope and the number of external networks to be advertised, network engineers can use manual route summarization on ASBRs to limit the amount of external routing information that can be advertised to the different OSPF areas.

1.6.7.6 Type 7 LSA: Not-So-Stubby Area LSA

As discussed earlier, an NSSA is a modification of the Stub Area concept that allows the importation of external routes into an OSPF Stub Area. Redistribution of external routes into an NSSA is done using a Type 7 LSA which can only be propagated in the NSSA in which the external routes are injected **[RFC3101]**. The Type 7 LSA provides a mechanism for propagating external routes within an NSSA. This LSA is originated by an NSSA ASBR and advertised only within that single NSSA. It is neither flooded into the Backbone Area nor into any other OSPF area by ABRs.

An ASBR connected to an NSSA (NSSA ASBR) generates the Type 7 LSA to advertise external routes in the NSSA, and an ABR attached to the NSSA (NSSA ABR) translates the Type 7 LSAs into a Type 5 LSAs which are then propagated outside to the rest of the OSPF autonomous system (Figure 1.28). As discussed earlier, NSSAs do not originate or receive Type 5 LSAs. Type 5 LSAs are not allowed to be injected into an NSSA, so the NSSA ASBR creates Type 7 LSAs for external routes instead, which can then only be propagated within the NSSA itself and not beyond. Upon reaching an NSSA ABR, the Type 7 LSA gets translated into a Type 5 LSA and then flooded into other OSPF areas.

The LSA header of a Type 7 LSA contains a "Propagate" (P) bit that OSPF routers in the NSSA would set to prevent the creation of propagation loops (i.e., recirculation of external LSAs) between the Backbone Area and the NSSA **[RFC3101]**.

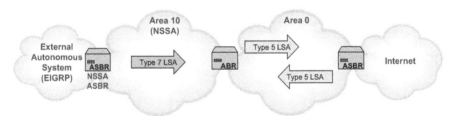

- Routers within a stub area must rely on a default route (0.0.0.0 or DefaultDestination in OSPF) originated/advertised (via Summary-LSAs (Type 3 LSAs)) by an ABR attached to the stub area to reach destinations external to the stub area and the OSPF autonomous system.
- A totally stubby area or totally NSSA does not allow summary LSAs (Type 3 LSAs) from entering the area, except Type 3 LSAs generated by an ABR used to describe a default route. A stub area or NSSA do allow the entry of Type 3 LSAs.
- To describe a default summary route (in all stub area types), the ABR will always set the Link State ID in the advertised Summary-LSAs to DefaultDestination (i.e., 0.0.0.0), the Network Mask to 0.0.0.0, and the metric to StubDefaultCost (as per area configurable parameter).
- The default summary routes are advertised only throughout the particular stub area, but not further.
- OSPF routers inside the stub area will use these summary default routes to reach any network destinations that are not explicitly reachable by intra-area or inter-area routes (including destinations that are external to the OSPF autonomous system).

FIGURE 1.28   Type 7 LSA in an NSSA

1.6.7.7 Other LSA Types

This section describes other specialized LSA types that have been defined for OSPF:

- **Type 6 LSA (Multicast-OSPF-LSA)**: This LSA type was originally defined for Multicast Extensions to OSPF (MOSPF) **[RFC1584]** which is an extension to OSPF to support IP multicast routing. However, MOSPF was not used and has been deprecated which leaves the Type 6 LSA to be reassigned in the future.
- **Type 8 LSA (Link-LSA for OSPFv3 [RFC5340])**: This is a link-local only LSA defined for OSPFv3, and is used by an OSPFv3 router to provide information about Link-Local Addresses and a list of IPv6 address prefixes on its attached link. The scope of flooding Link-LSAs is the local link only and flooding never goes beyond the link on which the LSAs are associated. An OSPFv3 router sends Link LSAs to provide its Link-Local Address to all other routers attached to the link, and to supply to these routers a list of IP address prefixes associated with the link.
- **Type 9 LSA**:
  o **Link-Local Opaque LSA in OSPFv2 [RFC5250]**: Opaque LSAs provide a mechanism for extending the capabilities of OSPF to allow for transmission of arbitrary information/data that are not necessarily specific or used by OSPF (e.g., MPLS Traffic Engineering **[RFC3630]**, OSPF Graceful Restart **[RFC3623]**, and advertising optional router capabilities **[RFC7770]**). An Opaque LSA contains a standard 20-byte LSA header followed by information/data that is specific to an application. The information carried may be parsed and used directly for OSPF extended capabilities not covered by the relevant routing standards (e.g., **[RFC2328]**), or by other applications that use such LSAs as a vehicle for exchanging information. These LSA types have a link-local scope and are not flooded beyond the local network or subnet.
  o **Intra-Area-Prefix LSA in OSPFv3 [RFC5340]**: Intra-Area Prefix LSAs are used by an OSPFv3 router to advertise IPv6 address prefixes that are associated with the OSPFv3 router itself, an attached transit network segment, or an attached stub network segment. Multiple Intra-Area-Prefix LSAs may be originated by an OSPFv3 router each with a unique Link-State ID for each connected router, transit, or stub network.
- **Type 10 LSA (Area-Local Opaque LSA in OSPFv2 [RFC5250])**: These LSA types have an area-local scope and are not flooded beyond the boundary of the area they are associated with.
- **Type 11 LSA (AS Opaque LSA in OSPFv2 [RFC5250])**: The LSA types are flooded throughout their OSPF autonomous system similar to the flooding scope of Type 5 (AS-External) LSAs. Type 11 LSAs are also flooded throughout all Transit Areas in the autonomous system but not into Stub Areas or NSSAs from the Backbone Area. These LSAs are also not originated by routers within a Stub Area.

1.7   SENDING AND RECEIVING LSAS

OSPF routers normally originate/advertise a single LSA for each known route (intra-area, inter-area or external) one at a time so that any change in a single route can be flooded without having to reflood the entire collection of routes it has learned. However, during the flooding process, a router can package many different LSAs in a single Link-State Update packet.

1.7.1   ORIGINATING OSPF LSAs

Whenever a router originates a new instance of an LSA, it will set its LS Age to 0, increment its LS Sequence Number, calculate its LS Checksum, add the LSA to its LSDB and then flood it out the appropriate interfaces. The following ten events can cause an OSPF router to originate a new instance of an LSA **[RFC2328]**:

1. **When the LS Age of One of the LSAs Originated by the Router Itself Reaches the Value *LSRefreshTime*:** When this happens, the router will originate a new instance of the LSA, even though the contents of the LSA (excluding the LSA header) has not changed. This is done to guarantee periodic refreshing/updating of all LSAs, a process which adds to the robustness to the OSPF link-state algorithm. However, LSAs that describe solely destinations that are unreachable will not be refreshed, but instead will be flushed from the routing domain.

   An important point to note here is, the router may not originate any two instances of the same LSA within the *MinLSInterval* time interval. This condition may require the router to delay the generation of the next instance of the LSA by up to *MinLSInterval*. The *MinLSInterval* is the minimum time a router must wait before originating another instance of a particular LSA. The default *MinLSInterval* value is 5 seconds.

   NOTE: The following three events which are mainly due to network/link-state changes may cause the contents of an LSA to change. These events will cause a router to originate a new LSA where the contents of the LSA is different:

2. **When the State of a Router Interface Changes:** In this case, the router will originate a new instance of a Router-LSA (Type 1 LSA) describing the change.

3. **When the Designated Router for a Network Segment Changes:** When a router that is directly attached to a router that is a Designated Router detects this change, it will originate a new Router-LSA describing the change. Also, if the router itself has now become the Designated Router for the network segment, it will originate a new Network-LSA (Type 2 LSA). However, if the router itself is no longer the Designated Router, all Network-LSAs that might have been originated by it have to be flushed from the routing domain.

4. **If the State of One of the Neighbor Routers Changes to/from the OSPF Neighbor State of Full State:** A router detecting this change in Neighbor State will originate a new instance of a Router-LSA. Also, if the router is itself the Designated Router, it will originate a new Network-LSA describing the change.

NOTE: The following four events describe what will cause an ABR to originate a new LSA:

5. **When an OSPF Router has Added/Deleted/Modified an Intra-Area Route
    in the IP Routing Table**: This may cause the ABR attached to the OSPF area
    (in which the router making the intra-area route changes belongs) to originate
    a new instance of a Summary-LSA (Type 3 LSA) for the affected route for
    advertisement to other areas in the OSPF autonomous system (including pos-
    sibly the Backbone Area).
6. **When an OSPF Router has Added/Deleted/Modified an Inter-Area Route
    in the IP Routing Table**: This may cause the ABR attached to the OSPF area
    (in which the router making the inter-area route changes) to originate a new
    instance of a Summary-LSA (for the route) for advertisement to each attached
    OSPF area (but not to the Backbone Area).
7. **When a Router Gets Newly Attached to an OSPF area (i.e., Assumes ABR
    Responsibilities)**: In this case, the router will start originating Summary-LSAs
    into the newly attached OSPF area for all relevant intra-area and inter-area
    routes it has maintained in the its IP Routing Table.
8. **When the State of One of the Configured Virtual Links on a Router
    Changes**: The router (ABR) detecting this change may originate a new Router-
    LSA into the Transit Area over which the Virtual Link passes through, as well
    as, originate a new Router-LSA into the Backbone Area.
    NOTE: The following two events will cause an ASBR to originate a new LSA:
9. **When An External Route Learned by an ASBR through an External
    Routing Protocol (such as BGP) Changes**: This will cause the ASBR to orig-
    inate a new instance of an AS-External-LSA to describe the external routing
    changes.
10. **When a Router Ceases to be an ASBR (Possibly after Restarting)**: When this
    happens, all AS-External-LSAs that were previously originated by the router
    will be flushed. A router can flush these LSAs from the routing domain via the
    premature aging procedure specified in **[RFC2328]** and as discussed below.

1.7.2 THE LSA FLOODING PROCEDURE

OSPF routers use Link-State Update packets as the main vehicle for flooding LSAs.
A router will create a Link-State Update packet that may contain several distinct
LSAs in it. Each LSA will then be flooded one (router) hop away from its node of
origination (i.e., to an immediate neighbor router). In order to make the flooding
process more reliable, each LSA that is received by a neighbor must be acknowledged separately. The neighbor transmits acknowledgments back to the sending
router using Link-State Acknowledgment packets. The neighbor may group many
separate acknowledgments to a particular sender together into a single Link-State
Acknowledgment packet.

1.7.2.1 First Part of the LSA Flooding Process

An OSPF router starts the flooding procedure when it receives a Link-State Update
packet. The router will make a number of consistency checks on the received packet

before it initiates the flooding procedure. Particularly, the Link-State Update packet will be checked to see if it is associated with a particular neighbor, and a particular OSPF area. If the neighbor is in a state lower than the Exchange State of the Neighbor State (see below), the packet will be dropped without further processing.

Router-LSAs, Network-LSAs, Summary-LSAs, and ASBR-Summary-LSAs, other than AS-external-LSAs, all have flooding scope that is associated with a specific OSPF area. However, LSAs have a format that does not include an OSPF Area field (Figure 1.18). This means a router that receives an LSA must deduce its associated area from the Link-State Update packet header. For each LSA contained in a received Link-State Update packet, the router will process it using the following steps:

1. The LSA's LS Checksum is validated and if it is found to be invalid, the LSA is discarded and processing proceeds to the next LSA in the Link-State Update packet.
2. The LSA's LS Type is examined and if it is unknown, the LSA is discarded and processing proceeds to the next LSA in the Link-State Update Packet.
3. Else, if the LSA is an AS-External-LSA and the receiving OSPF area has been configured as a Stub Area, the LSA is discarded and processing proceeds to the next LSA in the Link-State Update Packet. Recall that AS-External-LSAs are not injected into Stub Areas.
4. Else, if the value of the LS Age field in the LSA is equal to *MaxAge*, and the router's LSDB does not currently contain an instance of the LSA, and none of router's neighbors are in Exchange or Loading State, then the router will take the following actions:
   a. Send a Link-State Acknowledgment packet back to the sending neighbor acknowledging receipt of the LSA, and
   b. Discard the received LSA and then proceed to process the next LSA (if any) contained in the Link-State Update packet.
5. Otherwise, the router's LSDB is examined to see if it currently contains an instance of the received LSA. If the LSDB does not contain a copy, or the received LSA is more recent than the copy in the LSDB, the router will the perform the following steps:
   a. If the LSDB already contains a copy, and if that copy was received by way of flooding and installed in the LSDB for a period that is less than *MinLSArrival* seconds, the new LSA is discarded (without being acknowledged) and processing will proceed to the next LSA (if any) contained in the Link-State Update packet. The *MinLSArrival* is the minimum time that must elapse between reception of new LSA instances of any particular LSA during flooding. LSA instances that are received at a higher rate are discarded. The default *MinLSArrival* value is 1 second.
   b. Otherwise, the new LSA is immediately flooded out a subset of the router's interfaces. In some cases, the router will flood the LSA back out the receiving interface, for example, when the state of the receiving interface indicates it belongs to the Designated Router, and the LSA was received from

a router that is not the BDR. The router notes this occurrence for later use by the LSA acknowledgment process.

c. The router removes the current LSDB copy from the Link-State Retransmission Lists of all the neighbor routers.

d. The router installs the new LSA in the LSDB (replacing the existing copy). This situation may cause the router to perform the IP Routing Table calculation. In addition, the new LSA will be timestamped with the time it was received. The flooding procedure cannot cause the newly installed LSA to be overwritten until *MinLSArrival* seconds have elapsed.

e. The router may send a Link-State Acknowledgment packet back out the receiving interface to acknowledge the receipt of the LSA.

f. If the router finds that the new LSA is one that was previously originated by the router itself (i.e., a self-originated LSA), it must, either update the LSA or in some cases flush it from the routing domain.

6. Else, if the router detects that an instance of the received LSA already exists in the Link-State Request List of the sending neighbor router, then an error has occurred during the Database Exchange process. When this happens, the router will restart the Database Exchange process with the affected neighbor by generating a *BadLSReq* neighbor event and not process further the Link-State Update packet.

7. Else, if the router detects that the received LSA is the same instance as the copy in its LSDB (i.e., neither LSA is more recent), it will perform the following two steps:

a. If the received LSA is already listed in the Link-State Retransmission List of the sending neighbor, then the router is looking forward to an acknowledgment for this previously sent LSA. The router would treat the received LSA as an "implicit" acknowledgment (in the absence of an explicit one) and will then remove the LSA from the Link-State Retransmission List. The router notes this occurrence for later use by the LSA acknowledgment process.

b. The router may send a Link-State Acknowledgment packet back out the receiving interface to acknowledge the receipt of the LSA.

8. Else, the LSA copy in the router's LSDB is more recent. If the router detects that the LSDB copy has LS Age and LS Sequence Number equal to *MaxAge* and *MaxSequenceNumber*, respectively, it will simply discard the received LSA without sending an acknowledgment. When this happens, the LS Sequence Number of the LSA is wrapping, and this LSA (that has reached *MaxSequenceNumber*) must be completely flushed from the routing domain before any new LSA instance can be sent. Otherwise, as long as the router has not sent a copy of the LSA in its LSDB in a Link-State Update packet within the last *MinLSArrival* seconds, it must send a copy (encapsulated in a Link-State Update packet) directly back to the sending neighbor. When doing this, the router must not list the LSDB copy on the neighbor's Link-State Retransmission List, and must not acknowledge the received LSA instance which is deemed less recent.

1.7.2.2   Second Part of the LSA Flooding Process

When an OSPF router receives a new LSA or one that is more recent, it must flood it out some set of its local interfaces. We describe in section the second part of flooding process where a router selects the outgoing interfaces for an LSA, adds an LSA to neighbors' Link-State Retransmission Lists, and maintains the neighbors' Link-State Request Lists. The discussion equally applies to the flooding of LSAs that the router itself has just originated. Depending upon the LS Type of an LSA, the router will flood the LSA out only certain interfaces. The eligible interfaces for an LSA are defined as follows:

- **AS-External-LSAs**: These LSAs are flooded throughout an entire OSPF Autonomous System, with the exception of Stub Areas. The eligible interfaces on a router for AS-External-LSAs are all of the router's interfaces, excluding interfaces attached to Stub Areas and Virtual Links.
- **All Other LSA Types**: All other LSA types are flooded only within a specific OSPF area. The eligible interfaces in this case are all the interfaces attached to the specific OSPF area, and if that area is the Backbone Area, the flooding scope includes all of its Virtual Links.

The LSDBs associated with all adjacencies formed over the above eligible interfaces must be synchronized. Routers accomplish this by executing the steps described below for each eligible interface. A router following this procedure may decide not to flood an LSA out a particular interface, if it determines with high probability that the neighbor routers attached to the eligible interface have already received the LSA. The router, however, will still add the LSA to the Link-State Retransmission List of each adjacency so that it can ensure that the neighbor routers eventually do receive the LSA. For each eligible interface, the router will perform the following:

1. The router will examine each of the neighbors attached to the interface to determine whether they are eligible to be sent the new LSA. The router will execute the following steps for each neighbor:
    a. If the neighbor is in a state lower than the Exchange State (see below), it will move to examine the next neighbor since the neighbor does not partici-pate in LSA flooding in a lower Neighbor State.
    b. Else, if the router determines that the adjacency is not yet in the Exchange or Loading State, it will examine the Link-State Request List associated with it. If it finds that the list contains an instance of the new LSA, then this indicates that the neighbor already has an instance of the LSA. The router will then compare the new LSA to the copy in the neighbor's list:
        i. If it finds that the new LSA is less recent (than the copy), then it will examine the next neighbor (if any).
        ii. If it finds that the two LSA instances are the same, then it will delete the LSA instance on the neighbor's Link-State Request List, and then it will examine the next neighbor (if any).
        iii. Else, the new LSA will be considered to be more recent, and then it will delete the LSA from the neighbor's Link-State Request List.

    c. If the router determines that the new LSA was received from the neighbor currently being examined, then it will move to examine the next neighbor (if any).

    d. At this point, the router can conclude that the neighbor has already received an instance of this new LSA that is up-to-date. The router will then add the new LSA to the Link-State Retransmission List for the neighbor to ensure that the LSA flooding process is reliable. The router will retransmit the LSA at intervals of *RxmtInterval* until it receives an acknowledgment from the neighbor.

2. At this point, the router must determine whether to flood the new LSA out the router interface under consideration. If the router did NOT add the LSA to any of the Link-State Retransmission Lists in the previous step, then the router does not need to flood the LSA out the interface, and will examine the next interface.

3. If the router determines that the new LSA was received on the interface under consideration, and that it was received from either a Designated Router or the BDR, then it is highly probable that all the neighbor routers on the network segment have already received the LSA, and the router should examine the next interface.

4. If the router determines that the new LSA was received on the interface under consideration, and the state of the interface indicates Backup State (i.e., the router itself is the BDR), then the router will examine the next interface. This is because the Designated Router will be responsible for LSA flooding on this interface. However, if the Designated Router becomes unavailable, the router itself (being the BDR) will wind up retransmitting the updates.

5. If the checks performed reach this step, then the router must flood the LSA out the interface. This is done by sending a Link-State Update packet with contents (including the new LSA) out the interface. The router will increment the LS Age field value of the LSA by *InfTransDelay* (which must be greater than 0) when it is written into the outgoing Link-State Update packet. The LS Age field value is incremented by *InfTransDelay* until it reaches the maximum value of *MaxAge*.

   - On broadcast multiaccess networks, the Link-State Update packets are sent as multicasts and the destination IP address used for these packets depends on the state of the interface. If the state of the interface indicates Designated Router or BDR, then the destination IP address used is the "All OSPF Routers" address 224.0.0.5 (*AllSPFRouters*), otherwise, the "OSPF All Designated Routers" address 224.0.0.6 (*AllDRouters*) is used.
   - On non-broadcast networks, the router will send separate Link-State Update packets as unicasts to each adjacent neighbor (i.e., to neighbors in the Exchange State or greater) using the neighbor's IP address as destination IP address.

1.7.3 Determining Which LSA Is More Recent

Any time a router is presented with two instances of an LSA, it has to determine which one of them is more recent. This situation normally occurs when a router has to compare a received LSA with a copy already installed in its LSDB. The router also makes this comparison when it receives LSAs during the Database Exchange process when it is in the process establishing a full adjacency with a neighbor.

An LSA is uniquely identified by a combination of the LS Type, Link-State ID, and Advertising Router field values specified in it. When a router is presented with two instances of the same LSA, it uses the LS Age, LS Sequence Number, and LS Checksum field values to determine which instance is more recent.

The router will consider the LSA carrying the newer LS Sequence Number field value as more recent. However, if both LSA instances have the same LS Sequence Number values, then the following have to be considered:

- If the two LSA instances have different LS Checksum field values, then the router will consider the instance having the bigger LS Checksum value (when taken as a 16-bit unsigned integer) as the more recent LSA.
- Else, if only one of the LSA instances has its LS Age field value equal to *MaxAge*, then the router will consider the instance (with LS Age set to *MaxAge*) as the more recent one.
- Else, if the router finds that the LS Age field values of the two LSA instances differ by more than *MaxAgeDiff*, it will consider the instance having the smaller LS Age field value as the more recent LSA.
- Else, the router will consider the two instances to be identical.

1.7.4 WHEN A ROUTER RECEIVES LSAs IT HAS ORIGINATED

It is not unusual during the LSA flooding process for a router to receive LSAs that it has originated itself. A router detects a self-originated LSA when any one of the following happens [RFC2328]:

1. The Advertising Router field in the LSA is equal to the router's own Router ID, or
2. The LSA is a Network-LSA with a Link-State ID field that is equal to the IP address of one of the router's own interfaces.

However, if the router detects that the received self-originated LSA is newer than the last LSA instance that it had originated, it will take special action as follows. The reception of such a self-originated LSA indicates that the routing domain could be carrying LSAs that the router itself originated before the last time it was restarted. In such a case, the router must advance the LS Sequence Number of any new LSA it creates by one past the LS Sequence Number of the received LSA, and then originate a new instance of the received LSA.

The router may also decide that it no longer wishes to originate a new instance of the received LSA such as under the following example cases [RFC2328]:

1. The received LSA is a Summary-LSA (Type 3 LSA) or AS-External-LSA (LSA Type 5) and the router no longer advertises a route to the destination,
2. The received LSA is a Network-LSA (Type 2 LSA) but the router is no longer the elected Designated Router for the network segment, or

3. The received LSA is a Network-LSA with a Link-State ID field equal to the IP address of one of the router's own interfaces, but the Advertising Router field carries a value that is not equal to the router's own Router ID. This case, which is usually rare, indicates that the Router ID of the router has changed since it last originated the LSA.

For all these cases, instead of updating the received LSA and then sending a new instance of it, the router flushes the LSA from the routing domain by incrementing the LS Age value in the LSA to *MaxAge* and then reflooding it [RFC2328].

1.7.5 RETRANSMITTING LSAs AND THE USE OF LINK-STATE RETRANSMISSION LISTS

When an OSPF router floods LSAs out an interface to an adjacent neighbor, it also places them in the neighbor's Link-State Retransmission List. To ensure that flooding process is reliable, the router would retransmit these LSAs until they are acknowledged. The spacing between retransmissions (in seconds), *RxmtInterval*, is configurable on a per-interface basis. If the configured value is set too low for an interface, this can result in unnecessary retransmissions. If the value is set too high, the rate and amount of the LSA flooded can result in lost LSAs.

A router may pack several LSAs designated to be retransmitted, just enough to fit into a single Link-State Update packet. The router will send this single Link-State Update packet to the neighbor, and then prepare another update packet for retransmission whenever some of the LSAs are acknowledged, or upon the next triggering of the retransmission timer. The router always sends the Link-State Update packets carrying retransmissions directly to the neighbor, and on multiaccess networks, the retransmissions are sent directly as unicasts with destination address being the IP address of the neighbor.

The router increments the LS Age of each LSA by *InfTransDelay* when it is copied into the Link-State Update packet carrying the retransmissions. The LSA's LS Age value is not allowed to be incremented beyond the maximum value of *MaxAge*. The parameter *InfTransDelay* (which must be greater than 0) is the estimated length of time (in seconds) it takes for a Link-State Update Packet to be transmitted over a given router interface. The InfTransDelay value (which must be greater than 0) should be chosen to take into account the expected transmission and propagation delays of the router interface. A suggest value in [RFC2328] for a local area network is 1 second.

If an adjacent router goes down or is unreachable for some reasons, the OSPF router may still continue with retransmissions until the adjacency is terminated by the OSPF Hello Protocol. When the adjacency is terminated, router will clear the neighbor's Link-State Retransmission List of all its contents and stops retransmissions.

1.7.6 INSTALLING LSAs IN THE LSDB

When an OSPF router installs a new LSA in its LSDB, either as a result of flooding or a new LSA that the router itself has originated, this may cause the OSPF router to recalculate new routes for the IP Routing Table. To enable the router to decide what actions to take, the router will compare the contents of the new LSA to the old LSA instance in the LSDB, if present. If the two LSAs are the same, then there will be no need to recalculate routes for the Routing Table. When comparing a received LSA to a previously installed LSA instance, the router will consider the following as conditions under which the two LSAs are seen to have different contents:

- The Options field in any one of the LSAs has changed.
- The LS Age in one of the LSA instances has been set to *MaxAge*, while the other has not.
- The Length field in any one of the LSA header has changed.
- The contents of any one of the LSA (outside the standard 20-byte LSA header) has changed. This does not include changes in the LSA's LS Sequence Number and LS Checksum fields.

If the router determines that the two LSAs have different contents, then it has to recalculate the following for the Routing Table, but according to the LS Type field specified in the new LSA:

- **Router-LSAs and Network-LSAs**: For these LSAs, the OSPF router will recalculate the entire Routing Table, but by first performing the shortest path calculations for each area in the OSPF Autonomous System (not just the OSPF area whose LSDB has changed). The main reason why the shortest path calculations are not limited to the single OSPF area affected by the change is that ASBRs connected to the OSPF autonomous system may have interfaces belonging to multiple areas. This means link-state changes in the OSPF area currently providing the best route to an external destination (via the ASBR) may force routers in the affected area to use an intra-area route provided by a different area to that external destination (but by first going through the Backbone Area) as illustrated in Figure 1.29.

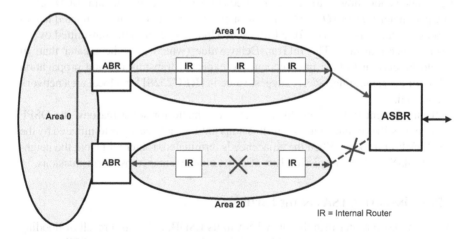

FIGURE 1.29 Routers in an OSPF area forced to use an intra-area route provided by a different area to an external destination

- **Summary-LSAs**: In this case, the router will recalculate the best route to the destination described by the Summary-LSA. If this destination is an ASBR, the router may also find it necessary to reexamine all received AS-External-LSAs.
- **AS-External-LSAs**: For this LSA type, the router will recalculate the best route to the destination described by the AS-External-LSA.

Furthermore, the router must take steps to remove any old instance of the LSA from its LSDB as soon as the new LSA is installed. The router must also remove the old instance of the LSA from the Link-State Retransmission Lists of all neighbor routers.

1.7.7 AGING THE LSDB

Each LSA carries an LS Age field with values expressed in seconds. An OSPF router increments the LS Age field value of an LSA while it is contained in the router's LSDB. The router, also, increments the LSA's LS Age by *InfTransDelay* when it is copied into a Link-State Update Packet to be flooded out a particular router interface.

The router, however, must not increments an LSA's LS Age past the maximum value of *MaxAge*. As a router ages LSAs in its LSDB, the LS Age of an LSA may reach the maximum value *MaxAge*. At this point, the router will attempt to flush that *MaxAge* LSA from its routing domain by reflooding it similar to what is done when flooding a newly originated LSA. LSAs with LS Age exceeding *MaxAge* are not used in the computation of the OSPF router's Routing Table.

When an OSPF router is in the process of creating a Database Summary List for a newly formed adjacency, it will add any *MaxAge* LSAs present in its LSDB to the Link-State Retransmission List of the neighbor instead of the neighbor's Database Summary List. The router must remove a *MaxAge* LSA immediately from its LSDB as soon as both of these two conditions are met:

a. The LSA is no longer listed in any neighbor's Link-State Retransmission List, and
b. None of the router's neighbors are in a Neighbor State of Exchange or Loading State.

When a router is in the process of aging LSAs in its LSDB, and the LS Age of an LSA reaches a multiple of *CheckAge*, the router must verify the LS Checksum of the LSA. If the LS Checksum is found to be incorrect, then a memory or program error could have occurred, and at the very least, it is recommended to restart the router itself.

1.7.8 PREMATURE AGING OF LSAS

We have seen above that a router can flush an LSA from the routing domain by setting its LS Age to *MaxAge* (while leaving its LS Sequence Number untouched), and then reflooding it. This action is similar to what a router will do when it wants to flush an LSA whose LS Age has progressively aged to the maximum value *MaxAge*. The action of setting the LS Age of an LSA to *MaxAge* is referred to as "Premature Aging".

A router applies premature aging to a self-originated LSA when it is time for its LS Sequence Number field to wrap around. At this point, the router prematurely ages the current LSA instance (with LS Sequence Number equal to the maximum value *MaxSequenceNumber*), and flushes it from the routing domain before originating a new instance of the LSA with Sequence Number equal to starting or initial value *InitialSequenceNumber*.

A router can also perform premature aging when, for example, it detects that one of the external routes it had previously advertised (via an AS-External-LSA) is no longer reachable. Under these circumstances, the router can flush that AS-External-LSA from the routing domain using premature aging. This action is preferable to the alternative, where the router originates a new AS-External-LSA for the external route carrying a routing metric of *LSInfinity*. *LSInfinity* is a metric value that indicates that the destination described by a Summary-LSA or AS-External-LSA is unreachable.

A router can also apply premature aging when it unexpectedly receives self-originated LSAs during the LSA flooding process. The router may only prematurely age LSAs it has originated itself and not LSAs that have been originated by other OSPF routers. Recall that a self-originated LSA is an LSA that has one of the following properties: (1) the Advertising Router field value of the LSA is equal to the router's own Router ID or (2) the LSA is a Network-LSA with a Link-State ID that is equal to the IP address of one of the router's own interfaces.

To summarize, an OSPF router can apply premature aging when it receives the following LSA types:

- A router that is no longer the Designated Router for a network segment must flush Network-LSAs that it had previously originated by prematurely aging the LSAs' LS Age to *MaxAge* and then reflooding them. These LSAs are no longer used in the router's Routing Table calculations.
- If a router advertises a Summary-LSA for a network destination which is no longer unreachable, it must flush that LSA from the routing domain by setting the LSA's LS Age to *MaxAge* and then reflood it.
- If a router advertises an AS-External-LSA for an external destination (i.e., external route) which is no longer unreachable, it must flush that LSA from the routing domain by setting the LSA's LS Age to *MaxAge* and the reflood it.

1.8   OSPF ROUTER IDENTIFIER

The Router ID is a 32-bit number that uniquely identifies a router in an OSPF autonomous system and must be available whenever OSPF runs. The OSPF Router ID is a parameter required in all OSPF messages and for the OSPF process in a router. When OSPF is started in a router, the system must be able to identify and use a unique Router ID for the OSPF process. As shown in Figure 1.11, the OSPF Router ID is a 32-bit value usually expressed in the dotted decimal IPv4 address format. As discussed in **[TEAREDIA15]**, at least, the router must be configured with one primary IPv4 address on an interface in the "Up/Up" State for it to be able to select a Router ID, otherwise, the router will log an error message, and the OSPF process will not start.

Cisco Systems routers use the following criteria to select the OSPF Router ID when the OSPF process is initialized (Figure 1.30) **[TEAREDIA15]**:

1. Use the unique 32-bit Router ID that has been manually specified for the router (which can be configured using the **router-id** *ip-address* command). An arbitrary value can be configured (in the IPv4 address dotted format) as long as the value is unique.
2. If no manually configured Router ID is available, use the highest IPv4 address among all of the IPv4 addresses assigned to all the active loopback interfaces in the router.
3. If no logical/loopback IPv4 address is set, then use the highest IPv4 address among all the IPv4 addresses assigned to all active nonloopback interfaces in the router (e.g., use 192.168.0.2 over 10.1.1.4).

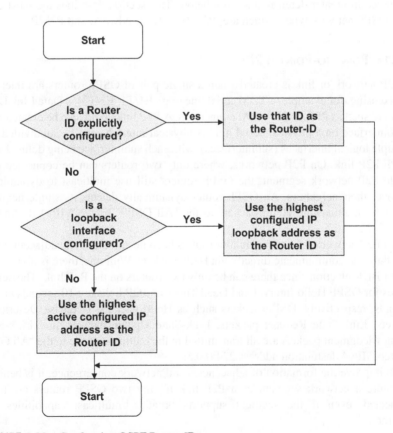

FIGURE 1.30   Configuring OSPF Router ID

Once the OSPF router has selected a Router ID, it will not change it even if the interface from which the ID was taken has changed its IP address or operational state. To change the Router ID, the OSPF process must be reset using the `clear ip ospf process` command, or the router must be reloaded [TEAREDIA15]. It is not recommended to change the Router ID while the OSPF router is running in a network since doing so requires all OSPF adjacencies to be reset, resulting in a temporary routing outage. This would also cause the router to send all new copies of all LSAs it had originated but identified with the changed new Router ID. Some routers [ALU7705SAR] allow the last 4 bytes of a router MAC address to be used when no other identifiers are available.

1.9   OSPF NETWORK TYPES

OSPF supports different network types which allows OSPF to be deployed over a wide variety of networks employing different architectures and/or underlying technologies. The operation of OSPF on each network type is slightly different with some network type-specific differences, which includes how the OSPF routers establish adjacencies, or if the network requires the election/use of a DR or BDR. DR/BDR are discussed in greater detail in a section below. This section describes the most common OSPF network types, which are, P2P, Broadcast, NBMA, and P2MP.

1.9.1   POINT-TO-POINT (P2P)

A P2P network or link is created when a single pair of OSPF routers are interconnected using, for example, a TDM serial line (e.g., E1/T1, E3/T3), a Data Link Layer protocol such as PPP, or P2P ATM connection. A P2P link could also be created from a subinterface on a physical serial link. A physical interface can be subdivided into multiple logical interfaces (subinterfaces), with each subinterface being defined as an OSPF P2P link. On P2P networks, where only two routers can be connected on a single P2P network segment, the OSPF routers still use multicast to dynamically discover their neighbors. An OSPF router dynamically detects its single neighbor router by multicasting Hello packets to the "All OSPF Routers" IPv4 destination address 224.0.0.5.

On P2P networks, the two neighbor routers become automatically adjacent whenever they can communicate directly via Hello packets. With this, there is no need for DR or BDR election since there can be only two routers on the P2P link. The default values for OSPF Hello Interval and Dead Timer on P2P links are 10 seconds and 40 seconds, respectively. OSPF packets such as Hello packets, Database Description packets, Link-State Request packets, Link-State Update packets, and Link-State Acknowledgment packets are all transmitted in the multicast mode to the "All OSPF Routers" IPv4 destination address 224.0.0.5.

To improve the formation of adjacencies and network convergence, it is better to configure a network segment as a P2P link if only two OSPF routers are to be connected, even if the segment supports broadcast/multicast capabilities like Ethernet.

1.9.2 BROADCAST

If multiple OSPF routers in a network segment are interconnect over a Data Link Layer protocol such Ethernet which provides built-in broadcast and multicast capabilities, then OSPF will exploit such capabilities to send OSPF packets. Any randomly chosen pair of routers on a broadcast multiaccess network can communicate directly with each other, and a single message is capable of reaching (broadcast or multicast to) all of the attached routers. In this network type, OSPF routers transmit Hello packets, Link-State Update packets, and Link-State Acknowledgment packets in multicast mode using the reserved IP multicast addresses (the "All OSPF Routers" address 224.0.0.5 and the "OSPF All Designated Routers" address 224.0.0.6).

However, OSPF Database Description packets and Link-State Request packets are transmitted in the unicast mode. In broadcast multiaccess network segments, the OSPF routers use OSPF's Hello Protocol to dynamically discover neighbors (via multicast Hello packets). In this case, the OSPF routers (via the Hello Protocol and packets) take advantage of the broadcast/multicast capabilities of the underlying network. The routers also perform a DR and BDR election in order to optimize the exchange of link-state information.

1.9.3 NON-BROADCAST MULTIPLE ACCESS (NBMA)

Multiple OSPF routers in a network segment can also be interconnected using Data Link Layer protocols such as Frame Relay and ATM that do not have built-in broadcast capabilities like Ethernet. These networks are generally referred to as NBMA networks. NBMA networks can also be used to interconnect more than two OSPF routers but without the broadcast capabilities found in Ethernet. NBMA networks are capable of supporting more than two routers, but it should be noted that they do that by only simulating the behavior and operations of a broadcast network because they do not have inherent broadcast capability.

An NBMA network is generally fully meshed (constructed using, for example, ATM Permanent Virtual Circuits [PVCs]) where any two OSPF routers in the network segment are reachable. It is a non-broadcast, multiaccess network where all OSPF routers are fully interconnected. If it is not possible/practical, in some cases, to connect any two routers, then it is generally preferable to configure the network as a P2MP (see below). If a router has only one peer in the NBMA network, then preferably a P2P link can be used to connect that router.

In NBMA networks, as routers cannot discover neighbors simply by broadcasting Hello packets, OSPF neighbors must be statically/manually configured, and after which DR/BDR election is performed. Although neighbor routers are still maintained using the Hello protocol, due to the absence of built-in broadcast capabilities, static/manual configuration is necessary to aid in the discovery of neighbors. In this network type, the OSPF routers transmit OSPF packets such as Hello packets, Database Description packets, Link-State Request packets, Link-State Update packets, and Link-State Acknowledgment packets in the unicast mode.

1.9.4 POINT-TO-MULTIPOINT (P2MP)

This OSPF network type is essentially a logical collection of P2P links all belonging to a common network segment, and with all P2P links having a common starting point. In this network type, OSPF routers discover neighbors dynamically using multicast transmission of Hello packets. Also, no DR/BDR election takes place in this network type. OSPF routers transmit Hello packets in the multicast mode, while the other OSPF packets such as Database Description packets, Link-State Request packets, Link-State Update packets, and Link-State Acknowledgment packets are transmitted in unicast mode.

Cisco has an extension to the P2MP network referred to as the "P2MP Non-Broadcast (P2MP-NB)" network that has the same characteristics except that OSPF routers do not discover neighbors dynamically. In P2MP-NB networks, neighbors must be statically configured and the routers use unicast for communication. P2MP-NB becomes very useful in P2MP networking scenarios where broadcast and multicast are not supported.

Table 1.1 summarizes the main characteristics of the different OSPF network types.

TABLE 1.1
OSPF Network Types and Their Main Characteristics

Network Type	Uses Dynamic Neighbor Discovery?	Requires DR/BDR Election?	Allows More Than Two Routers on Network Segment?	Hello Packet Transmission	Default Hello Interval/ Router Dead Interval (sec)
Point-to-Point (P2P)	Yes	No	No	Sent to the multicast address 224.0.0.5	10/40
Broadcast	Yes	Yes	Yes	Sent to the multicast address 224.0.0.5	10/40
NBMA	No	Yes	Yes	Sent as unicast to each neighbor	30/120
Point-to-Multipoint (P2MP) (Broadcast)	No	No	Yes	Sent to the multicast address 224.0.0.5	30/120
Point-to-Multipoint (P2MP) (Non-Broadcast)	No	No	Yes	Sent as unicast to each neighbor	30/120
Virtual Link	No	No	No	Sent as unicast directly to the other end of the Virtual Link	10/40

1.10 OSPF PACKET PROCESSING

OSPF routers send OSPF packets along adjacencies only (i.e., to neighbor routers only). Hello packets, which routers send out their active interfaces, are used to discover the adjacencies. Basically, all OSPF packets sent out a router interface travel only one IP router hop (to a neighbor), except packets sent over Virtual Links.

This section discusses some of the important processing requirements when OSPF routers send or receive packets. All OSPF packets start with a standard 24-byte packet header (Figure 1.11). We discuss below how an OSPF router fills in and verifies the various standard header fields.

1.10.1 SENDING OSPF PACKETS

When a router prepares an OSPF packet to be sent out one of its interfaces, it fills in the following OSPF packet header fields: Version Number which is set to 2; Packet Type; Packet Length; (originating) Router ID; Area ID; Checksum; Authentication Type; Authentication Data. The Area ID indicates the OSPF area to which the router interface belongs. The router will label all OSPF packets originating from each interface with its associated Area ID. These fields are described in Figure 1.11.

The IP destination address in the IP packet encapsulating an OSPF packet is selected as follows (see details in "network types" section):

Point-to-Point Networks:
- On P2P networks, OSPF packets (Hello, Database Description, Link-State Request, Link-State Update, Link-State Acknowledgment) are sent as multicasts with the IP destination always set to the "All OSPF Routers" address 224.0.0.5.

Broadcast Networks:
- Hello packets are sent as multicasts to the "All OSPF Routers" multicast address 224.0.0.5.
- Designated Router and BDR send Link-State Update and Link-State Acknowledgment packets as multicasts to the "All OSPF Routers" multicast address 224.0.0.5. All other routers on the network segment send their Link-State Update and Link-State Acknowledgment packets as multicast to the "OSPF All Designated Routers" address 224.0.0.6.
- Database Description and Link-State Request packets are always sent directly (as unicasts) to the neighbor's IP address (which is the IP address associated with the interface at the other end of the adjacency).
- Retransmissions of Link-State Update packets are always addressed directly (as unicasts) to the neighbor's IP address.

Non-Broadcast Multiple Access (NBMA) Networks:
- On NBMA networks, OSPF packets (Hello, Database Description, Link-State Request, Link-State Update, Link-State Acknowledgment) are sent as unicasts

with the IP destination always set to the IP address associated with the interface at the other end of the adjacency, that is, sent directly to the other end of the adjacency.

Point-to-Multipoint (P2MP) Networks:
- Hello packets are sent as multicasts to the "All OSPF Routers" multicast address 224.0.0.5.
- Database Description, Link-State Request, Link-State Update, and Link-State Acknowledgment are sent as unicasts with the IP destination always set to the IP address associated with the interface at the other end of the adjacency.

An OSPF router learns a Neighbor IP address when it receives Hello packets from the neighbor (i.e., from the source IP address field of the IP packet carrying the Hello packets). On Virtual Links, the OSPF router learns the Neighbor IP address during the IP Routing Table build up process [RFC2328]. The Neighbor IP address is the IP address of the neighbor router's interface sending the Hello packets, and attached to the common shared link or network segment. The router will use this IP address as the Destination IP address when it sends OSPF packets as unicasts along the adjacency. Also, this IP address is used as the Link ID in Router-LSAs (Type 1 LSAs) when they are sent over the attached network if the neighbor router has been selected to be Designated Router (see Figure 1.20).

The source IP address of the IP packet carrying an OSPF packet should be set to the IP address of the router interface originating the packet. Note that interfaces that are attached to unnumbered P2P links have no IP addresses associated with them. In such cases, the source IP address of a packet originated by these types of interfaces should be set to any of the other IP addresses assigned to the router. This means, the router must have at least one valid IP address assigned to it.

In many ways, Virtual Links behave in the same way as unnumbered P2P links, however, a Virtual Link usually has an IP address at the interface (which is discovered during the Routing Table build up process [RFC2328]). This discovered IP address is used as the source IP address in the packets being sent over the Virtual Link.

The IP network mask associated with the router's interface indicates the portion of its IP address that identifies the attached IP network/subnet, that is, the network address or prefix. Masking the IP address of the interface with its network mask yields the IP network prefix of the attached network. The interface network mask is not defined on P2P and Virtual Links because the link or interface itself may not be assigned an IP network address.

1.10.2 Receiving OSPF Packets

Whenever a router receives an OSPF packet, it marks it with the interface on which it was received. The router will then subject the packet through a number of tests which it must pass before it will be accepted at the IP level into the router. The router checks the validity of the IP header carrying the OSPF packet and the OSPF packet header itself. A packet that passes these tests will then be passed to OSPF for processing:

- The 16-bit IP header checksum in the IP packet carrying the OSPF packet must be verified to be correct.
- The IP destination address in the IP packet carrying the OSPF packet must be the IP address of the router interface receiving the packet, or one of the reserved OSPF IP multicast addresses: "All OSPF Routers" multicast address 224.0.0.5 or the "OSPF All Designated Routers" address 224.0.0.6.
- The 8-bit IP Protocol Number specified in the IP packet carrying the OSPF packet must be 89.
- OSPF packets that are originated locally by the router should not be passed on to OSPF. The router must examine the source IP address in the IP packet to make sure that it is not a multicast packet that the router itself has generated.

A packet that passes the IP level test will then have its OSPF packet header verified. The router will examine the fields specified in the OSPF header to make sure they match those configured for the receiving router interface. If the OSPF header field values do not match, the packet is discarded:

- The 8-bit OSPF Version Number field must have a value of 2.
- The 32-bit Area ID in the OSPF header must be verified. The packet will be discarded if the following two cases fail. The Area ID specified in the header must be one of the following:
  o **The 32-bit Area ID in the Packet Matches the Area ID of the Interface Receiving the Packet:** When this happens, the OSPF packet has been sent from a neighbor router only a single hop away and, therefore, the IP source address of the packet is required to be on the same IP network/subnet as the receiving router interface. The receiving router can verify this by masking the packet's source IP address and the interface's IP address with the interface's mask, and then comparing the resulting network addresses (prefixes). This comparison should not be used on P2P links, since on these network types, the interface IP addresses assigned to each end of the link is done independently (and not on the basis of them belonging to a common IP network/subnet).
  o **The 32-bit Area ID in the Packet Matches the Backbone Area ID of 0 or 0.0.0.0:** In this case, the OSPF packet has been sent from an ABR over a Virtual Link configured across a Transit Area. The receiving router must also be an ABR, and the Router ID specified in the OSPF packet (indicating the source router or ABR) must be the other end of the Virtual Link. The receiving interface must be that of an ABR attach to the Transit Area over which the Virtual Link is configured. The router will accept the packet if all of these checks succeed. After this the packet will be associated from now on with the configured Virtual Link and the Backbone Area.
- OSPF packets that have their IP destination addresses set to the "OSPF All Designated Routers" address 224.0.0.6 should only be accepted if the state of the interface receiving the packet indicates Designated Router or BDR.
- The Authentication Type specified in the received OPSP packet header must match the Authentication Type specified for the OSPF area associated with the receiving interface.

- The OSPF packet must be authenticated according to the setting indicated in the Authentication Type field in the OSPF header (Figure 1.11). The authentication type can be configured on a per-interface basis, and the corresponding authentication method may use one or more authentication keys. The authentication procedure may require the verification of the Checksum field in the OSPF packet header. The Checksum, when required, is computed according to the standard IP 16-bit one's complement checksum of the contents of the OSPF packet after excluding the OSPF packet header's 64-bit Authentication field. The router will discard the OSPF packet if the authentication procedure fails.

If the router receives a Hello packet type, then it will further process it according to the Hello Protocol. Some of the receive processing includes checking the values of the Network Mask, Hello Interval, and *RouterDeadInterval* fields in the received Hello packets (Figure 1.12) against the values configured for the receiving router interface. A router that finds any mismatches in neighbor configurations will discard the packet. However, on P2P and Virtual Links, the Network Mask in the received Hello Packet (Figure 1.12) is always ignored.

As stated above, all other OSPF packet types are sent/received only on router adjacencies (except Hello packets which are sent all OSPF-enabled interfaces to discover/maintain neighbors). When received, such OSPF packets must have been sent by one of the OSPF router's active neighbor routers. When a router receives a Hello packet it will attempt to match the source of the packet to one of the interface's neighbors:

- If the receiving interface connects to a network that is broadcast (as in Ethernet), P2MP, or NBMA, the sending/originating router interface is identified by the source IP address in the IP packet carrying the OSPF packet.
- If the receiving router interface connects to a P2P or Virtual Link, the sending interface is identified by the Router ID of the source router indicated in the OSPF packet's header.

The receiving interface will have associated with it a data structure that contains the list of active neighbor routers. The router will discard OSPF packets not matching any active neighbor. All received OSPF packets that are accepted by the router at this point will be associated with an active neighbor.

- When the router receives a Hello Packet from a neighbor on a broadcast, P2MP or NBMA network, it will set the Neighbor ID in the Neighbor Data structure equal to the Router ID in the OSPF packet header. For these network types, the router will also set the Router Priority, Neighbor's Designated Router, and Neighbor's BDR fields in the Neighbor Data Structure equal to the corresponding fields in the received Hello packet.
- When the router receives a Hello on a P2P link (but not on a Virtual Link), it will set the Neighbor IP address in the Neighbor Data Structure to the source IP address of the Hello packet.

After performing the above, the router will examine the rest of the Hello packet while generating events to be passed to the Neighbor and Interface State Machines [RFC2328].

1.11 NEIGHBOR DISCOVERY AND MAINTENANCE

For two OSPF routers to be able to establish a full adjacency, the interfaces forming the OSPF relationship must be in the same OSPF area. It must be noted that, in practice, any given interface on a router can only connect to or belong to a single OSPF area.

1.11.1 OSPF Neighbor States

To help understand how OSPF neighbor formation works, we explain the following data structures (lists) used by OSPF routers which are subsets of the area LSDB an OSPF router maintains:

- **Link-State Retransmission List**: This list contains the list of LSAs that a router has flooded on an adjacency and have not been acknowledged. The router will retransmit these LSAs at defined intervals until they are acknowledged by the neighbor, or until the adjacency is torn down.
- **Database Summary List**: This summary list contains the complete list of LSAs (i.e., LSA headers) that are in the area LSDB of an OSPF router at the time the router enters the Exchange State with a neighbor. The router sends this summary list to its neighbor in Database Description packets (Figure 1.13).
- **Link-State Request List**: This list contains the list of LSAs that an OSPF router needs to receive from a neighbor in order to synchronize its LSDB with that of the neighbor. The router creates this list as it receives Database Description packets from a neighbor, which allows it to determine if it needs to request missing LSAs by sending Link-State Request packets (Figure 1.14) to the neighbor. The router depletes the list as it receives appropriate Link-State Update packets (Figure 1.15) from the neighbor.

OSPF neighbor routers that want to form a full adjacency go through multiple Neighbor States as shown in Figure 1.31. A Neighbor State is the state of a conversation between a router and a neighbor router. This figure illustrates the states that an OSPF router interface goes through before establishing a full adjacency with another router interface. OSPF routers use the Hello Protocol for neighbor discovery and maintenance, and for ensuring two-way (or bidirectional communication) with neighbors. These Neighbor States are described as follows:

- **Down**: In this state (which is the initial state of a conversation with a neighbor), the router has not received any information from any other router on its interface. On NBMA networks, OSPF routers may still send Hello packets to neighbor routers that are in the Down State, but at a reduced rate determined by the *PollInterval*.

- **Attempt**: This state is only valid when the neighbors are connected to an NBMA network. It indicates that a router has not received any recent information from the neighbor, but is making a concerted effort to contact that neighbor. It does so by sending Hello packets to the neighbor at intervals indicated by the Hello Interval.
- **Init**: In this state, a router has detected a Hello packet on its interface sent from a neighbor router, but has not yet established bidirectional communication with that neighbor (i.e., its own Router ID is not written in Hello packets sent from the neighbor). An OSPF neighbor that is in this state (or higher) has its Router ID listed in the Neighbor field in Hello packets (Figure 1.12) received from the other neighbor.
- **2-Way**: In this state, the router has seen its own Router ID in Hello packets sent from a neighbor, and has established bidirectional communication with that neighbor. The election of one router to be the DR and another to be the BDR (if such capabilities are required in the network segment) will be performed in this state or after. Also, in the 2-Way State, depending on whether one of the routers has been elected a DR or BDR, or the link connecting the two neighbors is a P2P or a Virtual Link, the two routers will follow specific procedures in building the adjacency. If a decision has to be made to not continue with an adjacency, the state of the communication with the neighbor stops at the 2-Way State.
- **ExStart**: In this state, the two routers go through the process of establishing the initial Database Description packet Sequence Number that will be used in sequencing the routing information packets they exchange. The routers use the Sequence Number to ensure that the routing information they receive is always the most recent. The routers also decide in this state, which router will become the Master and which one will be the Slave. The Master will then start to exchange routing information with the Slave. OSPF neighbor conversations that are in this state or higher are referred to as adjacencies. In the ExStart State, the OSPF router (Master) will transmit Database Description packets that are empty to the Slave, but with each having the I (Initialize), M (More) and MS (Master) bits set (Figure 1.13). The Master will retransmit these packets every *RxmtInterval* seconds (which is the number of seconds the router has to wait before retransmitting Database Description and Link-State Request packets) until the next Neighbor State is entered. The "OSPF Database Description Packet" section above already provided an overview of the Master/ Slave relationship, and the Neighbor States.
- **Exchange**: In this state, each router describes its entire LSDB to its neighbor by sending Database Description packets. Each router includes its neighbor in the flooding procedure, and the two routers begin synchronizing their LSDBs. As shown in Figure 1.13, each Database Description packet carries information about the LSAs (i.e., the LSA headers) that are currently in the router's LSDB. An LSA entry may describe link-state information about a link or network connected to the router (see Figures 1.20 to 1.26). Each LSA carries among other pieces of information, the Link-State (LS) Type, Link-State ID, Router ID of the advertising router, LS Sequence Number, and the link cost. Each

router uses the Sequence Number to determine how current the received link-state information is.

Each Database Description packet which is assigned a Sequence Number must be explicitly acknowledged by the receiver. Also, at any given time, only one Database Description packet is allowed to be outstanding and unacknowledged. In the Exchange State, a router may also send Link-State Request packets to the neighbor asking for the more recent LSAs. All router interfaces that have adjacencies that have reached the Exchange State or greater, can participate in the flooding procedure. Router interfaces in this state are fully capable of receiving and transmitting all types of OSPF packets.

In Exchange State, the state of the router (Master or Slave) determines when a Database Description packet is sent:

o Master sends Database Description packets when either (a) the Slave has acknowledged the previous Database Description packet sent by repeating the corresponding Database Description Sequence Number, or (b) after *RxmtInterval* seconds has elapsed without receiving an acknowledgment from the Slave, in which case the router will retransmit the previous Database Description packet.

o Slave sends Database Description packets only in response to Database Description packets sent by the Master. If the Slave receives a Database Description packet from the Master that is new, it will send a new Database Description packet; otherwise, it will resend the previous Database Description packet.

- **Loading**: In this state, the routers are in the process of finalizing the exchange of link-state information (i.e., completing the synchronization of their LSDBs). As explained above, each router constructs a Link-State Request List and a Link-State Retransmission List where any missing or out-of-date link-state information in its LSDB, is placed in the Link-State Request List. Any LSA that a router sends is entered in the Link-State Retransmission List until it is acknowledged. In this state, the router will send Link-State Request packets to the neighbor requesting the more recent LSAs that it has learned (but has not yet been sent) during the Exchange State.

    To request these LSAs, the router will send a Link-State Request packet to the neighbor carrying the beginning of the Link-State Request List. When the neighbor sends the proper Link-State Update packet(s) in response to these requests, the router will truncate the Link-State Request List, and will send a new Link-State Request packet. The routers will continue this process until the Link-State Request List becomes empty. For LSAs in the Link-State Request List that have been sent, and corresponding Link-State Update packet(s) that have not yet been received, the router will package these in Link-State Request packets for retransmission after *RxmtInterval* seconds has elapsed. Only at most one Link-State Request packet should be outstanding at any given time.

    When the Link-State Request List finally becomes empty (with no more outstanding LSAs), and the neighbor is in the Loading State (i.e., the router has sent a complete sequence of Database Description packets to the neighbor and received also the proper response packets), the router will generate the neighbor event of Loading Done.

- **Full**: In this state, the two routers have completely established an adjacency and they are now considered to be fully adjacent. At this point, the adjacency will be listed in LSAs advertised by the two routers. Routers that have formed a full adjacency have similar/synchronized LSDBs. In this state, the adjacency established between the two routers will now show up in the Router-LSAs and Network-LSAs they advertised.

OSPF routers that are connected to Virtual Links, P2P, and P2MP networks automatically become adjacent. In all other networks, OSPF routers establish adjacencies only with some subset of their neighbors. Specifically, on broadcast and NBMA networks, all other routers form adjacencies with both the DR and the BDR (see discussion below). An OSPF router establishes a bidirectional adjacency with a neighbor when at least one of the following conditions are met:

- The underlying network type is P2P, P2MP, or a Virtual Link
- The router itself is the DR or BDR
- The neighbor router is the DR or BDR

1.11.2   OSPF Neighbor Adjacency Formation

As explained above, the process of establishing an OSPF adjacency involves a number of steps. Once two OSPF routers have established bidirectional communication, they follow a prescribed procedure as described in **[RFC2328]** to synchronize their LSDBs. Let us consider two neighbor routers, R1 (Router ID 1.1.1.1), and R2 (Router ID 2.2.2.2) connected over an Ethernet segment. We describe the adjacency formation by condensing the process as described next.

1. To establish a full neighbor relationship or adjacency, two OSPF routers start by exchanging OSPF Hello packets. Both routers start in the Down State (Figure 1.31) which is the initial state of a neighbor relationship that indicates that the routers have not yet received any Hello packets from the other. When a router receives a Hello packet from the other, but its own Router ID has not yet been listed in the received Hello packet, it will move to the Init State. In this state, the router will list the Router IDs of all known neighbors in Hello packets it sends to the neighbors (see Figure 1.12).

2. As soon as the router recognizes its own router ID in a Hello packet sent from the neighbor, it will transition to the 2-Way State. This state indicates that the two routers have established a bidirectional communication with each other. Recall that on broadcast networks, OSPF neighbor routers optimize the exchange of OSPF information by first determining which routers will play the DR and BDR roles.

3. Next, with bidirectional communication established, the routers start to exchange the contents of their LSDBs. However, to do this, the routers first

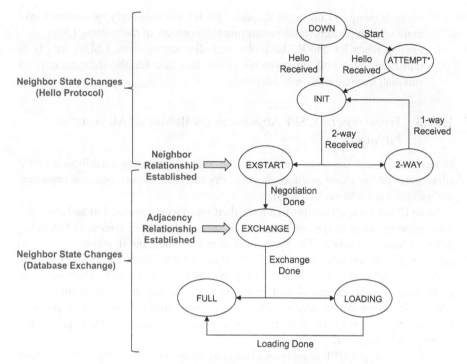

*Only valid for neighbors attached to NBMA networks

FIGURE 1.31　Neighbor State Changes

establish a Master/Slave relationship (to determine which becomes Master or Slave), and also choose the initial Sequence Number for the exchange of Database Description packets (see Figure 1.13). This is the ExStart State. The router with the higher Router ID (router R2) will become the Master, and will be the only router responsible for incrementing the Sequence Numbers. When a router receives the initial Database Description packet from the neighbor, it transitions its state to ExStart State.

4. With Master/Slave relationship established, the routers start exchanging their LSDBs. R2 will transition the Neighbor State of R1 to Exchange. In this state, R2 describes its LSDB to R1 by transmitting Database Description packets that carry the headers of all LSAs (see Figure 1.13) in its local Database Summary List. As described above, the Database Summary List contains all LSAs in R2's LSDB, but not its full content. The list contains the number of each type of LSA present in the router R2's LSDB, along with the total number of LSAs in it. To describe the content of its LSDB, R2 will send one or multiple Database Description packets to R1. R1 will then compare the content of its own Database Summary List with R2's list, and if it detects any missing LSAs, it will add these to its own Link-State Request List.

5. At this point, R1 and R2 both enter the Loading State. R2 will send a Link-State Request packet to R1 requesting the missing LSAs with their full

contents using its Link-State Request List. R1 will then reply by sending Link-
State Update packets to R2 containing full contents of the missing LSAs.

6. Finally, when R1 and R2 have obtained all contents of the LSDB, they both
transition to the Full State which signifies that their LSDBs are synchronized
and that, they are now fully adjacent.

1.11.3 FORMATION OF OSPF ADJACENCIES ON BROADCAST MULTIACCESS NETWORKS

In this section and the next, we discuss the main issues when establishing OSPF
adjacencies on broadcast multiaccess networks such as Ethernet, and non-broadcast
multiaccess networks such as ATM.

In an IP network, all routers that have their interfaces connected to and share the
same network segment (i.e., on the same IP subnet), have their interfaces belonging
to that common IP subnet. The router interfaces have the same IP address prefix and
subnet mask. When more than two OSPF routers form adjacencies on a broadcast
multiaccess network, each router will try to establish a full OSPF adjacency with
every other router on the shared common network segment. Establishing such
adjacencies on smaller broadcast multiaccess networks may not present a big problem
but could in larger networks with many router interfaces because of the large number
of adjacencies required.

The number of OSPF adjacencies required grows exponentially as the number
of routers in the network segment increases. In such a scenario, every OSPF router
is required to synchronize its LSDB with every other router in the network segment, and when the segment supports a large number of routers, this creates scalability issues, resulting in inefficiency in the formation and maintenance of
adjacencies.

Inefficiency arises when every OSPF router in a large network segment (with
many router interfaces) advertises all of its OSPF adjacencies to other routers in the
network. In a segment with full-mesh OSPF adjacencies, advertising adjacencies to
the other OSPF routers can lead to a situation where routers will be sent a large
amount of redundant link-state information since routers at both end of an adjacency
are likely to send the same information.

A solution that has been found to be scalable, effective and yet simple to implement is to have a central point (i.e., a specific OSPF router) responsible for handling
OSPF adjacencies and LSDB synchronization (see Figure 1.32). This central point
will also handle the advertisement of all relevant information to the other routers on
the segment. The central point (with which every other OSPF router in the segment
establishes an adjacency) will advertise the segment as a whole to other routers in the
network.

The routers on the network segment elect specific peer OSPF routers (a DR and
BDR) to which they delegate some key OSPF responsibilities, including
communications to handle adjacency issues on behalf of all routers connected to the
segment. The DR and BDR provide improvements to the establishment and maintenance of OSPF adjacencies as follows:

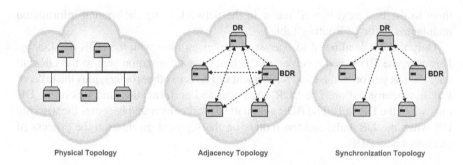

Physical Topology   Adjacency Topology   Synchronization Topology

FIGURE 1.32 OSPF Adjacency Formation and Link-State Synchronization on Broadcast Multiaccess Networks

- **Centralize Communications and Reduce Routing Information Traffic**: Each OSPF router on the network segment must establish a full adjacency with only the DR and the BDR. The DR/BDR pair serves as the centralized point with which other routers in the network exchange link-state information. Using the dedicated "OSPF All Designated Routers" IPv4 multicast address 224.0.0.6 (or equivalently the IPv6 multicast address FF00::6), each OSPF router sends its link-state information only to the DR and BDR instead of having to exchange link-state information with every other router on the segment. The DR communicates on behalf of all routers on the network segment by sending link-state information received from each one of them to the other routers in the network. The DR is responsible for flooding routing information, which significantly reduces such traffic on the segment. The DR's use of multicast further reduces the link-state traffic in the network. The DR uses IPv4 multicast address 224.0.0.5 (the "All OSPF Routers" address) to communicate with all other non-DR routers.
- **Managing the Synchronization of Link-State Information**: The DR and BDR receive link-state information from any router on the segment, and handle communications with other routers in the network to ensure that they all have the same link-state information originated by the sender. The presence of a DR and a BDR, thereby, reduces the amount of link-state information communicated that is prone to loss and retransmission, resulting in a reduction in the number of routing errors in the network.

The OSPF routers in the network segment send only link-state information (LSAs) to the DR/BDR and not end-user data packets (unless the DR or BDR itself is the target destination of the data packets). As noted above, each time a router sends LSAs, it sends them to the DR and BDR using the IPv4 multicast address 224.0.0.6. The DR will then re-advertise/flood the routing information to all other routers in the area using the IPv4 multicast address 224.0.0.5. This prevents all the routers on the segment from having to constantly update each other (in a full mesh) but rather obtain all their routing information from the centralized DR. However, when routing normal end-user packets to their destinations, the OSPF routers on the segment will forward them to the best next-hop IP router in the network using the routing information maintained in the IP Routing Table.

Only the DR is allowed to operate at any given time with the BDR on standby. However, the BDR also receives all the link-state information sent to the DR, but only the DR performs the re-advertisement of LSAs to other routers, plus the tasks of LSDB synchronization. The BDR takes over and performs the functions of the DR only when the DR fails. The BDR automatically takes over and becomes the working DR when the DR fails, and the routers on the segment go through the process of electing a new BDR.

1.11.4 ELECTION OF THE DESIGNATED ROUTER (DR) AND BACKUP DESIGNATED ROUTER (BDR)

When OSPF router wants to establish an adjacency with a neighbor, it will first send OSPF Hello packets to discover which neighbors are available and reachable on the common network segment. After the router and its neighbor have established a bidirectional communication and both are in the 2-Way Neighbor State (Figure 1.31), they will start the DR/BDR election process. The OSPF Hello packets that the routers exchange, contain among other fields, three specific fields as shown in Figure 1.12 (i.e., Designated Router, BDR and Router Priority) that are used for the DR/BDR election.

Each OSPF router in a broadcast multiaccess and NBMA network is assigned a Router Priority (Figure 1.12) which can take a value from 0 to 255. A value equal to 0 indicates that the router is ineligible to become a DR or BDR. A value of 1 means the router has the least chance of becoming a DR while a value of 255 means the router is always the DR unless there is another router with equal value.

When electing a DR and BDR, the OSPF routers on the segment examine the Router Priority value of other routers as they exchange Hello packets, and then use the following (condensed) rules to determine which router will become the DR and BDR. Details of the DR/BDR election process are given in Section 9.4 of **[RFC2328]**:

1. Among all the routers on the network segment sending Hello packets, the router that has the highest Router Priority will be elected the DR.
2. The router with the next highest Router Priority value will be elected the BDR.
3. In the event, multiple routers have equal Router Priority values (a tie), then the router with the highest OSPF Router ID will be elected the DR, while the one with the next highest OSPF Router ID will become the BDR. The OSPF Router ID, also carried in the Hello packet (Figure 1.12), is used as the tiebreaker in this case.

At the end of the DR/BDR election process, the Designated Router and Backup Designate Router fields of OSPF Hello packets sent over the segment will contain the routers that are DR and BDR. If a network segment first becomes active and no router is a DR or BDR on the segment, a BDR is first elected, and then the routers go through a second process of electing the DR. According to **[RFC2328]**, any OSPF router that has not been elected a DR or BDR is referred to as a DR Other (Figure 1.33).

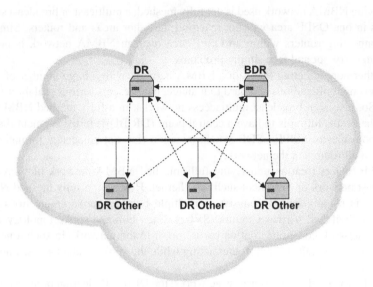

Network Segment

FIGURE 1.33 DR, BDR and DR Others

Once a DR and BDR are elected on a network segment, they will continue to function and not be preempted until there is a need for a new election. If an OSPF router with a higher Router Priority (which was the previous DR) comes online after the election has taken place, it cannot preempt the existing DR or BDR until at least either one fails. If the current DR becomes unavailable, the current BDR takes over as the new DR and a new election takes place to determine the new BDR. Even if the new DR fails and the previous DR is now available, the current BDR will still become DR.

This requirement makes the shared network segment more stable by preventing a DR/BDR election process from taking place whenever a new OSPF router becomes active on the segment. This means once two DR-eligible routers on the segment have been elected as DR and BDR, a new election will take place only when one of them fails.

1.11.5 OSPF Adjacencies on Non-Broadcast Multiple Access (NBMA) Networks

Establishing OSPF adjacencies on NBMA networks generally poses problems because of the requirement of full-mesh PVC connections between all routers in the network segment. Without these virtual connections, the routers in the NBMA network segment will not be able to synchronize their LSDBs directly among themselves.

From a general networking perspective, it is often very difficult to interconnect multiple OSPF areas over an NBMA network. This is because, if the topology of the underlying NBMA network used is not fully meshed, a multicast or broadcast sent by routers in one OSPF area will not reach all the other areas and routers. Similarly, interconnecting routers within an OSPF area over an NBMA network is not that trivial to carry out and face similar problems.

Furthermore, creating PVCs in a NBMA network with a large number of OSPF routers is not only tedious and hard to manage, but presents serious scalability problems. So, just as in broadcast multiaccess networks, in a fully meshed NBMA network the best solution is to have a central point (DR/BDR) in the segment that will be responsible for OSPF LSDB synchronization (but not adjacency formation) on behalf of all routers in the network.

OSPF routers treat an underlying fully meshed NBMA network like any other broadcast network environment such as Ethernet. However, a fully meshed NBMA network is usually built to consist of multiple hub-and-spoke connections using PVCs or Switched Virtual Circuits (SVCs). Each hub-and-spoke topology in the NBMA network constitutes only a partial mesh in the network. In such a network environment, the DR is a hub router acting while the spoke routers are just non-DR (DR Other routers).

Similar to broadcast multiaccess networks, the DR/BDR election process can also take place in a fully meshed NBMA networks. If the DR becomes unavailable, the BDR will immediately become the DR, and a new BDR will be elected. The DR/ BDR election process takes place when the network first becomes active, and also when the DR becomes unavailable. On NBMA networks, the DR and adjacent OSPF routers communicate using IP unicast addresses instead of multicast addresses.

A segment of an NBMA network may also be designed as a hub-and-spoke topology providing only a partial mesh between routers (Figure 1.34). Such a topology does not provide multiaccess capabilities for all OSPF routers as seen in broadcast and fully meshed NBMA network. In this hub-and-spoke environment, logically the hub router acts as the DR and the spoke routers simply as DR Other routers. In this case, it is recommended to configure an OSPF Router Priority value of 0 on the spoke routers so that they will not be able to participate in any DR/BDR election process nor eligible to be DR/BDR [TEAREDIA15].

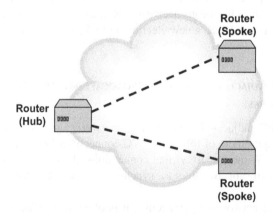

FIGURE 1.34 DR in a Hub-and-Spoke Topology

DRs/BDRs are not required on P2P links because the two OSPF routers on either end of the link must form a full OSPF adjacency, and also the link-state traffic exchanged between them cannot be further reduced or optimized.

1.12 OSPF LINK-STATE DATABASE SYNCHRONIZATION

When OSPF routers are initialized, they first go through an exchange process using the OSPF Hello protocol and packets. Figure 1.35 shows the exchange process that the OSPF routers go through. The upper portion of the figure shows when routers are establishing neighbor adjacencies, and lower portion shows when the routers are exchanging and synchronizing their LSDBs. We assume Routers 1 and 2 are both connected to a broadcast network (such as Ethernet), and Router 2 is the DR for the network. Router 2 has a higher Router ID than Router 1:

Establishing Neighbor Adjacencies (Figure 1.35):
- We assume that Router 1 is on a network segment and is in the Down State because it has not exchanged any information with any other OSPF router. Router 1 starts by transmitting a Hello packet through each of its OSPF-enabled interfaces even without the need to know the identity of the DR or DR Other routers it is connected to. The Hello packets are sent out with the destination address set to the "All OSPF Routers" multicast address 224.0.0.5.

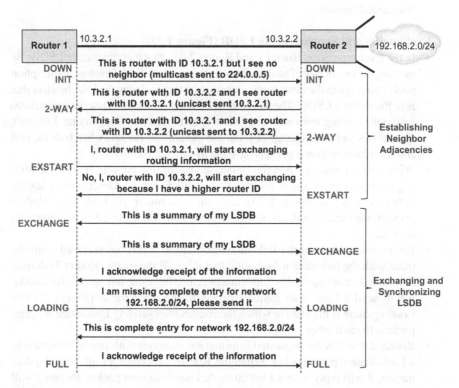

FIGURE 1.35 Link-State Database Synchronization

- All OSPF routers (including Router 2) that are directly connected to Router 1's interfaces receive the Hello packet and add Router 1 to their lists of neighbor routers (in the Neighbor Database). After Router 2 adds Router 1 to its neighbor list, it transitions to the Init State.
- Each OSPF router (including Router 2) that receives a Hello packet replies by sending a Hello packet as a unicast to Router 1 with its corresponding neighbor router list information. Router 2 writes all its neighbor routers including Router 1 in the Neighbor field in the Hello packet (see Figure 1.12).
- When Router 1 receives the unicast reply Hello packets from its neighbors (including Router 2), it will add all the OSPF routers that have listed its Router ID in their Hello packets to its own (local) neighbor list (database). After completing this process, Router 1 transitions to the 2-Way State. At this stage, any two OSPF routers that have listed each other in their local Neighbor Databases have established bidirectional communication. If the router links are attached to a network segment that requires the election of a DR and BDR, then this will be performed before the Neighbor State proceeds to the next phase. Router 2 indicates that it is itself the DR in the Hello packets sent to from Router 1.
- In the ExStart State, Router 1 and Router 2 will go through the process of establishing a Master/Slave relationship. Since Router 2 has the higher Router ID (10.3.2.2), it will act as the Master during the link-state exchange process. Router 1 transitions to Slave State and adopts Router 2's Database Description Sequence Number.

Exchanging and Synchronizing LSDB (Figure 1.35):
- In the Exchange State, Router 1 and Router 2 describe their respective LSDBs by exchanging one or more Database Description packets. Each Database Description packet conveys to the other router information that includes the LSA headers that is in the router's LSDB. The two routers exchange Database Description packets, with polls coming from Router 2 (Master) and responses from Router 1 (Slave). The routers end the exchange of Database Description packets when both the poll and associated response packets have the M-bit cleared.
- When either router receives the Database Description packets, it will acknowledge the receipt of these packets using still Database Description packets. However, when complete LSAs are sent to a router via Link-State Update packets, the receiver will acknowledge them using Link-State Acknowledgment packets.
- Each router compares the link-state information that it has received from the other with the link-state information that it has. If the router (Router 1) detects that it needs more up-to-date link-state or is missing some link-state information, it will send a Link-State Request packet to the other router (Router 2). The Loading State is the state in which the routers start sending Link-State Request packets to each other.
- Router 2 that has the requested information responds with that information in a Link-State Update packet. When the Router 1 receives a Link-State Update packet, it will reply with a Link-State Acknowledgment packet. Router 1 will add the new or missing link-state entries to its LSDB.

- When both Router 1 and Router 2 have received all the required information for all Link-State Request packets sent out, their LSDBs are considered to be synchronized. The Neighbor States of both routers are in the Full State, and their LSDBs are considered identical (or synchronized).

1.13 LSDB SYNCHRONIZATION ON MULTIACCESS NETWORKS

On multiaccess segments with built-in broadcast and multicast capabilities like Ethernet segments, the election of a DR and BDR allows OSPF to optimize the exchange of LSAs and the synchronization of LSDBs. The DR and BDR establish neighbor adjacencies with all the DR Other routers on the segment, while the DR Other routers, on the other hand, establish full adjacencies only with the DR and BDR.

A DR Other router exchanges its LSDB only with the DR, and the DR handles the synchronization of any changed or new LSAs with the rest of the routers on the network segment (Figure 1.36). Figure 1.36 illustrates the LSA exchange process between the DR and DR Other routers in the network segment. The following steps describe the process of LSDB synchronization on a multiaccess network segment:

Step 1: DR Other Router 2 detects a change in a link-state and multicasts a Link-State Update packet (which includes the updated LSA entry [Figure 1.15]) to the DR and BDR on the network segment using the "OSPF All Designated Routers" multicast address 224.0.0.6. The Link-State Update packet may contain several distinct LSA entries in it (Figure 1.15).

Step 2: The DR acknowledges receipt of the Link-State Update packet, and in turn floods it to the other DR Other routers on the network segment using the "All OSPF Routers" multicast address 224.0.0.5.

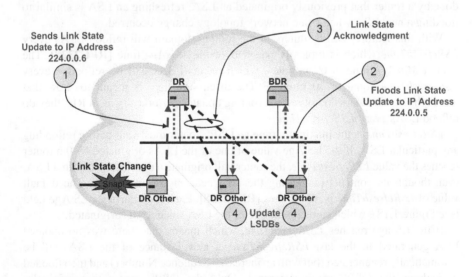

FIGURE 1.36 LSDB Synchronization on Multiaccess Networks

Step 3: Each DR Other router (Router 1) receives the reflooded Link-State Update packet and responds to the DR with a Link-State Acknowledgment packet.

Step 4: All DR routers on the network segment that receive the Link-State Update packet will update their LSDBs. This Link-State Update packet includes the changed LSA that was sent by DR Other Router 2.

1.14   ADVERTISING LSDB CHANGES AND OSPF "PARANOID" UPDATES

When a network topology change occurs, any affected OSPF router will generate an updated LSA to reflect the change. Each updated LSA carries a Sequence Number field (Figure 1.18) that is incremented so that other routers in the routing domain can distinguish an updated LSA from old ones. However, even in a stable OSPF routing domain (or area), OSPF routers still periodically reflood new instances of (existing) LSAs in order to synchronize their LSDBs. This periodic LSA update ensures that all the routers in the routing domain have the same view of the network topology by getting refreshed view of the LSDBs of neighbors.

This feature of periodic LSA update is commonly referred to as Paranoid Flooding or Paranoid Update **[TEAREDIA15]** because this LSA update process is only used to refresh the LSDB of routers. Note that the periodic updates each router performs are only on the LSA originated by the router itself and not the entire LSDB (i.e., not those LSAs originated by other routers). Thus, each OSPF router will periodically examine each LSA that it has originated itself, increment its Sequence Number, reset the LS Age and reflood it. Basically, each router refreshes a new instance or version of its own LSAs periodically via periodic reflooding. Note that performing Paranoid Updates does not involve resynchronization of an entire LSDB. Other than being done by a router that previously originated an LSA, refreshing an LSA is similar to flooding a new LSA when some network topology change occurred.

With this feature, OSPF routers in the routing domain will reflood LSAs every *LSRefreshTime* which is typically one half of the *MaxAge* time **[RFC2328]**. The default *MaxAge* time is 60 minutes, which means LSAs will be reflooded every *LSRefreshTime* equal to 30 minutes. The discussion here is meant to show that although OSPF routers do not refresh routing updates periodically as in RIP, they do reflood LSAs every *LSRefreshTime*.

LSRefreshTime is the maximum time an OSPF router must wait before reflooding any particular LSA. If the LS Age value of one of the LSAs originated by the router reaches the value *LSRefreshTime*, the router will originate a new instance of that LSA, even though its contents (excluding the LSA header) are unchanged. The default value of *LSRefreshTime* is 30 minutes **[RFC2328]**. Each LSA carries a LS Age field (see Figure 1.18), which sums up the age of the LSA since it was originated.

If the LS age reaches *LSRefreshTime*, which means that there was no updated LSA generated in the last *LSRefreshTime*, a new instance of the LSA will be automatically regenerated (but with an increased Sequence Number) and then flooded throughout the OSPF routing domain. Only the OSPF router that originally generated the LSA, will reflood the LSA every *LSRefreshTime*. It is important to note that nothing detrimental will happen if an LSA is not refreshed and the *LSRefreshTime* expires. This is because each flooded LSA will still be valid for another 30 minutes (given that the default MaxAge is 60 minutes).

MaxAge is the maximum age that any particular LSA originated by an OSPF router can reach. When the LS Age of an LSA reaches *MaxAge*, the OSPF router will reflood it with the goal of flushing that LSA from the OSPF routing domain. Routers will not use LSAs that have attained the age of *MaxAge* in their Routing Table calculations **[RFC2328]**. When a router determines that an LSA has reached a maximum age of *MaxAge* in the LSDB, it will remove it from the LSDB, and will perform a new SPF calculation. The router will flood the LSA to other OSPF routers in the routing domain, informing them to also flush the LSA from their LSDBs.

1.15   OSPF SHORTEST PATH COMPUTATIONS AND THE IP ROUTING TABLE

After establishing the LSDB, the OSPF router uses the SPF (based on the Dijkstra) algorithm to compute the best routes (shortest routes) to all known network destinations. The input information used for the SPF computations is the information maintained in the LSDB which is constructed through the exchange of LSAs between routers using different OSPF message types as described above.

A router maintains a separate LSDB for each OSPF area to which it belongs. Also, all routers belonging to the same OSPF area have identical LSDBs for the area after synchronization. A router always handles the LSDBs for each individual OSPF area separately and performs shortest path calculations separately for each area. Routers in an OSPF area flood LSAs from its LSDB throughout that area only. In addition to Router-LSAs (Type 1) and Network-LSAs (Type 2), the LSDB of an area contains Summary-LSAs (Type 3), and ASBR-Summary-LSAs (Type 4). Furthermore, all area LDSBs that are not Stub Area LSDBs will include AS-External-LSAs (Type 5).

Each time a network topology change occurs, OSPF routers use SPF to determine best paths to all known destinations. Each OSPF routers uses an SPT developed from its LSDB to calculate the best route to each network destination which is then installed in the IP Routing Table. As noted above, OSPF routers exchange appropriate LSAs to reflect the network topology in their LSDBs. The synchronized LSDBs are then used as input for the best path calculations. Each time there is an intra-area network topology change, OSPF routers within the area must rerun SPF algorithm to make best path selection.

Inter-area network changes, which are described in OSPF Type 3 LSAs (Summary-LSAs), do not cause OSPF routers to perform SPF recomputations because the routing information required for the best path computations remains unaltered. This is because OSPF routers (within an area) determine the best inter-area routes by calculating the best paths to an ABR. So, network topology changes that are described in Summary LSAs do not influence how router within an OSPF area reach an ABR, therefore, making SPF recalculation unnecessary.

Each OSPF router constructs an SPT where the router itself is the root and with the tree branching off to other routers (if any) and the destination networks. The router (i.e., the root) uses the SPF algorithm to calculate the cost along each branch (or path) to a destination. The router will then use the SPF algorithm to calculate the best path from itself (the root) to each known network destination. Note that the SPT is anchored on the router constructing the tree and calculating the best paths.

The SPT describes the entire path to any network destination including directly connected networks and hosts. However, each router will use only the next hop router (and its associated outgoing interface) leading to the destination for actual IP packet forwarding. The router selects the path with lowest cost as the best path to a destination. This path (having the lowest cumulative cost value between the router and the destination) will be installed in the router's IP Routing Table.

Cumulative path cost = Sum of the cost of all outgoing router interfaces along a route

Best route for a destination in the Routing Table = The route with the lowest total cost

Thus, the following steps give a high-level description of the best path computation process:

- Upon initialization (i.e., system power up or restart) or when a network change occurs (resulting in a change in routing information such as the cost of a link, a network being added or deleted), an OSPF router will generate an LSA that contains the collection of all link-states on the router.
- All routers in an OSPF area exchange LSAs by means of flooding. Each OSPF router that receives an LSA would store a copy in its LSDB and then will propagate it to other routers in the area.
- After the LSDB of each OSPF router in the area has been updated and synchronized, each router will calculate a SPT to all network destinations using the Dijkstra algorithm. Each router places itself at the root of a tree and calculates the shortest path (or lowest cost) to each destination based on the total cost (of all outgoing interface costs) required to reach that destination. Each router will construct its own SPT even though all the OSPF routers in the area will be using the same LSDB.
- After a router has constructed its SPT, it will populate the best routes in its IP Routing Table accordingly. Networks directly connected to the router will be assigned a cost of 0 while remote networks will be reached according to the cost calculated in the SPT. Each network destination address, its associated cost, and the next-hop router through which that destination can be reach, are added to the router's IP Routing Table.
- In the case where no network changes occur, the OSPF routers in the area do not perform any best path computations. However, the OSPF router will communicate any new network changes that occur by flooding new LSAs, and will use the Dijkstra algorithm to recalculate new best paths.

The SPT an OSPF router constructs cannot grow beyond the area the router belongs to. So, if the router has interfaces that are attached to multiple OSPF areas, it has to construct a separate SPT for each area. For the processing of external routes, each OSPF router (the root) will determine the next hop and cost/distance to the router (i.e., ASBR) advertising external routes.

The main difference between the LSDB and IP Routing Table in an OSPF router is that, the LSDB contains a complete collection of the raw LSAs that OSPF routers exchange, while the IP Routing Table contains the best routes to all known destinations reachable through specific interfaces (next-hops) of the router. The Routing Table contains all of the most relevant information a router will need to forward IP packets to their destinations.

Whenever an OSPF router receives an IP packet, it will identify the shortest path leading to the packet's destination in its local Routing Table. The IP packet is then forwarded to the next hop which is the adjacent router that is on the shortest path to the packet's destination. The Routing Table may also contain a default route (Destination ID = Default Destination, and with Network Mask = 0x00000000). The default route matches all IP destination addresses and, if it exists, is used when all other entries do not match a packet's IP destination address.

Three main routes are involved in inter-area routing of an IP packet from an Internal Router in a particular OSPF area. First, the Internal Router will identify the shortest or lowest-cost (intra-area) path to the Backbone Area via the nearest ABR, and the packet will be forwarded to that ABR. Second, the ABR will determine the shortest path through the Backbone Area to the destination OSPF area's ABR. Third, the destination area's ABR will determine the shortest (intra-area) path to the destination area's Internal Router. Once the Backbone Area routes the packet to the destination area's ABR, it will route the packet via the shortest path to the Internal Router which will then deliver it to its intended destination.

1.16   OSPF ROUTING METRICS

OSPF uses as its basic routing metric, a path cost that represents the cumulative costs of all outgoing interfaces from the router advertising the route up to the destination. The two-area routing abstraction supported by OSPF naturally leads to two types of routing; intra-area routing that determines routes within an area, and inter-area routing for routing between routers in different OSPF areas.

The relevant OSPF standards including ([RFC2328]) have not defined a specific method for computing a metric for OSPF (as done in RIPv2 [RFC2453] and EIGRP [RFC7868]). So, most vendors including Cisco [CISCID7039] [TEAREDIA15] and Juniper [JUNOSPFGUID] have defined their own metrics based simply on the configured interface speed (bandwidth) of the outgoing link on the router advertising a route.

This simple definition of OSPF routing metric allows network administrators to choose routing metrics suitable for their OSPF networks without resorting to complicated criteria or calculations to derive a routing metric. As common practice, Cisco routers define the interface cost to be inversely proportional to the outgoing interface bandwidth scaled by a factor which can be explicitly set up for the OSPF routers in an autonomous system (see Chapter 1 of Volume 1 this two-part book).

1.16.1 OSPF Metric Types and Routes

It is important to note that routing metrics used by different protocols are generally not comparable. Even metric used in the same routing protocol (e.g., OSPF) are only directly comparable when they are of the same type. OSPF network recognizes the following four types of routing metrics.

1. Intra-area metric
2. Inter-area metric
3. Type 1 External metric: This metric sums the external path cost advertised an ASBR and all the internal costs to that ASBR together to obtain the total cost to the external destination. Thus, this metric includes both the external path cost advertised and the sum of all internal path costs to the ASBR that advertised the external route. The internal cost is the cost of each link in the autonomous system to the advertising ASBR:
   a. An OSPF router uses this type of metric to avoid suboptimal routing when there are multiple ASBRs advertising a route into the same autonomous system.
   b. An OSPF router will increment the Type 1 External metric value of an external route according to its cost computation algorithm anytime it crosses the router's outbound interface.
4. Type 2 External metric: This metric considers only the external path cost from and ASBR to the external destination to be the OSPF cost. The value of the cost to the external destination is solely that of the external path cost and ignores all internal costs to reach the advertising ASBR. All route types except Type 2 External routes include the entire path's cost.
   a. An OSPF router uses this metric type if only one ASBR is advertising a route into the autonomous system.
   b. A Type 2 External metric remains unchanged no matter how many routers in the OSPF autonomous system it crosses.

When OSPF imports an external route, it includes the external cost or metric of the route. The difference between the Type 1 and Type 2 External metrics lies in how OSPF calculates the cost of the external route. Both Type 1 and Type 2 External metrics can exist in an OSPF autonomous system at the same time. When OSPF supports the two types of external metrics, then Type 1 External routes are always preferred over Type 2 External routes to the same destination. When all external routes are Type 2 external routes, then OSPF will always prefer the routes with the lowest advertised Type 2 External metric. *OSPF routers always prefer routes in the following order: intra-area, inter-area, Type 1 External, and Type 2 External* **[CISCID7039]**.

Considering the discussion above, the best path computation for internal and external destinations can be summarized as follows:

1. Each router in an OSPF area calculates the best paths to destinations (intra-area routes) and adds these to the IP Routing Table. This involves using OSPF Router-LSAs (Type 1) and Network-LSAs (Type 2).
2. Each router in an OSPF area calculates the best paths to the other OSPF areas (inter-area routes). This involves using Summary-LSAs (Type 3) and ASBR-Summary-LSAs (Type 4).
3. Each router in an OSPF area (except Stub Areas) calculates the best paths to the external destinations. This involves using AS-External-LSAs (Type 5) and either OSPF External Type 1 or External Type 2 routes.

Next, we will examine how OSPF routers determine the link costs that are used to calculate the best paths to network destinations. Once the OSPF routers have synchronized LSDBs in a network segment or routing domain (or area), each router will the determine individually the best paths through the network to all destinations.

The OSPF cost to a network destination is an indication of the communication overhead a router will incur to forward packets over one its interface to that destination. When an OSPF router uses the SPF algorithm to determine the best path to a known destination, it computes the total cost of each available path and then compares these total path costs to determine the path with the lowest cost. The router will then select the paths with the lowest costs as the best paths to that destination.

1.16.2 OSPF COST CALCULATION

OSPF cost based on interface bandwidth is computed using the following formula [CISCID7039] [TEAREDIA15] [JUNOSPFGUID]:

$$cost = \frac{Reference\ Bandwidth}{Interface\ Bandwidth}$$

In Cisco routers, the Reference Bandwidth has a default value of 100 Mb/s (i.e., 10^8 bits per second). With this definition, the interface cost value is a 16-bit positive value that takes a value from 1 to 65,535, where a lower value is a more preferable cost. This definition has some serious limitations in that, on high-bandwidth links (greater than 100 Mb/s), the cost computed has little meaning since all costs greater than 100 Mb/s are assigned a value of 1. This limits the ability of OSPF to optimally select the lowest cost path as it treats all the high-bandwidth interfaces/links as being equal. So, on high-bandwidth links, it is recommended to set the OSPF costs manually on each interface.

In an OSPF network, the cost is computed for all router links/interfaces along the path to a destination, and a router makes best path selection based on the total cost along a path. A router will compute a cost only on outbound paths/interfaces – inbound interface costs are not considered when making best path selection decisions. Routers recompute the OSPF cost to a destination after every network topology change, and they use the Dijkstra's algorithm to recompute the best path after adding all the outbound interface/link costs along a path to that destination.

It is important to note that the lowest cost route based on the interface bandwidth cost metric does not necessarily result in the true shortest route to a destination but is the best route in regards to the configured interface bandwidths. This metric does not account for the actual traffic loading and end-to-end latency along a route.

1.16.3 Cost of Intra-Area Routes

To compute the cost of intra-area routes, each OSPF router will first examine its LSDB to identify all (destination) networks within its area. For each available route, the OSPF router will compute the cost to reach a particular destination (network) by adding up the individual interface costs along that route. For each destination, the router will then select the route with the lowest total cost as the best route. In the scenario where two or more paths to a destination have the same lowest total cost, the router may select these routes as the best paths and install them in its Routing Table. The router can then perform equal-cost load balancing over these routes to the destination.

1.16.4 Cost of Inter-Area Routes

OSPF routers that are within an area receive only summarized routing information provided an ABR about inter-area routes. As a result, routers within an OSPF area do not calculate the cost of inter-area routes the same way as they do for intra-area routes. Routers calculate inter-area routes by examining Summary-LSAs advertised ABRs.

When an ABR connected to an OSPF area advertises routing information about inter-area routes by advertising Summary-LSAs (Type 3 LSAs), it will include also the lowest cost route to a specific destination in the LSA. An OSPF router within the area receiving these LSAs will add the cost to reach that specific ABR to the cost advertised in the Summary-LSA. The Internal Router in the OSPF area will then select the route with the lowest total cost as the best inter-area route.

To avoid a single point of failure when a single ABR is used for an area, best practice OSPF network design will use at least two ABRs for the area [**TEAREDIA15**]. However, using multiple ABR has some pitfalls and has to be used carefully. As a result of using multiple ABRs, an ABR can learn about an inter-area route to a specific network destination in another OSPF area, and also learn via the Internal Routers within the area, another inter-area route leading to the same destination from another ABR attached to the area.

With this, the ABR ends up learning about a direct inter-area route to the destination, plus an intra-area route pointing to another ABR that leads to the same network destination. In such a scenario, since OSPF routers prefer intra-area routes over inter-area routes to the same destination, the ABR (when forwarding traffic) may end up preferring the intra-area route going through the other ABR to the destination, rather than, the direct inter-area route. The direct inter-area route may have a lower cost to the destination, but the ABR will prefer the intra-area route over that inter-area route.

1.16.5 Cost of a Default Route in an OSPF Stub Area

A Stub Area or NSSA with a single ABR can be configured to advertise a default route into the Stub Area with a cost of 1. This can be a configurable cost for the summary default route that is generated by the ABR and advertised to routers within the Stub Area. When the Stub Area has redundant ABRs to the Backbone Area that are serving as primary and secondary exit points, the primary ABR can be configured to advertise a default route with a lower cost and the secondary ABR to advertise a default route with a higher cost.

The secondary ABR would advertise a higher cost default route and will attract traffic external to the Stub Area into the area only when the primary ABR fails. However, inter-area traffic originating from the Stub Area will follow the shortest path, that is, will travel via the primary ABR.

1.17 OSPF ROUTE SUMMARIZATION

We have seen above that an OSPF area is simply a logical collection of OSPF routers, links, networks, and hosts that have been assigned the same Area ID. Each router within an area maintains an identical LSDB for that area. Also, each router does not have to maintain detailed topology information about other OSPF areas or the network outside its local area which leads to a significant reduction in the size of its LSDB.

The creation of OSPF areas limit the scope of routing information distribution within the OSPF autonomous system. The LSDBs that routers within the same OSPF area maintain must be synchronized and be identical, which means, route summarization and filtering can only be done between different OSPF areas. The need to maintain identical LSDBs within an area means that it is not possible to perform route filtering when passing routing updates within an area. Thus, one of the benefits of dividing an OSPF autonomous system into areas is a reduction in the number of routes the ABRs will propagate, and that is, by allowing them to apply appropriate filtering and summarization of routes when needed.

Route summarization is the aggregation or consolidation of multiple routes (i.e., IP network address prefixes) into one single route (or address prefix) that can be advertised to other routers. Thus, an important feature of OSPF is it allows ABRs to summarize routes from within an area and advertise these summary routes at area boundaries (i.e., inter-area route summarization), Also, ASBRs can summarize external routes and advertise these summary routes into an OSPF autonomous system at the autonomous system boundaries (i.e., external route summarization) as illustrated in Figure 1.37. It is generally advantageous to summarize routes in the direction of the Backbone Area so that it will receive all the summary routes and in turn inject them into other OSPF areas.

1.17.1 Benefits of OSPF Route Summarization

One key benefit of route summarization is, it reduces the size of the LSDBs and the Routing Tables maintained by OSPF routers, which in turn reduces the memory size

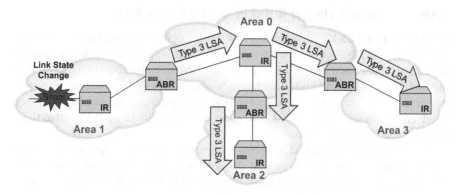

a) Inter-Area Route Summarization at ABRs

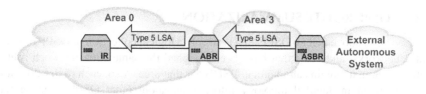

b) External Route Summarization at ASBRs

FIGURE 1.37 Route Summarization in OSPF: (a) Inter-Area Route Summarization at ABRs, (b) External Route Summarization at ASBRs

and protocol processing requirements on the routers. Another benefit is it reduces the amount and frequency of LSAs flooding by OSPF routers in the autonomous system. If a more-specific route of a summary route fails or flaps, or a network link behind a summary route fails, that network topology change will not be propagated into the OSPF Backbone Area or other areas. This protects the OSPF routers receiving routes from unnecessary Routing Table recalculations. All these factors allow OSPF networks to scale to very large sizes.

As discussed earlier, OSPF also defines a number of special area types such as Stub Areas, NSSAs, and Totally Stubby Areas. These stub area types allow default routes to be injected into them making the flooding of Type 3 LSAs (Summary-LSAs), Type 4 LSAs (ASBR-Summary-LSAs), and Type 5 LSAs (AS-External-LSAs) unnecessary. These stub area types are designed to prevent the flooding of certain LSA types and as a result, reduce the total amount of routing information flooded in the overall OSPF autonomous system. This in turn, reduces the size of the LSDBs and Routing Tables maintained by routers within the area.

Network designers are therefore encouraged to use stub area techniques wherever possible when designing their networks. Default routes reduce the size of the Routing Table maintained by routers in the Stub Areas, as well as, their memory and processing loads. By reducing memory and protocol processing requirements in routers, the use of stub area techniques can significantly improve the performance and scalability of OSPF networks, allowing them to scale to larger sizes.

1.17.2 Types of OSPF Route Summarization

This section describes the different types of route summarization used in OSPF networks. OSPF supports only manual route summarization which can be done on inter-area OSPF routes at the ABRs and OSPF external routes at the ASBRs. OSPF does not support any form of automatic summarization as done in other routing protocols like EIGRP.

Routers within each OSPF area generate Router-LSAs (Type 1 LSAs, see Figure 1.20) and Network-LSAs (Type 2 LSAs, see Figure 1.21) which then get translated into Summary-LSAs (Type 3 LSAs) by ABRs to be propagated in other areas in the OSPF autonomous system, as illustrated in Figure 1.37. When route summarization is configured, OSPF enables the following:

- ABRs and ASBRs receive multiple routes (from OSPF areas for ABRs and external networks for ASBRs) and summarize them into single advertisements for propagation to other areas.
  - o ABRs and ASBRs advertise multiple specific network address prefixes as only one summary network address prefix.
- ABRs consolidate multiple routes into Summary-LSAs (Type 3 LSAs) while ASBRs do so into AS-External LSAs (Type 5 LSAs).
  - o An ABR will summarize multiple internal routes into a single Summary-LSA (which describes the summary route) and propagate it into the Backbone Area. When a link failure occurs in an area, ABRs will not propagate the network topology change into the Backbone Area and then to other areas in the OSPF autonomous system. The ABRs will not flood LSAs describing the link failure outside the affected area.
  - o The ASBR will advertise each external route that is redistributed into the OSPF autonomous system from other routing protocol domains individually using AS-External-LSAs (Type 5 LSAs). Therefore, to reduce the size of the OSPF LSDBs, a summary route can be configured for external routes. With this, the ASBR will perform route summarization of external routes in Type 5 LSAs (carrying redistributed summarized routes) before injecting them into the OSPF autonomous system.

Without route summarization, an ABR will propagate all the IP address prefixes of internal routes from an OSPF area into the Backbone Area (Area 0) in Summary-LSAs as inter-area routes. Similarly, without route summarization, an ASBR will pass all the external network prefixes redistributed from an external autonomous system into the OSPF autonomous system.

Route summarization reduces unnecessary flooding of LSAs by ABRs and prevents routers outside an OSPF area affected by a network change from having to rerun their SPF algorithm, thereby, increasing the stability of the overall OSPF network. In such a case, injecting Summary-LSAs into an area does not cause the routers within that area to rerun the SPF algorithm. However, the routers within the OSPF area experiencing the network change will add or delete the routes being advertised in the Summary-LSAs from their Routing Tables and will perform the

SPF calculations. Because of the high demand placed on the OSPF routers' CPUs by the SPF calculations, proper route summarization forms an important part of OSPF network design.

A key requirement for route summarization is a good IP addressing plan for the OSPF autonomous system where IP subnets and addresses can be assigned based on the OSPF area structure. As much as possible, IP addresses in an area should be assigned contiguously to ensure that they can be aggregated into a minimal number of network prefixes. This IP addressing plan and area structuring facilitates IP prefix aggregation/summarization at the OSPF area borders and provides optimal route summarization results. The IP addressing plan will govern very much how effective OSPF route summarization will be and how well the overall network will scale.

1.18   OSPF VIRTUAL LINKS

As discussed earlier, and also discussed in greater detail here, OSPF Virtual Links provide a very useful tool for OSPF networking as highlighted by the following points:

- They allow OSPF areas to connect to other areas other than the Backbone Area
- They serve as a useful tool for repairing a discontinuous Backbone Area
- They can be used to create an optimal path to an OSPF area
- They can be used for path backup purposes

The two-level routing hierarchy used in OSPF dictates that if more than one OSPF area is required in an OSPF network, then one of these areas must be the Backbone Area (Area 0). All Non-Backbone Areas must be directly connected to the Backbone Area, and it must be contiguous to avoid making some areas of the Autonomous System unreachable. Also, all Non-Backbone Areas inject routing information into the Backbone Area which in turn distributes this information to other Non-Backbone Areas.

A Virtual Link is a logical or virtual connection that can be used to connect any two physically discontiguous or separate portions of the Backbone Area, or connect a detached Non-Backbone Area to the Backbone Area via a Transit Area. The two end-point routers of the (common) Transit Area are ABRs, and the (common) Transit Area through which the Virtual Link is configured, must support full routing information (i.e., cannot be a Stub Area or NSSA). The Virtual Link uses the intra-routing capabilities of the Transit Area. The Transit Area simply passes traffic from one adjacent area to another over the Virtual Link and does not originate or terminate any of this traffic.

It is recommended to use OSPF Virtual Links only in very specific cases, mainly as a temporary connection or as a backup connection that can be used when a main physical connection fails. Virtual links can be configured for redundancy in the event some network failures cause the Backbone Area to be partitioned into two parts. Generally, Virtual Links are not meant to be used as the primary method of connecting discontiguous parts of the Backbone Area.

1.18.1 CONNECTING A NON-BACKBONE AREA TO THE BACKBONE AREA THROUGH A VIRTUAL LINK

An OSPF Backbone Area interconnects the set of independent areas in an OSPF autonomous system into a single routing domain. All Non-Backbone Areas in the OSPF autonomous system must be directly connected physically or logically to the Backbone Area, and in the latter case through a simple logical connection or Virtual Link. In the cases where a physical connection is not practical or possible, a Virtual Link can be used to connect a Non-Backbone area to the Backbone Area.

In Figure 1.38, a Virtual Link is used to connect Area 5 to the Backbone Area through Area 3 (the Transit Area). The Virtual Link is configured between the OSPF routers, ABR 1 with Router ID 5.5.5.5 and ABR 2 with Router ID 3.3.3.3. The OSPF LSDBs in ABR 1 and ABR 2 treat the Virtual Link between the two as a direct link. To allow greater stability, the network designer can use the IP addresses of the loopback interfaces in the two routers as Router IDs, and the Virtual Link can be configured using these loopback addresses.

The performance of a Virtual Link depends heavily on the stability and reliability of the underlying routing within the Transit Area (i.e., intra-area routing stability). A Virtual Link cannot be configured through more than one OSPF area (or Transit Area), nor through a Stub Area, but rather only through a standard OSPF Non-Backbone Area. If a Virtual Link is required to connect a Non-Backbone Area to the Backbone Area across two Non-Backbone Areas (Transit Areas), then two Virtual Links have to be configured, one through each Non-Backbone Area.

OSPF routers send Hello packets over Virtual Links similar to the way they do over standard links, every 10-second intervals, that is, the default Hello Interval on P2P links and broadcast multiaccess networks. However, OSPF routers send LSA updates on Virtual Links in different manner. As discussed above, a router sends a new instance of an LSA (i.e., refreshes LSAs) every 30 minutes, that is, the default *LSRefreshTime* **[RFC2328]**. By default, an LSA expires after 60 minutes (the *MaxAge*) when it is not refreshed.

OSPF routers send LSAs over Virtual Links (demand circuits) with the *DoNotAge* bit set (also called *DoNotAge* LSAs) so that the LSAs will not be aged out **[RFC1793]**. *DoNotAge* LSAs allow routers to prevent excessive flooding of LSA updates over a

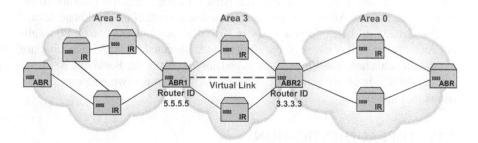

FIGURE 1.38 Connecting a Non-Backbone Area to the Backbone Area through a Virtual Link

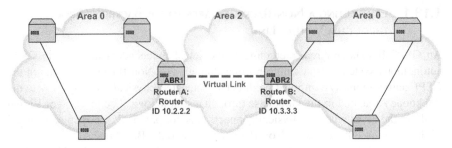

FIGURE 1.39   Interconnecting a Split Backbone through a Virtual Link

Virtual Link. A router will reflood a *DoNotAge* LSA only when there is a change in the LSA. The use of *DoNotAge* LSA over demand circuits reduces routing protocol traffic overhead over the link while allowing routers to flood any changed LSAs immediately.

1.18.2   INTERCONNECTING A SPLIT BACKBONE THROUGH A VIRTUAL LINK

Since the Backbone Area connects all the Non-Backbone Areas in the OSPF autono-mous system, it must be contiguous. Any two parts of a partitioned Backbone Area can also be connected together through a Non-Backbone Area using Virtual Links (Figure 1.39). A simple example is when two companies merge and have to merge their two separate OSPF networks into one bigger network that has a common Backbone Area. In this case, the two separate Backbone Areas have to be intercon-nected to create a single common logically contiguous Backbone Area.

The OSPF area through which the interconnecting Virtual Link is configured is also known as a Transit Area. The Transit Area must support full OSPF routing information and cannot be a Stub Area. If the Backbone Area is partitioned, for example, under router or link failure conditions, parts of the OSPF autonomous system can become unreachable. This may require Virtual Links to be configured to repair the partitioning.

A Virtual Link can be configured between the ABRs that connect to their respec-tive Backbone Areas through the Transit Area. In Figure 1.39, a Virtual Link is con-figured between the two ABRs, Routers A (Router ID 10.2.2.2) and Router B (Router ID 10.3.3.3), across a Non-Backbone Area, Area 2. Router A and B at each end of the Virtual Link are also ABRs and have interfaces that are part of the Backbone Area.

The Virtual Link here is similar to a standard adjacency between two OSPF neigh-bor routers, except that the neighbor routers interconnected by the Virtual Link are not directly attached. The Virtual Link is considered part of the Backbone Area and behaves as if it is an unnumbered P2P interface between the two endpoint routers (Routers A and B).

1.19   OSPF AUTHENTICATION

The header of an OSPF packet (Figure 1.11) includes an Authentication Type field and a 64-bit Authentication field that specifies the data to be used by the authentica-tion method indicated by the type field. As shown in Figure 1.11, OSPF supports Null

Authentication (Type 0), Simple Password (or Cleartext) Authentication (Type 1), and Cryptographic Authentication (Type 2) using, for example, MD5 and SHA algorithms. Cleartext is sometimes referred to as Plaintext.

OSPF authentication is an interface specific configuration. OSPF allows the authentication type to be configurable on a per-interface basis meaning, different interfaces on a router can support different authentication types **[RFC2328]**. However, any two interconnected neighbor interfaces must support the same authentication type. Also, for a given authentication type, the data used by that authentication method is also configurable on a per-interface basis. A router will accept routing information from another router only if the interconnecting interfaces are configured with the same authentication method.

1.19.1 NULL AUTHENTICATION

In Null Authentication, the OSPF packets (Figure 1.11) that OSPF routers exchange over their interconnecting interfaces are not authenticated. When an OSPF router uses this Authentication Type, it will fill the 64-bit Authentication field in the OSPF header with any type of data since this field is not examined when the OSPF packet is received **[RFC2328]**. Also, when Null Authentication is used, a router will compute a Checksum over the entire contents of each OSPF packet it receives (excluding the 64-bit Authentication field) in order to detect if any data in the packet has been corrupted.

1.19.1.1 Generating Null Authentication Messages

After an OSPF router has created the contents of an OSPF packet, it calls on the Authentication procedure configured for the sending interface (as indicated by the Authentication Type value) before sending the packet. For Null Authentication, the OSPF packet is modified as follows:

1. The router sets the Authentication Type field in the standard OSPF header to 0.
2. The standard IP Checksum of the entire contents of the OSPF packet (excluding the 64-bit Authentication field) is computed. The router then sets the Checksum field in the OSPF header to this value. The standard IP Checksum is calculated by first grouping the data in the packet (except the Authentication field) into 16-bit words. Next, the one's complement addition of all the 16-bit words in the packet is calculated, followed by the one's complement of this sum. If the length of the data in the packet is not an integral number of 16-bit words, the router pads the packet with redundant data consisting of bytes of all zeros before calculating the IP Checksum. For the purpose of computing the standard IP Checksum, the OSPF Checksum field value is set to zero (i.e., 16-bit word of all 0s).

1.19.1.2 Verifying Null Authentication Messages

When a router receives an OSPF packet on an interface, it will authenticate the packet first before further processing is carried out. The authentication procedure the router uses is indicated by the Authentication Type in the OSPF header, which must match the Authentication Type setting for the receiving router interface. If the OSPF packet passes the authentication, then the router can proceed to process the packet further, but if the authentication fails, the packet is discarded.

For Null Authentication, the router verifies the Checksum field in the OSPF header. The receiving router computes the OSPF Checksum on the received OSPF packet using the same algorithm as the sending router. The OSPF Checksum is computed as described above with the Authentication field also excluded. The packet may also be padded with bytes of all 0s before checksumming. If the computed OSPF Checksum yields a 16-bit word of all 0s, then this indicates that no error in the OSPF packet has been detected. Else, if the computed value is non-zero, then the packet is rejected.

1.19.2 SIMPLE PASSWORD AUTHENTICATION

In Simple Password Authentication, the 64-bit Authentication field in the OSPF packet carries essentially a simple "cleartext" 64-bit password. When this Authentication Type is used on an OSPF router interface, the data contained in the 64-bit Authentication field (i.e., the Authentication Key which is a cleartext password) can be configured on a per-network basis. OSPF uses the Authentication Key to generate and/or verify OSPF packets.

All OSPF packets that are sent out by routers on a particular interface must have this Authentication Key written in the 64-bit Authentication field of the OSPF header. Also, in this authentication method, the router must still compute a Checksum over the entire contents of each OSPF packet it receives (leaving out the 64-bit Authentication field) in order to detect if any data has been corrupted.

With Simple Password Authentication, a network engineer may decide to configure one or more passwords (or keys) per OSPF area, and where routers in the same OSPF area may be configured with the same authentication passwords. The authentication passwords do not have to be the same on all router interfaces in an OSPF area, but they must be the same on any two neighbor router interfaces sharing a link or network segment.

Generally, network operators use Simple Password Authentication as a way to isolate operator configuration errors (misconfigurations) related to adjacency formation. With Simple Password Authentication, it is hard for routers to inadvertently join an OSPF routing domain and establish adjacencies. This is because each router must first be configured with the correct passwords (that have been preselected for that routing domain) before it can participate in any routing activities. The main disadvantage of this authentication method, however, is that it is vulnerable to passive attacks where an attacker sniffing the link with a protocol analyzer could easily read the password off the OSPF packets.

1.19.2.1 Generating Simple Password Authentication Messages

For Simple Password Authentication, the router modifies the OSPF packet as follows:

1. The router sets the Authentication Type field in the standard OSPF header to 1.

2. The router computes the standard IP Checksum of the entire contents of the OSPF packet (excluding the 64-bit authentication field) and sets the Checksum field in the OSPF header to this value.
3. The router then sets the 64-bit Authentication field in the OSPF header to the 64-bit shared Authentication Key or password that has been configured for its sending interface.

1.19.2.2   Verifying Simple Password Authentication Messages

For Simple Password Authentication, the receiving router authenticates the received OSPF packet as follows:

1. The router verifies the Checksum field in the OSPF header as described above.
2. The router checks the 64-bit Authentication field in the OSPF header to see if the 64-bit shared Authentication Key carried is equal to the Key configured for the receiving router interface. The packet is accepted only if the two are equal.

1.19.3   CRYPTOGRAPHIC AUTHENTICATION

In Cryptographic Authentication (e.g., using the MD5 algorithm), a shared secret password or secret Authentication Key (or simply, Key), and Key-ID are configured on each OSPF router in the routing domain (e.g., OSPF area, network segment). A sending OSPF router will then apply an authentication algorithm on both an OSPF packet (to be transmitted) and the shared secret key to generate a one-way "message digest". The message digest is then *appended to the end* of the OSPF packet as authentication data. The Authentication Key is a shared secret password which OSPF uses for the generation (at sender) and verification (at receiver) of message digests that are appended to the OSPF packets.

The OSPF router transmits the message digest with the OSPF packet (appended to the end of packet), together with the Key ID and a non-decreasing Cryptographic Sequence Number. Specifically, to support Cryptographic Authentication, the 64-bit Authentication field in the OSPF packet header (Figure 1.11) has been reformatted or redefined to have subfields (Key ID, Authentication Data Length and Cryptographic Sequence Number) as shown in Figure 1.40.

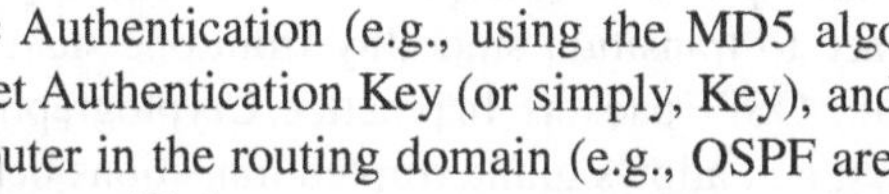

0	7	15	23	31
0		Key ID	Authentication Data Length	
Cryptographic Sequence Number				

Field	Meaning
Key ID (8 Bits)	Identifies the algorithm and secret key used to create the message digest appended to the OSPF packet. Key Identifiers are unique per-interface or, equivalently, per-subnet.
Authentication Data Length (8 Bits)	Specifies the length (in bytes) of the message digest appended to the OSPF packet
Cryptographic Sequence Number (32 Bits)	Specifies a non-decreasing sequence number used to guard against replay attacks

FIGURE 1.40   OSPF Authentication Field using Cryptographic Authentication

The message digest that is generated and attached to an OSPF packet is not considered to be part of the OSPF packet because it is not included in the Packet Length field in the OSPF header (Figure 1.11). However, the appended message digest is included in the Total Length field in the IP header of the IPv4 packet carrying the OSPF packet (Figure 1.10).

Unlike in the Simple Password Authentication method, the shared key is not carried in the OSPF packet being transmitted across the transmission medium, which provides protection against passive attacks. The sending OSPF router also includes a non-decreasing Sequence Number in a 32-bit Cryptographic Sequence Number field (Figure 1.40) in each OSPF packet it sends. This is to protect against replay attacks where OSPF packets can be captured, modified, and then retransmitted/re-advertised to other OSPF routers [RFC2328]. OSPF Cryptographic Authentication only allows the routing information carried in OSPF packets to be authenticated but does not provide confidentiality (i.e., does not protect the privacy of the OSPF routing information carried in the packets).

In Cryptographic Authentication, each shared secret key is identified by the combination of the interface running the authentication and the Key ID. A router interface may be configured with multiple shared secret keys active at any given time. This enables the router interface to transition smoothly from one shared key to another (key rollover) while sending OSPF packets. In practice, Cryptographic Authentication will support multiple keys, and allows uninterrupted transitions between keys when transmitting OSPF packets.

This feature (the key rollover mechanism) is very useful when an OSPF router needs to change the password (key) while sending OSPF packets without disrupting protocol packet exchange with neighbors [CISCID7039]. If a router interface is running with a preconfigured key and is then configured with a new key, the router will transmit multiple copies of the same OSPF packet, each authenticated with a different key (old and new). Once the OSPF router has detected that the newly configured key has been adopted by all of its neighbor routers, it will cease transmitting OSPF packets authenticated with the old key.

The Key ID allows an OSPF routers having multiple active keys to reference any one of them at any given time. This makes it easier and more secure for the router to migrate from one key to another. For example, a router can migrate from one key to another by simply configuring a new key under a different Key ID and then remove the old key. Similar to Simple Password Authentication, Cryptographic Authentication passwords do not have to be set the same in an OSPF area, but do need to be the same on neighbor router interfaces sharing a link or network segment.

Each Authentication Key has associated with it the following four time constants (expressed according to a local or network clock time-of-day):

- **KeyStartGenerate**: This specifies the time instant that the sending router interface will start using the given Authentication Key for OSPF packet generation.
- **KeyStopGenerate**: This specifies the time instant that the sending router interface will stop using the given Authentication Key for OSPF packet generation.

- **KeyStartAccept**: This specifies the time instant that the receiving router inter-face will start accepting OSPF packets that have been generated using the given Authentication Key.
- **KeyStopAccept**: This specifies the time instant that the receiving router inter-face will stop accepting OSPF packets that have been generated using the given Authentication Key.

1.19.3.1   Generating Cryptographic Authentication Messages

An OSPF router may have multiple Authentication Keys configured for an interface when using Cryptographic Authentication. In this case, the router chooses the Authentication Key that has the most recent *KeyStartGenerate* time among the multiple Authentication Keys that are valid for the interface to generate messages (i.e., Keys that have *KeyStartGenerate* ≤ current time < *KeyStopGenerate*). Using the chosen Authentication Key, router modifies the OSPF packet as follows:

1. The router sets the Authentication Type field in the standard OSPF header to 2.
2. The router does not calculate the Checksum field in the standard OSPF header, but instead sets it to 0.
3. The Key ID for the sending router interface is set to the Key ID associated with the selected Authentication Key.
4. The Authentication Data Length field (Figure 1.40) is set to the length of the message digest that is to be appended to the end of the OSPF packet. If the authentication algorithm of the sending interface is configured as MD5, then the Authentication Data Length is 16.
5. The 32-bit Cryptographic Sequence Number carried in the 64-bit Authentication field of the OSPF packet header (Figure 1.40) is set to a non-decreasing value.
6. The shared secret Authentication Key and the OSPF packet are used with an authentication algorithm to generate a one-way message digest that is appended to the OSPF packet. When MD5 is used as the authentication algorithm, the message digest is calculated as follows:
   a. A 16-byte MD5 message digest is computed and appended to the OSPF packet.
   b. As specified in **[RFC1321]**, a Trailing Pad field and Length field are then added.
   c. The router then runs the MD5 authentication algorithm over the concate-nated data made up of the OSPF packet, secret authentication key, Trailing pad field, and Length field, to produce a 16-byte message digest **[RFC1321]**.
   d. The router appends the MD5 message digest to the original OSPF packet. The Length field in the OSPF packet does not include the length of the mes-sage digest, but is included in the Length field in the IP packet in which the OSPF packet is encapsulated. Also, the router does not count or transmit any Trailing Pad field or the Length field beyond the appended message digest.

1.19.3.2   Verifying Cryptographic Authentication Messages

For Cryptographic Authentication, the receiving router interface authenticates the received OSPF packet as follows:

1. The receiving router checks if the Key ID of the configured Key of the receiving interface is equal to Key ID specified in the received OSPF packet (Figure 1.40). If the receiving router does not find that Key ID and its corresponding Authentication Key, or if the Key is determined to be not valid for receiving OSPF packets (i.e., current time $\geq$ *KeyStopAccept*, or current time < *KeyStartAccept*), then the packet will be discarded.

2. If the receiving router determines that the Cryptographic Sequence Number carried in the received OSPF header (Figure 1.40) is less than the Cryptographic Sequence Number recorded in the local Neighbor Data Structure it maintains for the sending router, the OSPF packet is discarded.

3. The receiving router verifies the message digest appended to the OSPF packet as follows:
   a. The received message digest that is appended to the OSPF packet is extracted (or precisely detached) from the OSPF packet and set aside.
   b. A new message digest is then computed on the OSPF packet, as described in the message generation steps described above.
   c. The received and locally computed message digests are compared to see if they match. The OSPF packet is discarded if there is a mismatch in the two digests. If these two digests match, then the OSPF packet has passed the authentication and is accepted. The receiving router sets the Cryptographic Sequence Number recorded in the local Neighbor Data Structure associated with the sending router to the Cryptographic Sequence Number carried in the OSPF packet header.

1.20   OSPF PROTOCOL DATA STRUCTURES AND PARAMETERS

Other than the LSDB, OSPF requires a number of other protocol data structures and configurable parameters in order to function. The settings of some of the OSPF parameters must be consistent among some groups of routers. For example, all routers in a given OSPF area must agree on the particular set of parameters to be used for that area, and all routers attached to a common shared network segment must have the same IP network address (or address prefix) and network mask. The IP address prefix and network mask must be consistent across all the routers attached to the shared network segment.

Other OSPF parameters may be determined by a router using other protocols and algorithms that are outside the OSPF specification (e.g., the Ethernet MAC address of an Ethernet based host connected to the router can be determined using the Address Resolution Protocol [ARP]). Yet, from OSPF's perspective, these other parameters are still considered to be configurable.

We discuss some of the top-level OSPF data structures and parameters in this section.

1.20.1   GLOBAL PARAMETERS

Two of OSPF global configuration parameters are described below:

- **Router ID**: This is a 32-bit value carried in the Router ID field in an OSPF packet header and uniquely identifies a particular router in the OSPF Autonomous System. One method of assigning a Router ID to a router is to choose the smallest or largest IP address belonging to the router. If the Router ID of an OSPF router is changed, it is generally recommended to restarted the OSPF protocol software in the router to allow the new Router ID to take effect. Before restarting the router (for it to assume the new Router ID), the router should flush all LSAs it had originated itself from the routing domain, or else they may persist for up to the maximum age of *MaxAge*.
- **RFC1583Compatibility**: This parameter controls the preference rules an OSPF router uses when choosing a single LSA from multiple received AS-External-LSAs advertising the same network destination. When the *RFC1583Compatibility* parameter is set to "enabled", the preference rules used are those specified in RFC 1583. When the parameter is set to "disabled", the preference rules used are those specified in RFC 2328 Section 16.4.1. The latter rules prevent routing loops from forming when an OSPF router receives AS-External-LSAs advertising the same destination but have been originated by different OSPF areas. The default setting of the parameter is "enabled".

  In order to minimize the possibility of routing loops being created, it is recommended that all OSPF routers in an OSPF autonomous system have the same *RFC1583Compatibility* parameter setting. If the OSPF network has routers that have not been updated to process the preference rules as specified in RFC 2328 Section 16.4.1, then the *RFC1583Compatibility* parameter should be set to "enabled" for all routers in the network. Otherwise, the RFC1583Compatibility parameter should be set to "disabled" for all routers to prevent all routing loops.

1.20.2 AREA DATA STRUCTURE AND PARAMETERS

An OSPF router maintains a separate Area Data Structure for each one of the OSPF areas it is connected to, including the Backbone Area (which is responsible for disseminating routing information to Non-Backbone Areas [i.e., inter-area routing information]). The Area Data Structure contains all the information an OSPF router in the area uses to run the basic OSPF protocol. Recall that routers in each area run a separate instance (copy) of the basic OSPF protocol for that area. This also creates the need for most configuration parameters for OSPF to be defined on a per-area basis.

Each router interface can only connect to a single OSPF area, and each adjacency established with a neighbor router also belongs to a single area. An area LSDB consists of the collection of Router-LSAs, Network-LSAs, Summary-LSAs, and ASBR-Summary-LSAs that have originated or injected by routers in the area (including ABRs and ASBRs). These LSAs are flooded/advertised throughout a single OSPF area only. The LSDB contains, in addition, the list of AS-External-LSAs by ASBRs connected to the OSPF autonomous system.

For routers belonging to an OSPF area to be able to form adjacencies with neighbor routers and exchange routing protocol and data traffic, they must agree on certain configuration parameters required for that area. When two neighbor routers disagree on these parameters, they will not be able to form adjacencies between them, resulting in a failure to exchange routing information and data traffic.

The following routing information and configuration parameters are required for an OSPF area: Area ID; List of IP address ranges (which comprises [IP Address, Network Mask], Status (i.e., set to *Advertise* or *DoNotAdvertise* an IP address range); Router interfaces connected to the area; List of Router-LSAs; List of Network-LSAs; List of Summary-LSAs; List of ASBR-Summary-LSAs; SPT; *TransitCapability*; *ExternalRoutingCapability*; and *StubDefaultCost*. All of these parameters have already been defined in appropriate sections above.

1.20.3 INTERFACE DATA STRUCTURE AND PARAMETERS

In OSPF, an interface is considered a connection between an OSPF router and a network or link. Note that an OSPF interface belongs to the OSPF area that contains the connected network/link of the interface. Each Interface Data Structure can have at most one IP address. All OSPF packets originated by a router interface are labeled with the Area ID associated with the interface. A router may form one or more router adjacencies over a given interface. The Router-LSAs a router originates describe the state of its interfaces/links to a particular area and their associated neighbor adjacencies.

An OSPF router maintains the following data items for each of its interfaces (in an Interface Data Structure). Some of these items are configuration parameters for the interface's connected network and must be the same for all the routers connected to that network: Link Type; Interface State (i.e., functional level of the interface); Interface's IP address; Interface's IP Network Mask; Area ID; Hello Interval; *RouterDeadInterval*; *InfTransDelay*; Router Priority; Hello Timer; Wait Timer; List of neighbor routers; Designated Router; BDR; Interface output cost(s); *RxmtInterval*; Authentication Type; and Authentication key. The Wait Timer is a timer (with length set equal to *RouterDeadInterval* in seconds) that causes a router interface to exit the Interface State "Waiting", and consequently select a Designated Router on the network **[RFC2328]**.

1.20.4 NEIGHBOR DATA STRUCTURE AND PARAMETERS

Each OSPF router communicates with its neighbor routers and each router-and-neighbor communication represents a conversation. An OSPF router describes each separate conversation with a neighbor in a Neighbor Data Structure. Also, the OSPF router associates each conversation with a particular router interface, and identifies it with either the neighbor router's Router ID or the Neighbor IP Address.

The Neighbor Data Structure an OSPF router maintains for a particular neighbor holds all the information relevant for forming or a formed adjacency with that neighbor. Note that a router does not necessarily form an adjacency with all neighbors. Thus, an adjacency can be considered as a highly developed conversation between an OSPF router and its neighbor.

A Neighbor Data Structure consists of the following items: Neighbor State (functional level of a conversation with the neighbor); Inactivity Timer (which indicates that the router has not seen a Hello packet from a neighbor recently); Master/Slave relationship (negotiated in the ExStart State); Database Description Sequence Number; Last received Database Description packet; Neighbor Router ID; Neighbor Router Priority; Neighbor IP Address; Neighbor Options; Neighbor's Designated Router; Neighbor's BDR; Link-State Retransmission List; Database Summary List; and Link-State Request List.

1.20.5 LIST OF EXTERNAL ROUTES

Routes to destinations external to an OSPF autonomous system are discovered by an EGP such as BGP, and advertised by an ASBR (via AS-External-LSAs) throughout the OSPF autonomous system (except Stub Areas). The ASBR floods the external routing information unaltered throughout the OSPF Non-Stub Areas in the autonomous system. An OSPF router maintains the external routes separately from its OSPF LSDB. However, a Non-Stub Router maintains a list of the AS-External-LSAs used to advertise the external routes in it LSDB (together with the Router-LSAs, Network-LSAs, Summary-LSAs, and ASBR-Summary-LSAs).

The advertising router (ASBR) can also tag each external route, to enable the passing of additional routing information between the external network and routers within the OSPF autonomous system. The 32-bit External Route Tag field value in an AS-External-LSA (Figure 1.26) is associated with each external route the ASBR advertises. OSPF does not use the External Route Tag itself for its operation but leaves it up to other processes in the router to use it for other non-OSPF routing related activities.

1.20.6 VIRTUAL LINK PARAMETERS

A Virtual Link is configured with an ABR at each endpoint. These two endpoint ABRs are connected to a common area, called Transit Area of the Virtual Link. The Virtual Link is identified by both the Router ID of the ABR at the other endpoint of the link, and the Transit Area through which the Virtual Link is configured. A Virtual Link relies on the intra-area routing capabilities of its Transit Area (which must not be a Stub Area) to forward OSPF packets between the endpoint ABRs.

The viability of a Virtual Link depends on the existence of an intra-area path through the Transit Area linking the two endpoint ABRs. A Virtual Link only becomes active (i.e., brought up) and becomes an interface to the Backbone Area when the SPTs for the Transit Area is constructed [RFC2328]. The cost and operational state of an ABR interface (*InterfaceUp* or *InterfaceDown*) connected to the Virtual Link is discovered during the SPT calculation through the Transit Area. Also, the cost of a Virtual Link between the two endpoint ABRs is equal to the cost of the intra-area shortest path through the Transit Area.

An OSPF router sets the IP source address of an OSPF packet to the IP address of the interface that sends it. However, router interfaces that are attached to unnumbered P2P links have no IP address associated with them. So, on these interfaces, a router sets the IP source address of packets it sends to any of the other IP addresses assigned to it. For this reason, the router must have at least one valid IP address assigned to it.

In many ways, a Virtual Link behaves very similar to an unnumbered P2P link. However, a Virtual Link does have an IP interface address which is discovered and set dynamically during the construction of the SPT and the IP Routing Table. The IP address of the local interface to the Virtual Link and the Virtual Link's neighbor IP address are both discovered during the SPT and Routing Table calculation process [RFC2328]. The endpoint ABRs use these IP addresses when they send OSPF packets over the Virtual Link.

The discovered source interface IP address is used as the source IP address when the router sends packets over the Virtual Link. Particularly, when an OSPF packet is sent over the Virtual Link, the receiving router must be an ABR, and the source Router ID specified in the OSPF packet (Figure 1.11) must be set to the Router ID of the sending ABR interface (at the sending end of the Virtual Link). The receiving ABR interface must also be connected to the Transit Area over which the Virtual Link is configured.

Note that on router interfaces attached to Virtual Links, the IP network mask is not defined [RFC2328]. Each Hello packet contains a Network Mask field that carries the IP network mask (see Figure 1.12) of the attached link or network, but on a Virtual Link, the Network Mask is always set to 0.0.0.0 and is ignored by all routers. Also, on Virtual Links, Hello packets are sent every Hello Interval (seconds) as unicasts and are addressed directly to IP address of the ABR interface at the other end of the link. For Database Description packets sent over Virtual Links, the Interface MTU field value (see Figure 1.13) is always set to 0 [RFC2328].

A Virtual Link appears in Router-LSAs sent over the Backbone Area as if it is just a normal router interface attached to the Backbone Area. As a result, a Virtual Link carries all of the parameters normally associated with a standard router interface attached the Backbone Area. For a Virtual Link, an OSPF router (i.e., the ABR) adds a link description to a Router-LSA it originates only when the neighbor router (endpoint ABR) over the Virtual Link is fully adjacent (a full virtual adjacency). For this LSA type, the router sets the Link Type field to 4 (indicating a Virtual Link), the Link ID to the Router ID of the neighbor ABR, the Link Data to the IP address of the local interface associated with the Virtual Link, and the link cost to the cost calculated for the shortest path over which the Virtual Link takes during the SPT and Routing Table calculation.

The cost associated with a Virtual Link is not a configured parameter but is the cost of the intra-area path between the two endpoint ABRs. This Virtual Link cost is also set dynamically and appears in the corresponding Routing Table entry for the Virtual Link. When the Virtual Link cost changes, the router will originate a new Router-LSA to be sent over the Backbone Area. Note that if the cost of the Virtual Link's path is greater than 0xffff, the maximum cost for an interface in a Router-LSA, the link is considered inoperational (i.e., treated as if the Virtual Link does not exist). For a Virtual Link, the spacing between link-state retransmissions, *RxmtInterval*, is a configured parameter and should be set to greater than the expected round-trip delay between the two endpoint ABRs. The Router Priority field as carried in all OSPF Hello packets (see Figure 1.12) is ignored and not used on Virtual Links.

1.20.7 NBMA NETWORK PARAMETERS

In OSPF, NBMA networks are treated much like broadcast multiaccess network such as Ethernet. So, given that an NBMA or a broadcast multiaccess network could contain many routers, a Designated Router must be elected for the network. The Designated Router then originates Network-LSAs which lists all OSPF routers attached to the network.

However, because an NBMA network lacks in-built broadcast capabilities, it may be necessary to configure some parameters to be used in the election of the Designated Router. These parameters only need to be configured in the OSPF routers that are eligible to be elected Designated Router (i.e., routers in the network that have non-zero Router Priority values), and only if there is no automatic procedure in the network for discovering neighbor routers. The NBMA network parameters include the following items: List of all other attached routers; and *PollInterval*.

An OSPF router adds a single link description to a Router-LSA it originates for each operational broadcast and NBMA interface as follows:

- If the Interface State is Waiting, the router adds a link description to the LSA and sets the Link Type equal to 3 (i.e., indicating a Type 3 link or stub network), Link ID to the IP address (prefix) of the attached network, Link Data to the network mask of the attached network, and the link cost to the configured outbound cost of the interface.
- Else, a Designated Router has already been elected for the attached network. If the router has already formed a full adjacency with the Designated Router, or if the router itself is Designated Router and has a full adjacency with at least one other OSPF router, the router adds a link description to the LSA, and sets the Link Type to 2 (indicating a transit network), sets the Link ID to the IP address of the interface of the network's Designated Router (which may be the router's own interface), sets the Link Data to the IP address of the router's own interface, and the cost equal to the configured outbound cost of the interface. Otherwise, the router adds a link description as if the Interface State was in the Waiting State.

1.20.8 POINT-TO-MULTIPOINT (P2MP) NETWORK PARAMETERS

A P2MP network has neighbor routers directly attached to its links, and it may be necessary to configure the set of neighbors that are directly reachable over these different links. Each attached neighbor router is identified by its attached interface IP address. Also, a P2MP network does not elect a Designated Router and BDR, therefore, the Designated Router eligibility of the attached neighbor routers is left undefined.

Alternatively, the (IP addresses of) neighbor routers on a P2MP network may be dynamically discovered using lower-level discovery/resolution protocols such as Inverse ARP (InARP) **[RFC2390]** used in ATM and Frame Relay networks. In these networks, a device uses InARP to find the IP address of another device by providing the device's Layer 2 address (such as the Data Link Connection Identifier [DLCI] in Frame Relay).

An OSPF router adds a single link description to a Router-LSA it originates for each operational P2MP interface as follows:

- The router adds a link description to the LSA, and sets the Link Type to 3 (indicating a stub network), the Link ID to the IP address of the router's own interface, the Link Data to the network mask 0xffffffff (indicating a host route), and the link cost set to 0.
- The router adds a link description to the LSA for each fully adjacent neighbor router associated with the interface, and sets the Link Type to 1 (indicating a P2P link), the Link ID to the Router ID of the neighbor router, the Link Data to the IP address of the router's own interface, and the link cost to the configured output cost of the interface.

1.20.9 Host Route Parameters

Each network in an OSPF autonomous system has associated with it an IP address and network mask. The IP address of a network combined with its network mask define the range of IP addresses supported by the network. The network mask indicates the number of IP hosts or end-user nodes in the network. An IP host that is connected directly to an OSPF router is referred to as host route and appears in the OSPF autonomous system as a stub network.

A host route is a considered to be a subnet that is directly attached to a router, and has a network mask that consists of "all ones" (i.e., 0xffffffff). The host route network mask of 0xffffffff indicates the presence of a single node in the subnet (directly attached to the router). Thus, a host route is always advertised with a network mask of 0xffffffff, indicating to other OSPF routers the presence of only a single host destination attached to the router.

In a Router-LSA, a host route has the Link Type set to 3 (same as a stub network or Type 3 link), the Link ID is set to the host's interface IP address, the Link Data are set to the network mask of 0xffffffff (indicating a host route), and the cost is equal to the configured cost of the host. A host route is classified and advertised as a link to a stub network having a network mask of 0xffffffff.

A host route may indicate either a router interface attached to a P2P network, a router looped interface (i.e., lookback interface), or an IP host that is directly connected to a router interface. An OSPF router maintains the following information which must be configured for each host directly connected to the router: Host IP address; cost of the router link to the host; and OSPF Area ID (of the host's area).

1.20.10 The Routing Table Structure

All routers in an OSPF area have identical LSDBs (when synchronized) which maintains the collection of LSAs originated by the routers belonging to the area. An OSPF router generates the best routes to network destinations for its IP Routing Table by calculating a SPT (from the topological information in the LSDB) with the router itself as the root. It should be noted that, the nature of the SPT depends on the particular router calculating the tree.

The SPT at a particular OSPF router (root) describes the entire path from the root to any network destination or host. However, the computing router (root) maintains only the next-hop router to the destination (and not the subsequent routers along the path) in its Routing Table which it uses in the packet forwarding process. The root also notes the best route to any router in the tree, and for the processing of external routes, the root calculates the distance from the next-hop to any router (ASBR) advertising external routes.

The IP Routing Table (which is a data structure) contains the routing information a router needs to forward an IP packet toward its destination. Each route entered into the Routing Table is indexed by a destination IP address, and contains among other pieces of information (see below), the path cost to the destination, and one or more paths over which packets can be forward to the destination. Each path is described by a Path Type (intra-area, inter-area, Type 1 External or Type 2 External, listed in decreasing order of preference) and next-hop router leading to the destination. If multiple paths exist to a particular destination, the Routing Table entry may describe the collection of best paths to that destination.

When a router is to forward an IP data packet to its destination, it searches its Routing Table for an entry (i.e., the longest matching prefix) that provides the best match for the packet's destination address. The best matching Routing Table entry then provides the IP address of the next-hop router and the outgoing router interface over which the packet should be forwarded on its way to the packet's destination. For example, a destination entry with address/mask pair of 10.15.1.0, 0xffffff00 is a more specific match than an entry with address/mask pair of 10.15.0.0, 0xffff0000. If a router finds no best matching entry in the Routing Table for the packet's destination, then the destination is considered unreachable. The packet will be discarded and an ICMP destination unreachable message will be sent to the packet's source.

The Routing Table may also provide a default route (with destination address equal to 0.0.0.0 (*DefaultDestination*), and network mask 0.0.0.0) in the event there is no matching entry in the Routing Table for the packet's destination address. A default route, when one exists, matches all IP destination addresses, is the least specific match, and serves as a gateway of last resort for a packet.

- The following set of items describes the destination in a Routing Table entry: Destination Type ("Network" entry or "Router" entry); Destination ID (for Network entry, the ID is the associated IP address, and for Router entry, the ID is the OSPF Router ID); Address Mask (defined only for Network entry, but for a host route, the mask is 0xffffffff); Optional Capabilities; OSPF Area.

A Network entry represents a range of IP addresses (i.e., IP address prefix) for a network to which the router forwards IP packets. A default route in the Routing Table is also a type of Network entry. A router uses only Network entries for forwarding packets. A Router entry describes a route to an ABR or ASBR. An OSPF routers uses Router entries primarily as intermediate steps in the Routing Table calculation process. A Router entry for an ABR is used when the router calculates inter-area routes, and also when it is maintaining a configured Virtual Link. A Router entry for an ASBR is used when the router calculates the external routes to destinations that are outside the OSPF Autonomous System.

- The following additional items (which form the rest of the items for a Routing Table entry) describe the set of paths available to the destination: Path Type; Cost; Type 2 Cost; Link-State Origin.

The Path-Type field specifies one of four possible types of paths (intra-area, inter-area, Type 1 External, or Type 2 External, listed in decreasing order of preference) a router can use to route packets to the destination. Intra-area paths are advertised via Router-LSAs and indicate paths to destinations that are inside an OSPF area. Inter-area paths are advertised via summary-LSAs and indicate paths to destinations that are inside other OSPF areas. AS external paths are advertised via AS-External-LSAs and provide paths to destinations that are external to the OSPF Autonomous System.

The Cost field applies to all path types except Type 2 External Paths and specifies the cost of the entire path to the destination. The Cost field value is calculated as the sum of the costs of the constituent router links that make up the complete path. Note that for *Type 2 External Paths*, the Cost field describes only the cost of the internal path component, that is, the path component that is *inside* the OSPF autonomous system (ignoring the outside path component).

The Type 2 Cost field applies and is valid only for Type 2 external paths and specifies the cost of only the external path component, that is, the path component that is *outside* the OSPF autonomous system. The Type 2 Cost is advertised by an ASBR, and represents the most significant part of the total path cost to the external destination. For example, an OSPF router will always prefer a Type 2 External Path with Type 2 Cost of 15 over a path with Type 2 Cost of 20, regardless of the costs of the internal path components of the two paths.

Link-State Origin field applies to only intra-area paths and indicates which LSA type directly references the network destination (Router-LSA or Network-LSA). For example, if the destination is a stub network, then the LSA that references this destination is a Router-LSA from the attached OSPF router.

- When multiple "equal-cost" paths (i.e., paths of equal Path Type and Cost) exist to a particular network destination, a router may store these paths in a single Routing Table entry. The following additional fields are used to distinguish each one of the "equal-cost" paths in the Routing Table: Next Hop; Advertising Router

Advertising Router field applies only to inter-area and AS external paths, and specifies the Router ID of the router (ABR or ASBR) that advertised the Summary-LSA or AS-External-LSA that in turn advertised the path. Each one of the equal-cost routes must be of the same Path-Type, Cost, and belong to the same OSPF area. However, a separate next-hop router and Advertising Router may be specified for each route.

1.21 HIGH-LEVEL OSPF ROUTER ARCHITECTURE, PROCESSES, AND DATABASES

In this section, we describe the processes, interfaces, and databases in a typical OSPF router implementation. These elements are linked by well-defined asynchronous interfaces as shown in the high-level block diagram of Figure 1.41. We describe these elements in greater detail below.

1.21.1 THE OSPF PROCESS

To allow all the OSPF routers in the OSPF autonomous system to make consistent routing decisions, each router must maintain the following major tables/databases (each of which comprises a number of data structures as explained earlier above):

- **OSPF Neighbor Table (also called the Adjacency Database)**: This holds a list of the immediate neighbor routers of a particular router.
- **OSPF LSDB (also called the Topology Table or Database)**: This describes the topology of all routers, links, and their attached networks plus their states in an area of the OSPF autonomous system network.
- **OSPF Routing Table or Database (also called the OSPF Routing Information Base [RIB])**: This holds the best routes to each network destination. Each entry of this database contains an IP network address, network mask, next-hop router IP address, outbound interface, and other information already discussed above in the "The Routing Table Structure" section.

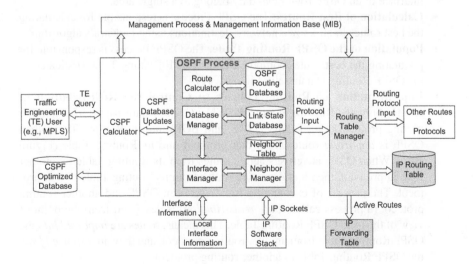

CSPF = Constrained Shortest Path First

FIGURE 1.41 OSPF Processes and Databases

The OSPF Process component in Figure 1.41 supports the following core protocol functions:

- **Management of the Router Interfaces and the OSPF Hello Protocol**: OSPF routers that share a common Layer 2 link or network segment can become neighbors on that link or segment. OSPF routers use the Hello Protocol to exchange Hello packets for neighbor discovery and as a keepalive mechanism. Each router uses the Hello Protocol to receive and periodically send Hello packets out each of its interfaces. The Hello packets that a router will send out an interface will list all of its known and active OSPF neighbors on its interfaces. Two OSPF routers become neighbors when each one sees itself listed in the Hello packet sent by the other router.
- **Maintenance of the OSPF LSDB**: The OSPF Process in each router is responsible for building adjacencies and the maintenance of the LSDB (sending and receiving LSAs).
- **Management of Neighbors and LSDB Synchronization**: OSPF routers establish and maintain neighbor relationships in order to be able to exchange routing updates. After two OSPF routers have become neighbors, they proceed to exchange LSAs in order to synchronize their LSDBs, a process that creates a full adjacency. The database in which a router maintains the neighbor relationship is called a Neighbor or Adjacency Database (Figure 1.41). The Neighbor ID field in the Hello packet (Figure 1.12) represents the Router ID of a neighbor router. Two OSPF routers can become neighbors if they share the same Area ID, are members of the same IP subnet (have the same IP network prefix), have the same network mask, and have the same authentication and timer settings (e.g., Hello Interval and Router Dead Interval). Any individual interface of an OSPF router can only belong to a single area.
- **Calculation of Best Routes**: The OSPF Process is responsible for calculating the best routes to all known network destinations using Dijkstra's algorithm.
- **Population of the OSPF Routing Table**: The OSPF Process is responsible for populating the best routes computed in the OSPF Routing Table (indicated as the OSPF Routing Database in Figure 1.41).
- **Implementation of Routing Policies to Control the Redistribution of Routes between Routing Protocols**: When OSPF installs routes from another routing protocol (e.g., EIGRP, IS-IS, and BGP) into its Routing Table, then OSPF is *importing* routes from that protocol into its Routing Table (Figure 1.42). When OSPF advertises active routes from its Routing Table to another routing protocol, then it is *exporting* routes from its Routing Table to that protocol. The process of exchanging routes between OSPF and another routing protocol (a process called *route redistribution*) is described from the point of view of the local OSPF Routing Table. This means, routes are *imported into* the OSPF Routing Table from another routing protocol, and they are *exported from* the OSPF Routing Table to another routing protocol.

  Routing policies serve as filters that are used to control the information a router running a particular routing protocol imports into its Routing Table and the information the router exports/advertises to its neighbors (Figure 1.42). A

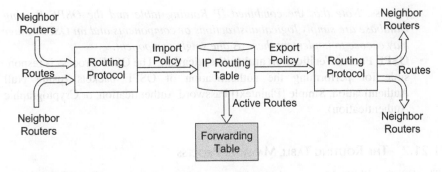

FIGURE 1.42 Importing and Exporting Routes in OSPF

routing policy can be created by defining a set of criteria against which routes are compared, and the action(s) to be performed if the criteria are met. Note that in OSPF, routing policies and filters can only be applied to routes installed in the OSPF Routing Table and not on the routing information carried in LSAs and meant to be installed in the LSDBs. This is because all OSPF routers in an area must have identical LSDBs and, therefore, must receive all LSAs flooded uninterrupted.

- **Distribution of Next Hop Routes to the Routing Table Manager and Traffic Engineering Link-State Information to the Constrained Shortest Path First (CSPF) Calculator (Figure 1.43):** The OSPF Process passes all the next-hop routes to the Routing Table Manager Process to be installed in the combined or integrated IP Routing Table. The combined IP Routing Table contains all the best routes computed by OSPF (along with their next hop IP addresses and outgoing interfaces) plus all imported routes into the OSPF

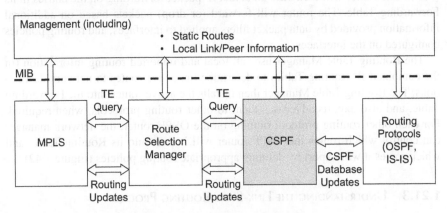

CSPF = Constrained Shortest Path First
MPLS = Multiprotocol Label Switching
TE = Traffic Engineering

FIGURE 1.43 CSPF Calculator Process

process. *Note that the combined IP Routing table and the OSPF Routing Database are simply logical abstractions or components and an OSPF router may choose to implement these as one single component.*

• **OSPF Packet Verification and Authentication**: The OSPF Process is responsible for performing the authentication of OSPF packets (using Null Authentication, Simple [Plaintext] Password Authentication, or Cryptographic Authentication).

1.21.2 THE ROUTING TABLE MANAGER PROCESS

The Routing Table Manager determines an active route (or multiple active routes when multipath load balancing is used) for each network destination and copies this information to the Forwarding Table(s). We show in Figure 1.41 that the Routing Table Manager Process supports other functions. In many practical implementations, routing protocols in a router store the routes that they have learned in their respective protocol-specific Routing Tables.

When a router determines the best route to a particular destination, this information is passed to the Routing Table Manager. The Routing Table Manager may also receive more than one best route to a particular destination from different routing information sources or protocols other than OSPF (e.g., static routes, RIP, EIGRP, IS-IS, and BGP). However, given that the routing metrics used by different routing protocols are not the same and/or comparable, the Routing Table Manager uses the Administrative Distance (also called the Route Preference) associated with each protocol to select the overall best route among the multiple candidate best routes.

The Routing Table Manager will select the route with the lowest Administrative Distance as the best route to be installed in the combined IP Routing Table and subsequently, the Forwarding Table (also called the Forwarding Information Base [FIB]). The router will then base all actual IP packet forwarding on the entries in its Forwarding Table. The router will forward (or drop) packets based on additional information provided by data packet filters applied to interfaces, and routing policies configured on the interfaces.

The Routing Table Manager uses all local and imported routing information (in the combined IP Routing Table) to determine the active routes to network destinations. The Routing Table Manager then installs the active routes into the Forwarding Table, and also exports/advertises them to other routing protocols, when required. For each other routing protocol running on the OSPF router, the network manager can control which routes the OSPF router will import into its Routing Table, and which routes it will export by defining appropriate routing policies (Figure 1.42).

1.21.3 UNDERSTANDING THE LINK-STATE ROUTING PROCESS

In OSPF, a link in a network is simply an interface on an OSPF router and the information about the state of a link is referred to as a link-state. All routers in an OSPF area follow the general link-state routing process described below to exchange routing information and synchronized their LSDBs.

1.21.3.1   Link and Link-State

The first step in the OSPF Routing Process is that each OSPF router will determine which one of its interfaces/links is in the up state (i.e., OSPF enabled), as well as, the networks the interface is directly connected to. Before it can participate in the routing process, each router interface must be configured with an IP address and network/subnet mask, and the interface, when in the up state, becomes part of the configured network (which is specified the network mask).

When the OSPF router boots up, it will load its saved startup configuration file and learn/discover the networks that are directly connected to its configured and active interfaces. A router interface/link must be in the up state before OSPF can learn about the networks attached to it. The router will then add these directly connected networks as entries in its Routing Table (a process that applies to all routing protocols, RIP, EIGRP, etc.).

The link-state information associated with each active router interface includes:

*   The IPv4 address and network mask of the interface
*   The type of network connected to the interface, such as P2P link, multiaccess broadcast (Ethernet), and Virtual Link.
*   The routing metric or cost for the link
*   Neighbor routers on network segment attached to the link

Cisco routers specify the OSPF cost of a link based on the bandwidth of the outgoing router interface.

1.21.3.2   Exchange of Hello Packets and Neighbor Discovery

The second step in the OSPF Routing Process is that each router will exchange Hello packets with it directly connected neighbor routers to detect and establish full adjacencies with these neighbors. OSPF routers use the Hello Protocol to discover neighbors attached to their links. Each router sends Hello packets out its active interfaces to discover if there are any active OSPF neighbors attached to the interfaces.

Any two OSPF routers go through a prescribed neighbor formation process [RFC2328] to discover each other and to form a full adjacency. After establishing a full adjacency, OSPF routers continue to exchange Hello packets with neighbors. The exchange of Hello packets serves as a keepalive mechanism used by a router to monitor the state of the neighbor and to maintain the adjacency. If a router does not receive Hello packets from a neighbor after a specified time period (i.e., RouterDeadInterval), it will declare that neighbor as unreachable and the adjacency will be terminated.

1.21.3.3   Building the Link-State Advertisements

The third step in the OSPF Routing Process is that each router will construct one or more LSAs containing the state of each of its directly attached links and networks. The router records in them all the relevant link-state information about the links and each connected neighbor (e.g., Link Type and Neighbor ID).

This step is performed after a router has established adjacencies with its neighbors. The router will construct LSAs that contain the pertinent link-state information about these links and directly connected networks.

1.21.3.4   Flooding the LSAs

The fourth step in the OSPF Routing Process is that each router will flood the constructed LSAs to all neighbors in the OSPF area who will store a copy of each received LSA in their local LSDBs. The neighbor routers will in turn flood the received LSAs to their neighbors until all routers in the OSPF area have received the flooded LSAs.

In addition to link-state information, each LSA carries other information such as, LSA authentication data, LSA aging information (LS Age), and sequence numbers that are used to manage the LSA flooding process. The aging and sequence number information are used by each router in the OSPF area to determine if it has already received a particular LSA via another router, or if the recently received LSA carries more up to date routing information than the information already entered in its LSDB. This process of screening received LSAs flooded by other routers allows each router in the OSPF area to maintain only the most current information in its LSDB.

When a router in the OSPF area receives an LSA from a neighbor, it will immediately transmit that LSA out all of its interfaces except the interface on which that LSA was received. This process of propagating LSAs from routers results in the flooding of LSAs throughout the OSPF area. A router will almost immediately flood LSAs after they are received without performing any intermediate routing calculations. This allows OSPF networks to converge more quickly than RIP. OSPF uses the SPF algorithm only after the flooding of LSAs is complete and LSDBs are synchronized.

Recall that OSPF routers do not depend on periodical routing updates as in RIP. A router only sends:

- During initial router startup or restart of the OSPF Routing Process;
- Whenever a network topology change occurs (e.g., a router interface/link comes up or goes down, an existing neighbor adjacency is broken, and a new neighbor adjacency is established).

1.21.3.5   Building the Link-State Database

Eventually, all routers in the OSPF area receive new LSAs from every other OSPF router and store these LSAs are in their LSDBs. As a result of the LSA flooding process, each router will learn the link-state information from all other routers in the OSPF area. Note that each router will also include its own link-state information in its LSDB.

After all the LSDBs in an OSPF area have been synchronized, the final step in the OSPF Routing Process is that each router will use its synchronized LSDB or topology table (which is a complete map of the network topology) and the SPF algorithm to compute the best paths to each network destination. The network topology map represents a complete map of the network and all destinations and the routes to reach them.

1.21.3.6 Building the Shortest Path Tree (SPT)

Each router in an OSPF area uses its LSDB (which maintains the link-state information obtained from all other routers) and the SPF algorithm to construct a SPT (with the router itself as root). To begin, a router will use the SPF algorithm to interpret the LSA in its LSDB to identify network destinations and the associated costs to them. The router will then use the SPF algorithm to calculate the shortest paths to reach each individual destination, resulting in a SPT for all the destinations. Each router in the OSPF area constructs its own SPT in parallel and independently from all other routers.

1.21.3.7 Adding OSPF Best Routes to the Routing Table

Each OSPF router will then add the best paths determined by the SPF algorithm to its Routing Table. The router will also include in its Routing Table all directly connected networks and routes obtained from other routing information sources such as static routes. With the Routing Table properly populated, the router will now forward packets according to the entries in the Routing Table. When forwarding an IP packet, the Routing Table entry that provides the best match (i.e., the longest matching prefix) for the packet's IP destination address is used to forward the packet. The next-hop IP router and its outgoing router interface associated with the best matching Routing Table entry is used to toward the packet.

An OSPF router may employ an Equal Cost Multipath (ECMP) routing (if such a feature is supported) when there are multiple equal-cost routes (or multiple equal-cost next-hops) to the same network destination in its Routing Table. With ECMP routing, a router potentially has multiple next hop routers available toward a specific destination. The Routing Table in this case contains a collection of best routes to a particular destination. When ECMP routing is used, the router may use a hash algorithm to select one of the next-hop addresses in the ECMP to which a given flow of packets will be forwarded. A load-balancing routing policy will have to be defined to schedule flows across the equal-cost paths.

Recall that when multiple routing protocols in the router provide routes to the same network destination, the Administrative Distances (or Route Preferences) can be used to select which route to be installed in the Routing Table. The route with the lowest Administrative Distance will be selected for the Routing Table.

1.21.4 THE CSPF CALCULATOR PROCESS

Traffic engineering (TE) provides routers in a network with the ability to establish routes in the network according to specific criteria and constraints in order to meet specific traffic requirements rather than, letting them rely solely on the route selection criteria of the conventional routing protocol they are running (e.g., OSPF, IS-IS) [RFC3630]. Traffic engineering enables a network manager to define criteria and constraints that will allow a network to make the best use of its limited resources to satisfy different traffic flows, thereby reducing over- and underutilization of its resources. Besides source-based routing, a network must implement the following in order to support traffic engineering:

- Compute a path from the source to the destination while taking into account administrative and network constraints such as bandwidth and delay requirements.
- Distribute the information describing the network topology and link attributes to all nodes in the network once the path is computed.
- Set up the path and reserve network resources for it.

1.21.4.1 Traffic Engineering Protocols

OSPF-TE **[RFC3630] [RFC7471]** is an extension to OSPF that can be used for traffic engineering over IP and non-IP networks (e.g., Generalized Multi-Protocol Label Switching [GMPLS] **[RFC4203]**). OSPF-TE, via the use of OSPF Opaque LSAs carrying Type-Length-Value (TLV) elements, can be used for exchanging traffic engineering information for a specific network. For example, a network administrator can use MPLS-TE and OSPF-TE to establish Label-Switch Paths (LSPs) in an MPLS network. To implement traffic engineering, the following protocol components are needed:

- **Protocol for Building a Traffic Engineering Database (TED)**: OSPF-TE generates Opaque LSAs that describe the traffic engineering network topology (including administrative and network constraint parameters such as bandwidth) and distributes this information within a given OSPF area. OSPF-TE distributes link information so that every router in the network can build a local TED. IS-IS **[RFC5305] [RFC8570]** and BGP **[RFC7752] [RFC8571]** can also be used to distribute traffic engineering information.
- **Protocol for Building MPLS LSPs**: Resource Reservation Protocol – Traffic Engineering (RSVP-TE) **[RFC3209]** is a Transport Layer protocol that can be used as an MPLS label distribution protocol for traffic engineering purposes. RSVP-TE was designed for distributing MPLS labels and allows MPLS Label Switched Paths (LSPs) to be established while taking into consideration a number of network constraint parameters such as the available bandwidth and the explicit number of hops to a network destination. After a TED is created, RSVP-TE uses the information contained in it to establish LSPs that satisfy a set of traffic engineering constraints. The Label Switch Routers (LSRs) in an MPLS network with Traffic Engineering use RSVP-TE to communicate the MPLS labels used to forward traffic between and through the LSRs. RSVP-TE enables an LSR to learn the MPLS label mappings of other peer LSRs.

MPLS is often used when transit traffic is routed through an IP network. Particularly, MPLS traffic engineering requires the following components:

- Setup of MPLS LSPs for forwarding packets.

- Distributing information about the network topology and link attributes using IGP extensions (OSPF-TE and IS-IS TE). For traffic engineering across multiple routing domains or autonomous systems, BGP Link-State (BGP-LS) is more suitable (see Chapter 3).
- Path computation and path selection using CSPF.
- Establish the forwarding state in nodes along the path and reserve resources along the path using RSVP-TE.

LSPs can be created manually, or through the use of signaling protocols – LDP (Label Distribution Protocol) **[RFC5036]** and RSVP(-TE). These signaling protocols are used within an MPLS network to establish LSPs for traffic crossing a transit network.

The extensions in OSPF-TE allow OSPF-TE to be used completely outside the data plane (out-of-band) of non-IP networks such as GMPLS **[RFC4203]**. When used in GMPLS networks, OSPF-TE sends OSPF Opaque LSAs with TLVs that can be used to describe the network topology over which GMPLS LSPs are to be established. Once GMPLS has learned the full network map, it will use its own signaling protocols for path setup and forwarding protocols for traffic forwarding

1.21.4.2 Traffic Engineering Database (TED)

To implement traffic engineering, a detailed knowledge of the network topology is required as well as information about the dynamics of the network, link loading, and state. Figure 1.43 shows an example of some of the information distribution component and extensions to OSPF required to implement traffic engineering. As discussed above, some link attributes have been included as part of each router's LSA for the purpose of OSPF Traffic Engineering **[RFC3630] [RFC7471]**.

These OSPF extensions are implemented using OSPF Opaque LSAs **[RFC5250]**, whereas IS-IS extensions are implemented via the definition of new TLVs. The standard flooding mechanisms used by OSPF and IS-IS ensure that link attributes used for traffic engineering are distributed to all the routers in the required routing domain. Some of the traffic engineering extensions that can be included in an OSPF LSA include current bandwidth reservation, maximum reserved link bandwidth, and maximum link bandwidth.

For the purpose of MPLS traffic engineering, each router has a local TED in which it maintains the network topology information and link attributes. The router uses the TED exclusively for computing explicit paths on which MPLS LSPs can be constructed through the network. Typically, a router will maintain a separate TED so that its traffic engineering computations will be independent of the normal OSPF Routing Process and its LSDB. This allows OSPF to continue its operation without modifications from the traffic engineering operations, as well as, allowing it to perform the traditional shortest path calculation based on information contained in its LSDB.

1.21.4.3 Constrained Shortest Path First (CSPF)

After a router has received and placed in its TED the network topology information and link attributes that are flooded by OSPF, it will use the TED to calculate the paths over which a set of LSPs can be created across the routing domain. The router applies a CSPF algorithm to the information in its TED to determine the physical path for each LSP. Standard OSPF **[RFC2328]** uses the SPF algorithm and the OSPF costs to compute the best paths to network destinations. The CSPF algorithm extends the SPF algorithm so that other network constraints can be taken into account during path computations.

With OSPF-TE, the CSPF algorithm takes into account the LSPs to be set up by MPLS. OSPF-TE is used to generate the OSPF Opaque LSAs that carry traffic engineering parameters. The generated traffic engineering parameters are then used to populate the TED. Recall that the TED is used exclusively for computing explicit paths over which MPLS LSPs can be established across the overall network. While taking into account path constraints, the CSPF algorithm uses the TED to compute the explicit paths over which the MPLS LSPs are to be created. RSVP-TE then uses the path information computed by the CSPF algorithm to set up the LSPs and to reserve bandwidth for them.

Constraint-based routing (CR) allows network engineers to implement traffic engineering by considering network resource availability and traffic resource requirements, rather than, simply relying on the shortest path calculations provided by the conventional routing protocols (OSPF and IS-IS). Typically, the constraints and their associated criteria are determined a priori and are implemented at the edge of the network, at the edge router. The criteria can include factors such as the required bandwidth values or required explicit paths through the network.

The CSPF algorithm was developed to be used with link-state routing protocols such as OSPF and IS-IS as illustrated in Figure 1.43. The CSPF algorithm is a modification of the SPF algorithm (used in OSPF and IS-IS) to take into account specific restrictions/constraints when routers compute the shortest paths across the network-to-network destinations. Some of the input information used by the CSPF algorithm include:

- Network topology and link-state information learned from OSPF-TE (or IS-IS-TE) and maintained in the router's TED. OSPF-TE and IS-IS-TE are both IGPs extended for Traffic Engineering (IGP-TE).
- State of network resources described by attributes such as available link bandwidth, reserved link bandwidth, and total link bandwidth that are carried by OSPF LSAs and extensions and stored in the TED.
- Attributes required for administrative purposes and obtained from user configuration to support the traffic to be sent over a candidate LSP to be created (such as requirements regarding maximum hop count, bandwidth, and administrative policy).

When a CSPF router considers each candidate link and node (next router) for placing a new LSP, it will accept/reject a specific link/router based on the availability of resources, or whether that particular link/router violates the defined policy constraints. The router will output from the CSPF computations an explicit route that lists the sequence of next-hop router addresses that provides the shortest path through the network to a particular network destination that meets the required constraints.

The router will then pass this explicit route to its signaling component which will use the appropriate signaling protocol (RSVP-TE) to establish the forwarding state in the routers along the path for the proposed LSP.

Figure 1.43 shows the place of the CSPF Calculator Process in the overall IP routing architecture (for both OSPF and IS-IS). The CSPF Calculator Process generally will be designed to respond to all traffic engineering queries by finding the best route to a destination that meet some prescribed or specified network constraints. The CSPF Calculator Process implements the appropriate algorithms and associated databases and has access to the local copy of the LSDB.

The CSPF Calculator Process has all information in a format optimized for its calculations, and uses this to generate traffic engineering solutions for each query. A distributed router architecture may choose to distribute multiple instances of the CSPF Calculator Process to the line cards (distributed processing) for scalability and redundancy.

1.22   SUMMARY OF OSPF FEATURES

OSPF advertises and uses link-states rather than distance vectors for best route computations to network destinations. OSPF advertises LSAs and not Routing Table updates as in RIP, and uses the SPF algorithm to calculate the shortest path to each known destination. Each OSPF router in an area supports an identical LSDB which contains at a minimum, a list of the router's interfaces and corresponding reachable neighbor routers.

Some of the most important features of OSPF are summarized below:

- **VLSM and CIDR**: Similar to RIPv2 and EIGRP, OSPF supports VLSM and CIDR.
- **Triggered Updates**: OSPF sends routing updates only when network changes occur instead of periodically as in RIP. OSPF only sends periodic link-state refresh updates ("paranoid updates") at long time intervals of 30 minutes (default) but does not depend on these updates to function. Routers send these LSAs periodically to refresh the network topology information before they are aged out of the LSDBs. Also, OSPF recalculates best routes to network destinations when the network topology/state changes, using the Dijkstra algorithm, and uses triggered updates (and not periodic updates) to minimize the routing protocol traffic that it generates and to ensure better use of network bandwidth.
- **Optimization on Multiaccess Networks**: OSPF optimizes the exchange of LSAs and synchronization of LSDBs on multiaccess network segments like Ethernet by electing a Designated Router and BDR. When OSPF routers form adjacencies with neighbor routers on a multiaccess segment, the election of a Designated Router and BDR takes place when the routers are in the 2-Way State. The OSPF router on the network segment with a highest OSPF Priority field value, or the highest Router ID when a tie occurs, will be elected the Designated Router. The router that has the second highest OSPF Priority or Router ID will be the BDR.

- **Multilevel Routing Hierarchy**: OSPF provides a multilevel (specifically, a two-level) routing hierarchy referred to as "area routing", where routing information about the topology within a specific area of an OSPF autonomous system is hidden from routers that are outside this area. The two-layer routing hierarchy divides networks into a Backbone Area (Area 0) and Non-Backbone Areas. This multilevel routing hierarchy makes routing in a large autonomous system more manageable, and provides routing protection and isolation, and a reduction in the overall routing protocol traffic. All routers within an OSPF area maintain the same LSDB contents, and the flooding of LSAs and the calculation of best paths, are limited only to changes within the area. OSPF has better scalability than the 15 hops limit of RIP.
- **Route Summarization**: The summarization of routes is an important feature for scaling OSPF networks because it reduces the frequency of LSA flooding in an OSPF autonomous system, as well as, the size of OSPF LSDBs and Routing Tables. An ABR summarizes routes advertised from a specific OSPF area before propagating them (via the Backbone Area using Summary [Type 3] LSAs) to other areas in the OSPF autonomous system. Route summarization reduces the amount of bandwidth consumed by OSPF LSAs, and the memory and processing resources used in OSPF routers. Without route summarization, Router-LSAs and Network-LSAs generated within an area will be propagated into the OSPF Backbone Area and to other OSPF areas, causing unnecessary router and network traffic overhead.
- **Equal Cost Multipath (ECMP) Routing**: OSPF supports ECMP routing where the Routing Table it creates contains multiple next-hop addresses with equal cost to the same network destination. All the equal cost routes have the same Administrative Distance and metric values.
- **Authentication**: OSPF packets can be authenticated to allow only trusted routers to participate in the exchange of routing information in an autonomous system.

OSPF version 2 (OSPFv2) whose features are discussed in this chapter was designed for use with IPv4. OSPFv3 (defined in RFC 5340) was created by updating OSPFv2 for compatibility with IPv6 and IPv6's 128-bit address space. The OSPFv3 Address Families feature allows it to support routing for both IPv6 and IPv4. Some differences between OSPFv2 and OSPFv3 include the following:

- OSPFv3 supports various packet and LSA formats with new LSAs created to carry IPv6 addresses and prefixes.
- OSPFv3 supports protocol processing on a per interface or link basis and not on a per-subnet basis.
- OSPFv3 supports the addition of LSA flooding scope which can be defined as link-local scope, area-local scope, or autonomous system wide scope.
- OSPFv3 supports the running of multiple instances of OSPFv3 on a link. Unlike OSPFv2, which has a built-in authentication mechanism, OSPFv3 relies solely on IP Security (IPsec) protocol suite to provide all needed authentication functionality [RFC4552].

REVIEW QUESTIONS

1. How are OSPF messages sent over IP?
2. What is an OSPF Link-State Advertisement (LSA)?
3. What is the purpose of LSA flooding in an OSPF area?
4. Explain the difference between the OSPF LSDB and the IP Routing Table.
5. What causes OSPF to send triggered updates?
6. What are the main functions of the OSPF Backbone Area or Area 0?
7. What is the main difference between an ABR (Area Border Router) and an ASBR (Autonomous System Boundary Router)?
8. What is an OSPF Stub Area? Explain the benefits of creating OSPF Stub Areas (whenever possible) in an OSP network.
9. Explain the use of Virtual Links in OSPF.
10. What is a Transit Area in OSPF?
11. What are the functions of the Designated Router in OSPF?
12. Explain briefly how the Designated Router and Backup Designated Router are elected on an OSPF network segment.
13. What is the purpose of the "All OSPF Routers" IPv4 Multicast Address 224.0.0.5 (AllSPFRouters Address)?
14. What is the purpose of the "OSPF All Designated Routers" IPv4 Multicast Address 224.0.0.6 (AllDRouters Address)?
15. Explain the main difference between the OSPF Router-LSA (Type 1 LSA) and Network-LSA (Type 2 LSA).
16. Explain the main difference between the OSPF Summary-LSA (Type 3 LSA) and ASBR Summary-LSA (Type 4 LSA).
17. What is the purpose of the OSPF AS-External-LSA (Type 5 LSA)?
18. What is the purpose of the Not-So-Stubby Area LSA (Type 7 LSA)?
19. What are the main functions of OSPF Hello packets?
20. What are the main functions of OSPF Database Description packets?
21. What are the main functions of OSPF Link-State Request packets?
22. What are the main functions of OSPF Link-State Update packets?
23. What are the main functions of OSPF Link-State Update packets?
24. Explain the use of the Link-State Request List and Link-State Retransmission List in an OSPF router.
25. What is the OSPF Router ID (Identifier)?
26. What is OSPF Paranoid Flooding (also called Paranoid Updates)?
27. What happens to an OSPF LSA when its LS Age reaches *MaxAge*? What actions do OSPF routers in the area take?
28. Explain the difference between an OSPF Type 1 External metric and a Type 2 External metric.
29. What is a host route?
30. What are the main benefits of OSPF route summarization?
31. Explain why routing policies and filters cannot be applied to OSPF LSAs and the LSDB but instead to the Routing Table.
32. Explain what is required in the IP Routing Table to implement Equal Cost Multipath (ECMP) Routing.

33. Explain the main differences between Null Authentication, Simple (Plaintext) Password Authentication, and Cryptographic Authentication in OSPF.

REFERENCES

[ALU7705SAR].   Alcatel Lucent, 7705 SAR OS Routing Protocol Guide, OSPF Chapter, page 159.

[CISCID6208].   Cisco Systems, "OSPF Not-So-Stubby Area (NSSA)", Document ID: 6208, August 10, 2005.

[CISCID7039].   Cisco Systems, "OSPF Design Guide", Document ID: 7039, March 28, 2005.

[CISCOSPFCOMD19].   Cisco Systems, *Cisco IOS IP Routing: OSPF Command Reference*, August 19, 2019.

[ISO/IEC10589].   ISO/IEC 10589:2002 – Information technology – Telecommunications and information exchange between systems – Intermediate System to Intermediate System intra-domain routing information exchange protocol for use in conjunction with the protocol for providing the connectionless-mode network service (ISO 8473), November 2002.

[JUNOSPFGUID].   Juniper Networks, "OSPF Feature Guide", September 23, 2018.

[MOYJT1998].   John T. Moy, *OSPF: Anatomy of an Internet Routing Protocol*, Addison-Wesley Professional, 1998.

[RFC1191].   J. Mogul and S. Deering, "Path MTU Discovery", IETF RFC 1191, November 1990.

[RFC1321].   R. Rivest, "The MD5 Message-Digest Algorithm", IETF RFC 1321, April 1992.

[RFC1584].   J. Moy, " Multicast Extensions to OSPF", IETF RFC 1584, March 1994.

[RFC1793].   J. Moy, "Extending OSPF to Support Demand Circuits", IETF RFC 1793, April 1995.

[RFC2328].   J. Moy, "OSPF Version 2", IETF RFC 2328, April 1998.

[RFC2390].   T. Bradley, C. Brown, and A. Malis, "Inverse Address Resolution Protocol", IETF RFC 2390, September 1998.

[RFC2453].   G. Malkin, "RIP Version 2", IETF RFC 2453, November 1998.

[RFC3101].   P. Murphy, "The OSPF Not-So-Stubby Area (NSSA) Option", IETF RFC 3101, January 2003.

[RFC3209].   D. Awduche, L. Berger, D. Gan, T. Li, V. Srinivasan, and G. Swallow, "RSVP-TE: Extensions to RSVP for LSP Tunnels", IETF RFC 3209, December 2001.

[RFC3623].   J. Moy, P. Pillay-Esnault, and A. Lindem, "Graceful OSPF Restart", IETF RFC 3623, November 2003.

[RFC3630].   D. Katz, K. Kompella, and D. Yeung, "Traffic Engineering (TE) Extensions to OSPF Version 2", IETF RFC 3630, September 2003.

[RFC4203].   K. Kompella, Ed. and Y. Rekhter, "OSPF Extensions in Support of Generalized Multi-Protocol Label Switching (GMPLS)", IETF RFC 4203, October 2005.

[RFC4552].   M. Gupta and N. Melam, "Authentication/Confidentiality for OSPFv3", IETF RFC 4552, June 2006.

[RFC5036].   L. Andersson, I. Minei, and B. Thomas, Eds., "LDP Specification", IETF RFC 5036, October 2001.

[RFC5250].	L. Berger, I. Bryskin, A. Zinin, and R. Coltun, "The OSPF Opaque LSA Option", IETF RFC 5250, July 2008.
[RFC5305].	T. Li and H. Smit, "IS-IS Extensions for Traffic Engineering", IETF RFC 5305, October 2008.
[RFC5340].	R. Coltun, D. Ferguson, J. Moy, and A. Lindem, "OSPF for IPv6", IETF RFC 5340, July 2008.
[RFC7471].	S. Giacalone, D. Ward, J. Drake, A. Atlas, and S. Previdi, "OSPF Traffic Engineering (TE) Metric Extensions", IETF RFC 7471, March 2015.
[RFC7752].	H. Gredler, Ed., J. Medved, S. Previdi, A. Farrel, and S. Ray, "North-Bound Distribution of Link-State and Traffic Engineering (TE) Information Using BGP", IETF RFC 7752, March 2016.
[RFC7770].	A. Lindem, Ed., N. Shen, J. P. Vasseur, R. Aggarwal, and S. Shaffer, "Extensions to OSPF for Advertising Optional Router Capabilities", IETF RFC 7770, July 2007.
[RFC7868].	D. Savage, J. Ng, S. Moore, D. Slice, P. Paluch, and R. White, "Cisco's Enhanced Interior Gateway Routing Protocol (EIGRP)", IETF RFC 7868, May 2016.
[RFC8201].	J. McCann, S. Deering, J. Mogul, and R. Hinden, Ed., "Path MTU Discovery for IP version 6", IETF RFC 8201, July 2017.
[RFC8570].	L. Ginsberg, Ed., S. Previdi, Ed., S. Giacalone, D. Ward, J. Drake, and Q. Wu, "IS-IS Traffic Engineering (TE) Metric Extensions", IETF RFC 8570, March 2019.
[RFC8571].	L. Ginsberg, Ed., S. Previdi, Q. Wu, J. Tantsura, and C. Filsfils, "BGP – Link State (BGP-LS) Advertisement of IGP Traffic Engineering Performance Metric Extensions", IETF RFC 8571, March 2019.
[TEAREDIA15].	Diane Teare, Rick Graziani, and Bob Vachon, Implementing Cisco IP Routing (ROUTE) Foundation Learning Guide: (CCNP ROUTE 300-101), Chapter 'OSPF Implementation', February 3, 2015.

2 Intermediate System-to-Intermediate System (IS-IS) Protocol

2.1 INTRODUCTION

Intermediate System-to-Intermediate System (IS-IS) is a link-state routing protocol and an IGP similar to OSPF, and is used for distributing routing information within an autonomous system. IS-IS also uses the Shortest-Path First (SPF) algorithm (a variant of the Dijkstra algorithm) to calculate loop-free routes to network destination. Compared to RIP, IS-IS was designed to respond quickly to network topology changes just like OSPF (both exhibit the same convergence properties), but IS-IS uses relatively fewer routing protocol messages and generates less protocol traffic when compared to OSPF.

IS-IS was originally designed as a routing protocol for the Connectionless Network Service (CLNS) environments, but was extended to include routing over IP **[ISO10589:2002]**. The extended version that has routing capabilities for CLNS only, IP only, or mixed CLNS and IP environments is sometimes referred to as Integrated IS-IS or Dual IS-IS **[RFC1195]**. In the dual mode, Integrated IS-IS can carry CLNS routing information in addition to IP routing information and network address prefixes.

This chapter describe the IS-IS protocol and its structures. The chapter begins with the basics of Open System Interconnection (OSI) routing, and then focuses on IP routing as supported by Integrated IS-IS **[RFC1195]**. Differences between the Integrated IS-IS protocol as described in **[RFC1195]** and the protocol as deployed in practice for IP routing, are described in **[RFC3719] [RFC3787]**. Reference **[RFC5308]** describes extensions to the IS-IS protocol for IPv6 Routing. Multi-Topology (MT) concept for IS-IS or M-ISIS (which allows a number of independent IPv4 and IPv6 topologies to be run in a single IS-IS domain at the same time) is defined in **[RFC5120]**. The section "Summary of IS-IS Features" at the end of the chapter lists other extensions and applications of the IS-IS protocol.

2.2 OVERVIEW OF IS-IS

Routers running a link-state routing protocol (IS-IS or OSPF) exchange network topology information with their nearest neighbors, and this information in turn gets flooded by other neighbor routers throughout the network. This allows every router in the network to have a complete picture or map of the network topology. Each router maintains a full topology map of the network (i.e., which routers and end systems are connected to which other routers and end systems). Routers then use this map to compute the best path to each network destination. The best path information to a destination includes the best adjacent router or next-hop to which a packet will be forwarded on its way to the destination.

Each IS-IS router uses Link-State Packets (LSPs) to distribute information about its local state to other routers. The local state information includes the router's usable interfaces, neighbors that are reachable over each interface, and the routing cost associated with each interface. The IS-IS routers receive the LSPs and use the information they contain to build identical Link-State Databases (LSDBs) that describe the topology of the routing domain. An IS-IS routing domain is a collection of IS-IS areas (each consisting of End Systems [ESs] and Intermediate Systems [ISs]), and is equivalent to an autonomous system in OSPF. An IS-IS routing domain is any portion of an OSI internetwork that is under the control of a common administrative authority.

After all the LSDBs have been synchronized and are identical, each router will use the SPF algorithm to calculate the best routes to all known network destinations. These best routes will then be installed in the router's local IS-IS Routing Database (also known to as the IS-IS Forwarding Database in [ISO10589:2002]). Each entry of the IS-IS Routing Database contains at a minimum, a known destination address, and its corresponding next-hop IP address and outgoing interface. Each IS-IS router updates its LSDB and recalculates best routes whenever network topology changes occur. Any network change will cause LSPs to be flooded, LSDBs to be updated, and then each router to run the SPF algorithm over its map (or LSDB) to calculate the shortest path to all possible destinations. As discussed in Chapter 2, OSPF messages are directly carried in IP and use the IP Protocol number 89, while IS-IS messages are directly encapsulated in Layer 2 (Data Link Layer) protocols.

One important advantage of a link-state routing protocol is that, by providing the routers in the network a complete knowledge of the network topology, they can calculate best routes that satisfy particular criteria, for example, in traffic engineering applications. Extensions to OSPF **[RFC3630] [RFC7471]** and IS-IS **[RFC5305] [RFC8570]** for traffic engineering purposes, allow routes to be calculated subject to constraints to meet particular quality-of-service (QoS) requirements.

The main disadvantage of a link-state routing protocol is that, the requirement of developing and maintaining a consistent, identical, and synchronized complete map of the network (i.e., LSDB) makes link-state routing protocols relatively more computationally intensive and requiring more memory than distance-vector routing protocols. As more routers are added to the routing domain, the size of the network topology database (or LSDB) increases. The size and frequency of the topology updates also increases, as well as, the length of time it takes for routers to calculate the best routes to all network destinations. The advantage of link-state routing protocols, however, is they make better path decisions that are less prone to routing loops.

IS-IS and OSPF have similar features such as the use of triggered updates, support of Equal-Cost Multi-Path (ECMP) routing, Variable Length Subnet Mask (VLSM), Classless Inter-Domain Routing (CIDR), two-level hierarchical routing, and various authentication mechanisms. IS-IS supports authentication so that all protocol exchanges can be authenticated. This allows only trusted IS-IS routers to join in the routing information exchanges in the IS-IS routing domain.

TABLE 2.1

OSPF and IS-IS Terminology

OSPF	IS-IS
Host	End System (ES)
Router	Intermediate System (IS)
Link	Circuit
Packet	Protocol Data Unit (PDU)
Designated Router (DR)	Designated IS (DIS)
Backup DR (BDR)	N/A (no Backup DIS is used)
Link-State Advertisement (LSA)	Link-State PDU (LSP)
Hello Packet	IS-IS Hello PDU
Database Description (DBD)	Complete Sequence Number PDU (CSNP)
Area	Subdomain (Area)
Non-backbone Area	Level 1 Area
Backbone Area	Level 2 Subdomain (Backbone)
Area Border Router (ABR)	Level 1/Level 2 IS (Router)
Autonomous System Boundary Router (ASBR)	Any Level 2 Capable IS

One method used to reduce the memory and processing requirements on routers, as well as, the size of topology updates, the LSDB, and the IS-IS Routing Databases, is to divide a network into smaller routing domains (or subdomains) called areas. Similar to OSPF, IS-IS supports a two-level routing hierarchy. IS-IS also uses the concept of "area routing" where an autonomous system is divided into smaller sub-domains and then organized in a two-level routing hierarchy. The use of subdomains allows information about the topology details within a subdomain (or area) to be hidden from routers outside that area providing a number of benefits as discussed in Chapter 1. Chapter 1 discusses extensively the advantages of creating areas in an autonomous system that uses a link-state routing protocol. Table 2.1 compares OSPF and IS-IS terminology.

2.3 BASIC OSI TERMINOLOGY AND OSI NETWORK LAYER PROTOCOLS

The International Organization for Standardization (ISO) is an international, independent, non-governmental body (made up of representatives from various national standards organizations) that sets standards for various areas ranging from technology, manufactured products, agriculture, healthcare, and others such food safety. Part of the responsibilities of the ISO is to develop standards for data networking. The OSI standardization program of the ISO developed a suite of international network reference models and protocols that were geared toward facilitating multivendor communication equipment interoperability. In this section, we explain the basic OSI terminology and the type of Network Layer protocols used in OSI.

The OSI networking program provided international standards (network reference models and protocols) to facilitate communication and interoperability between the hardware and software systems developed by different manufacturers and vendors, despite differences in their underlying architectures. The OSI networking suite has two major components:

- **The OSI Reference Model**: This is a seven-layer model which is an abstract model that characterizes and standardizes the functions of data communication systems, without placing emphasis on their underlying internal structure and technology.
- **OSI Protocol Suite**: The represents a collection of standard networking protocols developed at each of the seven OSI reference model layers. Numerous standard protocols were developed as part of the OSI protocol suite; protocols at the seven OSI reference model layers – Physical, Data Link, Network, Transport, Session, Presentation, and Application Layers are supported.

A Network Service Access Point (NSAP) is a conceptual point located at the boundary between the OSI Transport and Network Layers; it is the location at which the Network Layer provides services to a Transport Layer entity. Each Transport Layer entity is assigned a single NSAP which also has an individual NSAP address in the OSI internetwork. Addressing in at the OSI Network Layer is implemented using two types of hierarchical addresses: NSAP addresses and Network-Entity Titles (NETs) which are a specific subset of NSAP addresses **[ISO/IEC8348:2002]**.

Similar to IP, Network Layer addressing is required for end-to-end communication. NSAP addressing is used in OSI networks, both in connectionless and connection-oriented mode services. An NSAP represents the address of a particular network service on a particular node in the OSI network. Unlike addressing in TCP/IP networks, an NSAP address is not assigned to individual interfaces on a network node, but rather to the entire network node. In common setups, a single OSI node is assigned only one NSAP address regardless of how many network interfaces it has.

IS-IS's NSAP addressing, scheme, unlike OSPF, avoids dependencies on IP addressing. Using IS-IS, IP connectivity between routers is not required for an IS-IS router to be able to share routing information. IS-IS sends routing updates (including updates when Integrated IS-IS is performing IP-only routing) via CLNS instead of IP.

The OSI protocol suite supports the following two types of Network Layer services (that are available to the Transport Layer):

- **Connectionless Network Service (CLNS)**: This type of service provides datagram transport (similar to IP datagram service) and does not require the establishment of a connection or circuit before a Transport Layer entity can transmit data. An entity can send data without having to establish a connection first. The *Connectionless Network Protocol (CLNP)* is an OSI Network Layer protocol that is used to carry data from an OSI upper-layer, and provides error indications over the underlying service provided by CLNS. CLNP provides the interface between the OSI Transport Layer and CLNS.

  A Transport Layer entity requests CLNSs through CLNP. Paths through a network are determined independently for each packet, meaning CLNS does not perform connection setup or termination. CNLS uses routing protocols to exchange routing information so that end-user packets can be routed from source to destination. CLNS provides no guarantees that data transmitted from a source will not be lost, corrupted, misordered, or duplicated; CLNS provides only best-effort data delivery. CLNS relies on a Transport Layer protocol or an upper-layer protocol to perform error detection and correction.

- **Connection-Mode Network Service (CMNS):** This type of service requires the explicit establishment of a connection or circuit between any two Transport Layer entities before they can start to exchange data. In general, this type of network service provides some level of data delivery guarantee, whereas the CLNS does not. The *Connection-Oriented Network Protocol (CONP)* is an OSI Network Layer protocol that is used to carry data from an OSI upper layer, and to provide error indications over connection-oriented paths.

  CONP is a Network Layer service and provides the interface between the OSI Transport Layer and the CMNS. When a Transport Layer entity requests connection-oriented network services through CONP, CMNS performs a number of functions related to the explicit establishment of paths, such as, connection setup, maintenance, and termination. CMNS also provides a mechanism for a Transport Layer entity to request a specific QoS.

CLNS and CMNS are Network Layer *services*, while CLNP and CONP are Network Layer *protocols*.

2.4   OSI ROUTING HIERARCHIES

In an OSI network, an ES is any non-routing node or host that sends and receives packets (e.g., a workstation, laptop, and email server). An Intermediate System (IS) is a node that performs routing of data (i.e., relay packets) but can also send and receive packets; it is router as in an IP network. Routing in the OSI networks is based on the concept of hierarchies, or levels, depending on the scope of the network in which the routing is performed.

2.4.1   OSI ROUTING LEVELS

OSI defines four levels of routing: Level 0, Level 1, Level 2, and Level 3 routing.

2.4.1.1   Level 0 Routing

Level 0 routing refers to routing between two ESs on the same link, or between an ES and the closest IS. Level 0 routing deals with how an ES discovers its nearest gateway (an IS), as well as, how an IS discovers which ESs are connected to it. The discovery is accomplished by allowing ESs and ISs to send periodic Hello messages advertising their presence. Hello messages sent by ESs are called ES Hello (ESH) messages, while Hello messages sent by ISs are called IS Hello (ISH) messages.

Level 0 routing is sometimes referred to as ES-IS routing **[ISO9542:1988]**. Level 0 routing is analogous to the Address Resolution Protocol (ARP) in IPv4 **[RFC826]**, and Neighbor Discovery in IPv6 **[RFC4861]**; routers in IPv6 networks send Router

Advertisement messages (allowing end-stations to detect their presence), and hosts send Router Solicitation messages.

2.4.1.2 Level 1 Routing

Level 1 routing refers to routing between ESs via one or more ISs within a single area of an OSI domain. Level 1 routing deals with intra-area routing, that is, routing between ESs that are members of the same area. An area is an administrative subdomain of an OSI domain.

Using IS-IS link-state routing, ISs in a given area in the domain will have the same detailed and complete picture of the entire area's topology. At Level 1 routing, ISs collect the lists of all ESs that are directly attached to them, and advertise these lists to each other to allow all ISs to discover the location of all ESs in the area. Thus, for routing, Level 1 ISs are required to keep track of the ESs within their own local IS-IS area.

2.4.1.3 Level 2 Routing

Unlike Level 1 routing which refers to routing within an IS-IS area, Level 2 routing refers to routing between IS-IS areas. Level 2 routing refers to routing between ESs located in different areas of an OSI domain. At Level 2 routing, ISs do not advertise to each other the list of ESs that are connected within IS-IS areas. Instead, ISs exchange Area Address prefixes to allow them to learn and determine how to reach particular areas in the routing domain. Thus, Level 2 ISs are required to keep track of the paths to destination IS-IS areas.

If a Level 1 IS determines that the destination ES of a packet is located in a different area in the OSI domain, it will forward the packet toward the closest Level 2 routing-capable IS, irrespective of the destination area. The packet will then be forwarded by one or more Level 2-capable ISs until it reaches the destination area, where it will again be forwarded by one or more Level 1 ISs to the specified ES. Thus, Level 1 routing can be viewed as routing according to the System IDs (assigned to nodes), while Level 2 routing can be viewed as routing according to the Area Address prefixes (assigned to OSI areas). Level 2 routing provides communication between individual areas of the domain and constitutes the backbone of the OSI domain.

2.4.1.4 Level 3 Routing

Level 3 routing refers to routing between ESs located in different OSI domains. Level 3 routing deals with routing between different OSI domains (i.e., interdomain routing). This is analogous to inter-autonomous system routing as provided by the Border Gateway Protocol (BGP) **[RFC4271]** in IP internetworks. Before BGP became the preferred and only routing protocol for routing between autonomous systems, the Inter-Domain Routing Protocol (IDRP) **[ISO10747:1994]** was defined for interdomain routing in OSI internetworks.

However, with BGP capable of multiprotocol interdomain routing, as well as, capable of carrying information about NSAP addresses, today's multiprotocol internetworks including OSI networks, use BGP and not IDRP.

2.4.2 OSI INTRADOMAIN ROUTING

The OSI protocol suite supports two protocols at the Network Layer for intradomain routing (Figure 2.1):

- **End System-to-Intermediate System (ES-IS) Discovery Protocol**: This is used to locate systems on directly connected network segments. It enables "routing" between an ES and an IS, and is analogous to the ARP in IPv4 **[RFC826]**.
- **IS-IS Routing Protocol**: This performs hierarchical (Level 1 and Level 2) routing between ISs. IS-IS is a link-state routing protocol used in ISO CLNS environments. ISs can exchange network reachability information with other ISs using the IS-IS protocol. Similar to IP, static routes can also be created in an CLNS environment.

There is no ARP, or Internet Control Message Protocol (ICMP) for CLNS, so, the ES-IS protocol **[ISO9542:1988]** provides the same kind of functions for ESs and ISs. ES-IS and IS-IS protocols allow routers to learn about the network topology and perform the function of routing data packets from one ES to another within the OSI domain.

As mentioned above, OSI uses hierarchical routing and the concept of areas to simplify network, and router design and operation (Figure 2.1). Each ES resides in a particular area and listens to ISH packets to discover the nearest IS. An ES that wants

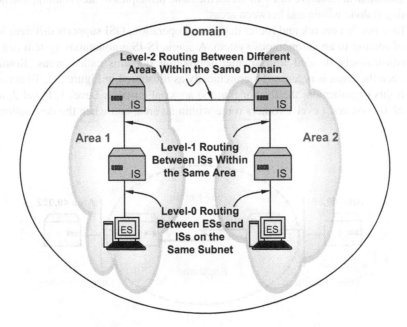

FIGURE 2.1 OSI Hierarchical Routing

to send a packet to another ES, sends it to any randomly selected Level 1 IS on it directly attached network in the local area (Level 0 routing). In some cases, to achieve optimal routing, the IS will send a redirect message to the sending ES informing it about a more suitable IS.

The Level 1 IS then performs a lookup in it IS-IS Routing Database using the packet's destination address, and forwards the packet on the best route leading to the destination. An IS discovers ESs by listening to the ES Hello packets they send. ISs send ISH packets to ESs. By listening to ES Hello packets, the sending Level 1 IS can determine if the destination ES is on the same subnetwork, so, that it can forward the packet accordingly. The Level 1 IS may also send, if necessary, a redirect message back to the source ES advising it that a more direct route is available.

If the destination ES on another subnetwork in the same area, the routing information provided to the Level 1 IS, allows it to determine the correct route (Level 1 routing) to the destination. If the destination ES is an in different area, the Level 1 IS will send the packet to the nearest Level 2 capable IS in the local area (to perform Level 2 routing). Level 2 ISs forward the packet until it reaches a Level 2 IS in the destination ES's area. Within the destination area, Level 1 ISs forward the packet along the best path until it reaches the destination ES.

2.5   IS-IS HIERARCHICAL ROUTING

This section describes IS-IS concepts and capabilities, and the different types of IS-IS routers, along with their roles in an IS-IS autonomous system. We describe also the hierarchical structure of IS-IS areas, the basic principles of area routing, and how routing is done within and between areas.

To simplify network and router design and operation, OSI supports different levels of routing in an autonomous system. A single IS-IS autonomous system can be administratively divided into smaller subrouting domains called areas. Routing between the areas is organized hierarchically as illustrated in Figure 2.2. To accomplish this organization, the ISs (or routers) are configured as Level 1, Level 2, and Level 1-2 routers. Level 1 routers route within an area, and when the destination is

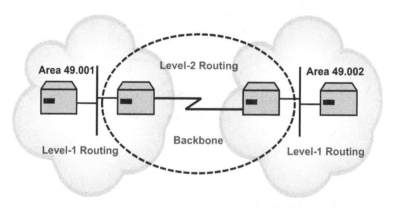

FIGURE 2.2   IS-IS Hierarchical Routing

outside the area, they forward the traffic to a Level 2-capable router, which provides routing between areas.

The scope of propagation of link-state information is determined by the routing level and area in which the IS-IS routers belong. Each router within a routing level (Level 1 or Level 2) and area maintains a complete LSDB of all the link-state information provided by the routers in the same routing level and area. Note that Level 1 routing in an IS-IS routing domain is provided collectively by all the Level 1 areas in the domain. Each router in an IS-IS area then uses the SPF algorithm (or the Dijkstra algorithm) to determine the shortest path from the local router to other IS-IS routers and ESs as represented in its LSDB.

2.5.1 WHAT IS INTEGRATED IS-IS?

Integrated IS-IS **[RFC1195]** is an adaptation of the IS-IS protocol that can carry and propagate routing information for ISO CLNS as well as IP. Integrated IS-IS can distribute routing information for ISO CLNS only, IP only, or mixed CLNS and IP (called the dual mode) environments. Integrated IS-IS can be used as an alternative to OSPF in an IP environment, but also allows the mixing of ISO CLNS and IP routing in one protocol.

Some of the important features supported by Integrated IS-IS include the following:

- VLSMs and allows network masks and address prefixes to be sent in routing updates.
- Redistribution of routes between IS-IS and other IP routing protocol (RIPv2, EIGRP, OSPF, BGP).
- Summarization (or aggregation) of IP routes.

OSPF and Integrated IS-IS are both link-state protocols and have similar features.

- Routing information based on network link-states
- Support more comprehensive routing metrics than RIP
- Flood routing updates only when network changes occur
- Use various mechanisms for aging routing information
- Routers use identical LSDBs (after routing information synchronization) and SPF algorithms for computing best paths to network destinations

Volume 1, Chapter 2 of this two-part book discusses in detail the general characteristics of link-state routing protocols. In the rest of this chapter and in others, Integrated IS-IS is referred to as simply, IS-IS.

2.5.2 IS-IS AREAS

An OSI domain can consist of one or more IS-IS areas. An IS-IS area is a logical routing entity and is formed by a group of contiguous IS-IS routers and their interconnecting links. An area is a routing subdomain in which the IS-IS routers within maintain detailed routing information about its internal composition or structure, and also maintain identical routing information which allows them to reach each other and other areas. All IS-IS routers in the same area exchange routing information about all the IS-IS routers and ESs that they can reach in the area. The Level 2 capable routers in the OSI domain are connected in a contiguous fashion to form a backbone for the OSI domain. All routers that form the backbone know how to reach all areas in the domain. An IS-IS routing domain (equivalent to an OSPF autonomous system) supports a two-level routing hierarchy (see Figures 2.3 and 2.4).

In Figure 2.3, IS-IS Level 2 and Level 1-2 routers are chained together to form a Level 2 backbone. Area 1 contains two routers, one being a Level 1-2 router which borders Area 2. The other router is a Level 1 router and is contained entirely within Area 1. Area 2 has five routers with two routers specified as Level 1 only routers. These Level 1 only routers perform routing internal to Area 2 only (and to the exit points). Three Level 1-2 routers form a chain across Area 2 linking to the neighbor areas, Area 1, Area 3, and Area 4.

Although the middle router of these three Level 1-2 routers in Area 2 does not have a direct link to another IS-IS area, it supports Level 2 routing so that the IS-IS backbone is contiguous. If this middle router fails, the remaining two Level 1-2 routers (although have a physical path through one of the Level 1 routers in the area) will no more be contiguous and the backbone would be broken. The two Level 1 routers in Area 2, although are physically connected to the remaining Level 1-2 routers, cannot perform Level 2 functions. Router redundancy can be implemented to ensure backbone continuity (i.e., redundant Level 2 capable routers and links). The network

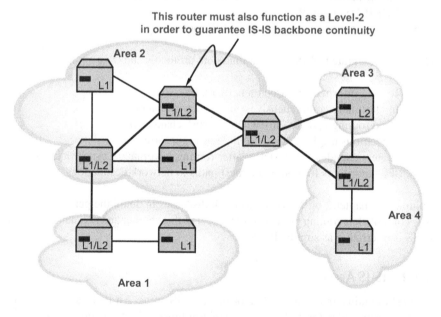

FIGURE 2.3 IS-IS Backbone – IS-IS Level 2 and Level 1-2 Routers Forming a Level 2 Backbone

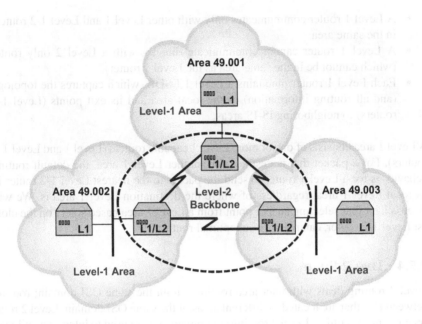

FIGURE 2.4 IS-IS Areas

operator also must ensure that area design and partitioning is done in such a way so that the backbone cannot be easily partitioned when Level 2 capable router failures occur.

Area 3 contains only one Level 2 router that connects to Area 2 and Area 4, but has no IS-IS neighbors within the area. If another Level 1 router were to be added to Area 3, the Level 2 router would revert to a Level 1-2 router. We see from Figures 2.3 and 2.4 that the border between the IS-IS areas lies on the links between Level 2 capable routers (in contrast to OSPF, where the border exists within an Area Border Router [ABR]). Detail characteristics of Level 1, Level 2, and Level 1-2 routers are given below.

2.5.3 LEVEL 1 ROUTER

A Level 1 router is responsible for routing to and from ESs within an IS-IS area. This is similar to an internal non-backbone router in an OSPF Totally Stubby Area as discussed in Chapter 1. Level 1 routers enable ESs in an IS-IS area to communicate with each other, and to communicate via a Level 1-2 router with the rest of the OSI domain. The following are properties of a Level 1 router:

- Level 1 routers share only intra-area routing information; exchange topology information for the local area (i.e., keep track of paths within the local area).
- A Level-1 router has IS-IS neighbors only in the same area. A Level 1 router will not form an adjacency with a router that does not share a common Area Address. When two routers have successfully exchanged Hello PDUs and have established an adjacency, they are considered to be adjacent.

- A Level 1 router communicates only with other Level 1 and Level 1-2 routers in the same area.
- A Level 1 router cannot communicate directly with a Level 2 only router (which cannot be in the same area as the Level 1 router).
- Each Level 1 router maintains a Level 1 LSDB, which captures the topology (and all routing information) of the local area and its exit points (Level 1-2 routers) to neighboring IS-IS areas.

A Level 1 area consists of one or more Level 1 capable routers (Level 1 and Level 1-2 routers). For a packet that is destined for another Level 1 area, the default routing behavior is for a Level 1 router to send the packet to the nearest Level 1-2 router in its local Level 1 area, regardless of where the destination Level 1 area is. We will discuss later that selecting an exit point from a Level 1 area based solely on the closest Level 1-2 router, can result in suboptimal routing.

2.5.4   LEVEL 2 ROUTER

Level 2 routing deals with inter-area routing within the same OSI domain; routing between ESs that are located in different areas of the same OSI domain. Level 2 routers are routers within a Level 2 routing subdomain and are used to interconnect Level 1 areas. Level 2 routers share inter-area routing information (i.e., keep track of the paths to Level 1 areas) and perform routing between different areas only. This is similar to an internal router in an OSPF Backbone Area as discussed in Chapter 1. The following are properties of a Level 2 router:

- Level 2 routers provide routing between Level 1 areas and form an intradomain routing backbone. Routing between areas is performed based on the Area Address.
- Level 2 routers can communicate directly with Level 2 and Level 1-2 routers in the same or different areas but only a Level 2 adjacency can be established. Level 2 routers advertise their assigned NSAP addresses to the other Level 2 capable routers making up the Level 2 backbone.
- A Level 2 router may have neighbors in the same or other areas.
- Level 2 routers do not need to learn the network topology within any Level 1 area, except if the Level 2 router is also Level 1 capable (i.e., is a Level 1-2 router) within a single area.
- Each Level 2 router maintains a separate LSDB that contains only the inter-area topology information (i.e., a Level 2 LSDB that contains all inter-area routing information). The LSDB maintains information about other Level 2 routers and the local Level 1-2 routers in each area.

Level 2 routers can also announce external routing information, corresponding to network addresses that are reachable by routers in other autonomous systems or routing domains. Links to other routing domains (external destinations) must be from Level 2 routers.

IS-IS does not support the concept of a dedicated Area 0 that serves as a backbone as in OSPF. Instead, an IS-IS backbone consists of a chain of Level 2 capable routers that may pass through different IS-IS areas. The IS-IS backbone is a set of Level 2 capable routers in distinct and interconnected IS-IS areas. In IS-IS, no single area functions strictly as the backbone as in OSPF. As discussed in Chapter 1, the OSPF backbone is a separate OSPF area (i.e., Area 0).

An IS-IS backbone consists of a contiguous set of Level 2 and Level 1-2 routers. The IS-IS backbone is a path composed of a chain or connection of Level 2 capable routers that must be contiguous, but can pass through different IS-IS areas; all individual areas making up the backbone must also be contiguous. This means Level 2 adjacencies are contiguous and exist independent of the contributing areas. A Level 2 backbone can be viewed as a virtual IS-IS area consisting of Level 2 capable routers. IS-IS hierarchical routing simplifies the IS-IS backbone design by allowing Level 1 routers to know only how to get to the nearest Level 2 capable router (i.e., a Level 1-2 router).

2.5.5 LEVEL 1/LEVEL 2 (OR LEVEL 1-2) ROUTER

A Level 1-2 router performs routing between an IS-IS area and the IS-IS backbone (which consists of a contiguous collection of Level 2 capable routers each of which can be in a different IS-IS area). Level 1-2 routers participate in the Level 1 intra-area routing as well as the Level 2 inter-area routing. A Level 1-2 router is equivalent to an ABR in OSPF. A Level-1-2 router may have IS-IS neighbors in any area.

Level 1-2 routers maintain two separate LSDBs (i.e., Level 1 and Level 2 LSDBs) which allows them to function as if they are two separate IS-IS routers:

- A Level 1-2 router can act as both a Level 1 and Level 2 router, sharing intra-area routes with other Level 1 routers in its area, and inter-area routes with other Level 2 routers in other areas. A Level 1-2 router exchanges topology information within its local area and between IS-IS areas; performs routing within and between areas.
- A Level 1-2 router supports a Level 1 function that allows it to communicate directly with other Level 1 routers in the area in which it is located. It maintains Level 1 LSP information in a Level 1 LSDB. It also informs other Level 1 routers in the local area that it has an exit point from the area.
- A Level 1-2 router also supports a Level 2 function to communicate with the rest of the Level 2 capable routers that form the IS-IS backbone. It maintains a Level 2 LSDB that is separate from its Level 1 LSDB.

If a destination address is in the same area (i.e., the Area IDs are equal), a Level 1-2 router uses the Level 1 LSDB to forward a packet, based on the System ID, to the Level 1 router that advertises the destination address. If the destination address is in another IS-IS area (i.e., the Area IDs are not equal), the Level 1-2 router uses the Level 2 LSDB to forward the packet based on the Area ID. The Level 1-2 router located at the boundary between subdomains also has the following properties:

- Routes in the Level 1 LSDB are added to Level 2 LSPs so that they appear as leaves off the Level 2 router. The Level 1-2 router does not inject topology information from a Level 1 area into the Level 2 routing subdomain, only reachability information.
- By default, a Level 1-2 router does not inject routing information from the Level 2 routing subdomain or other areas into the local Level 1 area. Instead, when connected to the Level 2 routing subdomain, it sets the ATT (Attached) bit in its LSPs so that the Level 1 routers in the local area can know that it is attached to the Level 2 routing subdomain. Level 1 routers receiving LSPs with the ATT bit set, will see the advertising Level 1-2 router as a default route leading outside the area. IP routers in the local area will install a default route pointing to the Level 1-2 router sending the LSPs with the ATT bit set. An IS-IS Level 1 area is analogous to an OSPF Not-So-Stubby Area (NSSA) (see Chapter 1).

If a Level 1-2 router is running Integrated IS-IS and its Level 1 area has IP subnets, it will leak all these IP subnets from the Level 1 area into the Level 2 routing subdomain. These IP subnets can be summarized (by the Level 1-2 router) when desirable.

2.5.6 CHARACTERISTICS OF IS-IS AREAS

An IS-IS router can be a Level 1 router, a Level 2 router, or a Level 1-2 router. A Level 1-2 router advertises itself at Level 1 when presenting an exit point to a Level 1 area. Figure 2.5 (top diagram) shows the physical view of an example IS-IS network with Level 1 and Level 1-2 routers. R1 and R4 are Level 1 routers while R2 and R3 are Level 1-2 routers. Each Level 1-2 router physically connects to a Level 1 router in its area and possibly, to Level 2 capable routers that make up the backbone. Level 1-2 routers, R2 and R3, belong to their respective Level 1 areas (Area 1 and Area 2) and have a physical connection between them directly, or through a Level 2 router.

In IS-IS, the boundary between IS-IS areas is on the link between Level 2 capable routers; that is, between two Level 1-2 routers, or between a Level 1-2 router and a Level 2 router. The boundary is not inside a router (ABR) as in OSPF, but on the link that connects two IS-IS routers that are in different areas. This difference is because an IS-IS router generally has one NSAP address (that contains the router's Area ID), while an IP router generally can have multiple IP addresses. Figure 2.5 (bottom diagram) shows the logical view of the network in the top diagram. R2 and R3 are Level 1-2 routers, but they provide in addition, an entry point to the Level 2 backbone that interconnects the two Level 1 areas.

2.5.7 OSPF AREAS VERSUS IS-IS AREAS

It is discussed in Chapter 1 that OSPF uses the concept of a central backbone (Area 0) to which all other OSPF areas must be physically or logically attached (via Virtual Links) (see example network in Figure 2.6). In OSPF, the border between

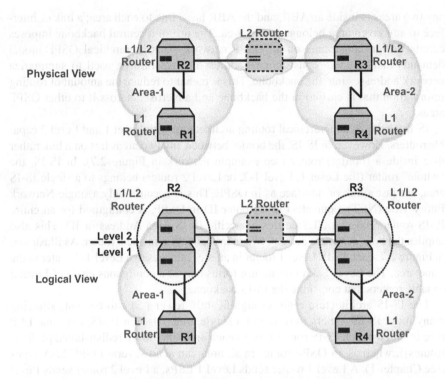

FIGURE 2.5   Physical and Logical Views of an IS-IS Area Configuration

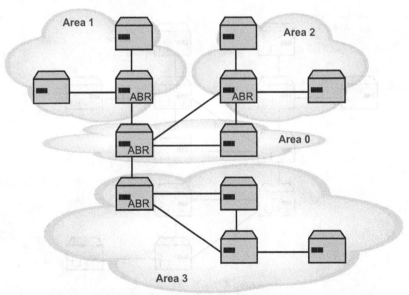

FIGURE 2.6   OSPF Areas: Area Borders are Within Routers

any two areas is inside an ABR, and the ABR has a link to each area; a link or inter-face to any given area belongs to that area. The use of a central backbone imposes certain design constraints on an OSPF network. This hierarchical OSPF model demands that, a good, consistent IP addressing structure be used to summarize network addresses into the backbone. This is meant to reduce the amount of routing information that is carried in the backbone and advertised across it to other OSPF areas.

IS-IS also uses a hierarchical routing architecture with Level 1 and Level 2 capa-ble routers. However, in IS-IS, the border between any two areas lies on a link rather than inside a (border) router (see example network in Figure 2.7). In IS-IS, the "whole" router (the Level 1, Level 1-2, or Level 2 router) belongs to a single IS-IS area, not just a link or interface as in OSPF. This is because only a single Network Entity Title (NET), equivalent to a Router ID in OSPF, is configured for an entire IS-IS router, and that NET carries a specific IS-IS Area Address or ID. This also implies the boundary between any two areas cannot lie inside a router. As illustrated in Figure 2.7, each IS-IS Level 1 router in an area attaches to a Level 1-2 router in the same area. The Level 1-2 router in turn forms part of a contiguous chain of Level 2 capable routers that constitute the IS-IS backbone.

The IS-IS architecture requires significantly fewer LSPs to be sent, allowing many more IS-IS routers to reside in a single area. Also, in IS-IS, only one LSP type is sent by each IS-IS router type in each area (including redistributed prefixes [routes]), whereas an OSPF router in an area can send several OSPF LSA types (see Chapter 1). A Level 1 router sends Level 1 LSPs, a Level 2 router sends Level 2 LSPs, and a Level 1-2 sends both Level 1 and Level 2 LSPs. Each router creates

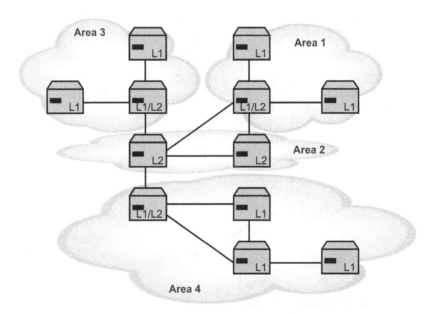

FIGURE 2.7 IS-IS Areas: Area Borders are on Links Between Routers

an LSP and floods it to its neighbors. This means, with regard to the sending and processing of routing updates and router CPU processing overhead, IS-IS is more efficient than OSPF. IS-IS routers have fewer LSPs to process, and the mechanisms IS-IS uses to install and withdraw network prefixes are less processor intensive.

The IS-IS hierarchical architecture provides more flexible and allows the IS-IS backbone to be extended by simply adding more Level 2 routers. This flexibility allows IS-IS to be more scalable than OSPF. Extending the backbone in IS-IS is less complex than in OSPF. However, OSPF supports some special features that IS-IS does not have, such as stub areas, and NSSAs, route tags, and so on.

2.6 NSAP ADDRESSES

The Network Layer addresses that identify systems in an OSI network are based on NSAP addresses (sometimes called ISO addresses) **[ISO/IEC8348:2002]**. Each NSAP address identifies a point of connection in an OSI network, such as a router. IS-IS uses NSAP addresses to identify individual IS-IS routers, the area memberships of routers, and their adjacencies. NSAP addressing does not support the notion of per interface addressing as in IP networks.

This section describes the hierarchy and addressing structures in OSI networks. A number of NSAP address formats have been defined for different systems, and each OSI routing protocol uses a different NSAP representation. NSAP addresses usually are expressed in hexadecimal format and have a variable length of up to 40 hexadecimal digits (or 20 bytes).

IS-IS Hello PDUs, LSPs, and other routing PDUs use OSI NSAP address formats. This means every IS-IS router requires an OSI NSAP address even if it is performing only IP routing. IS-IS uses the NSAP address in LSPs to identify the source router, construct the LSDB as well as the underlying IS-IS Routing Database.

2.6.1 NSAP ADDRESSING FORMAT

Figure 2.8 shows the general format of an NSAP address. IS-IS uses a two-layer addressing architecture, which combines the Initial Domain Part (IDP) and High-Order Domain Specific Part (HODSP) fields as the Area Address or ID for Level 2 routing, leaving the remaining System ID for Level 1 routing. The NSAP is divided into three parts: 1 byte for NSAP Selector (NSEL), 6 bytes for the System ID, and from 1 to 13 bytes for the Area Address field. An NSAP address has a variable length and can be from 8 bytes to 20 bytes long.

The NSAP address is usually longer than 8 bytes to allow some granularity for allocating IS-IS areas. All nodes within an IS-IS area must have the same Area Address. The exact format and length of the IDP and the Domain Specific Part (DSP) depend on the actual application in which the NSAP addressing is used, making them variable to a large extent. This results in NSAP addresses having variable lengths. In a single message instance, an NSAP address contains information about a node's autonomous system, area within that autonomous system, unique identifier, and even the upper-layer service associated with the message.

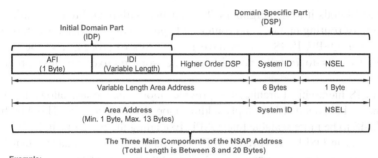

FIGURE 2.8   NSAP Address

2.6.2   INITIAL DOMAIN PART (IDP) AND DOMAIN SPECIFIC PART (DSP)

The IDP consists of two fields as shown in Figure 2.8: The Authority and Format Identifier (AFI) and the Initial Domain Identifier (IDI). The AFI (a 1-byte value in the range of 00 to FF) indicates the format of the remaining NSAP address fields. It identifies the format of the IDI and DSP. The IDI is a variable length field and its structure depends on the address format indicated by the AFI; it is even omitted in some cases. Together, the AFI and IDI in the NSAP address indicate the routing domain (or autonomous system) to which an OSI node belongs.

The DSP consists of a variable-length HODSP field that identifies the area within the IS-IS domain in which a node is located. The format of the DSP depends on the particular address format indicated by the AFI. The DSP can be further structured into the System ID and the NSAP Selector (NSEL) subfields. The System ID is a unique identifier of the node in its IS-IS area. The System ID field can be from 1 to 8 bytes long, but all current implementations fix the length of this field to 6 bytes. The System ID throughout the domain must be of the same length.

2.6.3 NSEL Field in NSAP Addressing

The NSAP address of a device in the OSI network contains the following device-specific information (Figure 2.8):

- The OSI address of the device itself (i.e., System ID)
- The identifier of the higher-layer process (called the N-Selector [NSEL], Service Identifier or Process ID)

The 1-byte NSEL field identifies the particular service in or above the OSI Network Layer on the destination node that should receive the message. This is analogous to the Protocol Number or Identifier in IP (i.e., an upper-layer protocol, for example, a Transport Layer protocol like TCP or UDP). The entire NSAP address including the Area Address can be viewed as equivalent to the combination of an IP address and an upper-layer protocol (TCP or UDP port number) in an IP header.

An ISO ES can have multiple NSAP addresses, where the addresses differ only by the NSEL (the last byte). Each NSAP address represents a particular service that is available at that ES. In addition to having multiple services (NSAPs), a single ES can belong to multiple IS-IS areas.

If the NSEL field is 00 (in hexadecimal), the NSAP address refers to the OSI device itself. An NSEL value of 00, means no particular service in the destination node is being addressed; the entire NSAP address identifies the destination node itself without referencing any particular service on that node. An NSAP with an NSEL equal to 00 is equivalent to the OSI Layer 3 address of the device. An NSAP address with the NSEL set to 00 is known as the NET. The NSAP addresses for all IS-IS routers have their NSEL equal to 0.

The NET is the address that is configured on an IS-IS router, and is the address a router uses to identify itself in the LSPs it sends. The NETs form the basis for node identification during the IS-IS best route calculations. IS-IS uses the NET (a subset of the NSAP) to represent each router during route calculation (using the Dijkstra algorithm), while OSPF uses the Router IDs. NETs and NSAPs must start and end on a byte boundary and are usually expressed in all hexadecimal digits.

Official NSAP address prefixes are required for ISO CLNS routing. NSAP addresses that start with an AFI value of 49 are considered to be private NSAP addresses. The AFI of 49 is used to define private address spaces analogous to private IP addresses in [RFC1918]. This group of addresses are routable by IS-IS, however, they are not to be advertised to other CLNS networks. In typical IS-IS deployments using an AFI of 49 in network addressing, the length and meaning of the HODSP field are entirely up to the network administrator, making this addressing format flexible and adaptable for different environments.

The following are example of NETs:

- 49.0001.1234. **00a0.c96b.c490**.00
  Area Address = 49.0001.1234
  System ID = **00a0.c96b.c490**
  NSEL = 00
- 49.0001.**1111.1111.1111**.00
  Area Address = 49.0001
  System ID = **1111.1111.1111**
  NSEL = 00
- 1234.5678.**90ab.cdef.0001**.00
  Area Address = 1234.5678
  System ID = **90ab.cdef.0001**
  NSEL = 00.
- 39.0f01.0002.**0000.0c00.1111**.00
  Area Address = 39.0f01.0002
  System ID = **0000.0c00.1111**
  NSEL = 00

The easiest way to determine that the Area Address is 39.0f01.0002 in the last example, is to start from the right-hand side of the NSAP address, and work toward the left. The last two digits of the NET are the NSEL; which are always set to 00. The next 12 digits (separated into three groups of four digits) are the System ID. Note that in most IS-IS implementations, the System ID is always 6 bytes long and anything else to the left of the System ID is the Area Address (which is of variable length from 1 to 13 bytes).

The following are not valid router NETs:

- **2**.49.0000.00c0.1234.00 is not a valid NET because the first digit in the Area Address must be at least one byte (two digits in hexadecimal) long. Note 2.49 is only one and half byte long (when examining the address from right to left). The NSAP address must be a multiple of a whole byte. Recall that a NET ranges from 8 bytes to 20 bytes in length.
- 40.0000.00c0.1234.**56** is not a valid NET because the last two digits, the NSEL, must always be 00.

Each node in an IS-IS area must have a unique System ID. It is customary to create the System ID based on either a MAC address on the router, or the IP address of a loopback interface on the router. The Area ID (or Area Address) can be from 1 to 13 bytes, however, using a 1-byte field for the Area ID limits the number of areas that can be defined. Thus, in most implementations, the Area ID consists of 3 bytes, with the AFI taking 1 byte, and the remaining 2 bytes used for the actual Area ID **[CISCTEAPAQ00]**. For example, in the NSAP address, 49.0001.0000.0c12.3456.00, the AFI is 49, and the remaining 2 bytes are 0001, resulting in an effective Area ID of 49.0001.

Unlike in IP networks, individual interfaces in an OSI network are not assigned their own Network Layer addresses. However, the Layer 2 addresses in the OSI network are used in the same way as in IP networks. When an IS-IS packet is to be sent out a router interface, it is encapsulated in a Layer 2 frame that is addressed to the next directly attached IS-IS hop. The Layer 2 frame is identified by the receiving interface's Layer 2 address.

The Layer 2 address of an interface in OSI networks is called a Sub-Network Point of Attachment (SNPA). The SNPA is similar to the Layer 2 address used in IP and can be taken from an Ethernet MAC address, or the Virtual Circuit ID of an Asynchronous Transfer Mode (ATM) connection. A link is the path between two IS-IS neighbor routers, and it is defined as being "up" when the two neighbors' SNPAs are able to communicate with one another.

Several techniques exist for creating unique System IDs in an IS-IS routing domain:

- The naïve approach where systems are numbered 1, 2, 3, 4, and so on.
- Basing the System ID on the Media Access Control (MAC) address of the IS-IS router.
- Taking a system's loopback IP address and converting it into a System ID as illustrated in Figure 2.9.

With the introduction of "Dynamic Hostname Exchange Mechanism for IS-IS" [RFC5301], the practice of modifying a loopback IP address and using it as the System ID of an IS-IS router is now considered outdated. Reference [RFC5301] defines a new Type Length Value (TLV Code 137) that can be used to map the router's hostname to the System ID.

Figure 2.10 shows three different formats of an ISO NET [CISCISISWP02]:

(a) An 8-byte Area ID/System ID format
(b) An OSI NSAP format
(c) A Government OSI Profile (GOSIP) NSAP format

2.6.4　SYSTEM IDENTIFICATION IN IS-IS

IS-IS associates an Area ID or Area Address with a specific IS-IS routing area. An IS-IS router can only belong to only one IS-IS Level 2 routing subdomain (IS-IS backbone); meaning, it can be a member of only one Level 2 backbone. Recall

Router R1
Interface Loopback IP Address 192.168.3.25

Router IS-IS
NET Address: 49.0001.1921.6800.3025.00

Converting an IP Address to System ID:

192.168.3.25　→　192.168.003.025　→　1921.6800.3025　→　49.001.1921.6800.3025.00

FIGURE 2.9　Coding an IP Address into a System ID – An Example

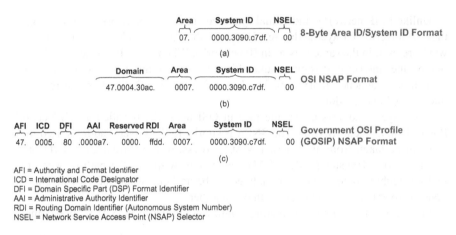

AFI = Authority and Format Identifier
ICD = International Code Designator
DFI = Domain Specific Part (DSP) Format Identifier
AAI = Administrative Authority Identifier
RDI = Routing Domain Identifier (Autonomous System Number)
NSEL = Network Service Access Point (NSAP) Selector

FIGURE 2.10   Three NSAP Formats

that, because the Area ID uniquely identifies an IS-IS routing area in the OSI domain, and the System ID uniquely identifies each IS-IS node in an area, an IS-IS router can belong to only one area.

The following are characteristics of IS-IS areas and System IDs **[CISCTEAPAQ00]**:

- All IS-IS routers within an area must use the same Area ID. The shared Area ID the routers use in OSI PDUs actually defines the area.
- An ES is considered adjacent to a Level 1 router only if they both share the same Area ID. This means, an ES recognizes only IS-IS routers (and other ESs on the same subnetwork) that share the same Area ID.
- Routing within an area (Level 1 or intra-area routing) is based on System IDs. Therefore, each device (ES and IS-IS routers) within the same area must have a unique System ID, and all System IDs must have the same length (e.g., a 6-byte System ID).
  - o Although, each router must have a unique System ID within an area, it is generally recommended to have unique System IDs across an IS-IS domain to avoid conflicts in System IDs at Level 1 or Level 2 if a device is moved from one area into a different area in the domain.
- All Level 2 IS-IS routers become aware of all other IS-IS routers in the Level 2 backbone. Therefore, each of these routers must have a unique System ID within the backbone area.

As discussed above, the 1-byte NSELs in the NSAP addresses are set to 00, indicating that these are NETs (i.e., NSAP address that refer to the OSI device itself). Recall that IS-IS routers use NETs to identify themselves in the LSPs they send and, these also form the basis for the OSI routing calculation. The NET is similar to the Router ID used in OSPF (see Chapter 1).

Figure 2.11 shows examples of NETs for IS-IS routers in an IS-IS domain: the 3-byte Area IDs are unique to each area (all IS-IS routers in the same area must have the same Area ID), the 6-byte System IDs are unique across the domain, and the

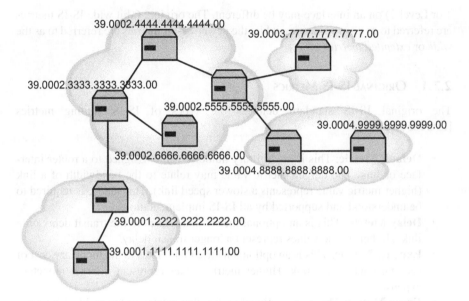

FIGURE 2.11   NSAP Addresses in an IS-IS Network

1-byte NSELs are all set to 00. The NETs must all be a multiple of whole byte (cannot have fractional byte parts).

An IS-IS router can be configured with multiple NETs, but the router cannot be in more than one area. When multiple NETs are configured on a router, this causes the areas in the NETs to merge into a common area, and the Level 1 LSDBs to be leaked into each other **[CISCISISWP02]**. The only reasons for configuring multiple NETs on IS-IS routers are for the purposes of merging, splitting, or renumbering areas in a routing domain. Multiple NETs are mostly used during periods of network transition. In Cisco routers, the number of configurable NETs is limited to three per router **[CISCISISWP02]**.

A Level 1 router will not form an adjacency with a neighbor router if their Area Addresses do not overlap. However, if a Level 1 router has Area Addresses X, Y, and Z, and a neighbor has area addresses W and X, then the Level 1 router can form an adjacency with the neighbor since they have the overlapping Area Address X.

A Level 2 capable router can form an adjacency with a neighbor Level 2 capable router, regardless of their Area Addresses. If their Area Addresses do not overlap, they can only form a Level 2 only adjacency, and only Level 2 LSPs would flow on the adjacency.

2.7   IS-IS ROUTING METRICS

Each circuit emanating from an IS-IS router is assigned one or more metric values by the system management. IS-IS associates metrics with the outgoing interface or link pointing toward the neighbor IS router. The metric for each IS-IS routing level (Level

1 or Level 2) on an interface may be different. The original 6-bit wide IS-IS metrics are referred to as *narrow metrics*, while the newer 24-bit metrics are referred to as the *wide* or *extended metrics*.

2.7.1 ORIGINAL IS-IS METRICS

The original IS-IS standard defines four types of IS-IS routing metrics **[ISO10589:2002]**:

- **Default Metric**: This is an arbitrary numeric value assigned to a router inter-face or link. The default metric value may relate to the bandwidth of a link (higher metric value represents a slower speed link). This metric is required to be understood and supported by all IS-IS implementations.
- **Delay Metric**: This is an optional metric that relates to the transit delay on a link. Higher metric values represent a longer transit delay.
- **Expense Metric**: This is an optional metric that relates to the monetary cost of sending data over a link. Higher metric values represent a larger monetary expense.
- **Error Metric**: This is an optional metric that relates to the residual bit error probability or rate of a link. Higher metric values represent a larger probability of undetected errors on the link.

Every IS-IS router must be capable of calculating routes based on the default metric, and it is not mandated to support any or all of the other optional metrics. If a router system supports the calculation of routes based on a particular metric, it may adver-tise that metric value for the associated link in its LSPs. When a router calculates paths using one of the optional routing metrics, it only utilizes LSPs carrying a value for that metric. If a router's circuit has a metric value that is not associated with any one of the optional metrics, then it must not calculate routes based on that metric.

IS-IS routing metrics are assigned to individual router interfaces (links) in the routing domain. An IS-IS router uses these metrics to calculate a path's cost when running the SPF algorithm. A metric may be used only in the SPF calculation for Level 1 (intra-area) routing, or for Level 2 (interarea) routing; each routing level has its own SPF calculations. Most IS-IS implementations today support only the default metric. When the default metric is used, most IS-IS implementations assign all inter-faces the default value of 10, regardless of the bandwidth of the interfaces.

Unlike in OSPF implementations, most IS-IS implementations do not automati-cally take an interface's bandwidth into account when calculating the IS-IS metric. Typically, OSPF calculates the link metric automatically based on bandwidth, while EIGRP uses bandwidth and delay. In IS-IS, the administrator takes up the responsi-bility of configuring different interface metrics if necessary, using appropriate router per-interface commands.

The original IS-IS standard defined what is normally referred to as narrow metrics for any single interface (link). A narrow metric is 6 bits wide, resulting in a per-interface metric values in the range of 1–63 ($=2^6 - 1$). The value 63 is the IS-IS *MaxLinkMetric* parameter, and is the maximum value of a narrow routing metric assignable to a link. A complete path metric is defined to be 10 bits wide, resulting in a total path metric (*MaxPathMetric*) of 1,023 (=2^{10} − 1). The total cost from a source to a destination along a particular path is the sum of the costs on all outgoing interfaces along that path. The router selects the least-cost path as the best path to that destination.

Each of the four routing metrics above may be described as an Internal metric or an External metric. Internal metrics are used to describe links/routes to destinations that are within (internal) to the IS-IS routing domain. External metrics are used to describe links/routes to destinations that are outside (external to) the IS-IS routing domain. Internal and External metrics are not directly comparable, except that, internal routes are always preferred over external routes. This means, routers will always select an internal route even if an external route with lower total cost exists.

2.7.2 Extended or Wide Metrics

The small range of the narrow metric is seen to be insufficient for large networks (with many hops), and does not provide much granularity for new applications such as IS-IS with traffic engineering and other applications, especially, when using IS-IS over high-bandwidth interfaces. To address the limitations of the narrow metrics, 24-bit metrics, called wide metrics have been defined. This new metric style provides a maximum metric value of 16,777,215 (= 2^{24} − 1) and a total path metric of 4,294,967,295 (= 2^{32} − 1) **[CISCISISWP02] [CISCTEAPAQ00]**.

Running different metric styles (narrow or wide) within a single IS-IS area can cause serious routing problem. Link-state routing protocols like OSPF and IS-IS are able to calculate loop-free routes by requiring all routers (within one area) to calculate best routes for their Routing Databases based on identical LSDBs. This requirement is violated if some routers use the old-style narrow metric and others the new-style wide metrics. However, if all routers within an area are required to use the same metric style, the SPF algorithm will be able to compute loop-free routes. Whenever available and supported, it is strongly recommended to use wide metrics. However, all routers in an IS-IS area must use the same type of metrics (narrow or wide).

2.7.3 Maintaining Per-Metric LSDBs in an IS-IS Router

IS-IS requires that each router supports the default (narrow) metric plus a corresponding "default metric" LSDB. If a router supports any of the three optional narrow metrics, it must maintain a LSDB for routes based on that optional metric. If a router supports all four narrow metrics, then it will end up having four separate LSDBs, one for each metric. This is because IS-IS does not support any form of a combined or composite metric created from the individual metrics as in EIGRP, which means the router has to maintain a separate for any optional metric supported.

When an IS-IS router supports the wide (new-style) metric, it will maintain only one LSDB. Notice that the default (narrow) metric and wide metric are both dimensionless metrics that the network administrator assigns to each link. Each metric value is an arbitrary numeric value assigned to a link, and does not translate directly into a network performance parameter (e.g., bandwidth, delay, and packet loss). Both metrics serve simply as a mechanism for ranking/weighting the importance a particular link plays in routing and the SPF calculations. For example, a higher bandwidth link will be considered more important than a lower bandwidth link and as a result, will be assigned a lower metric value.

2.8   IS-IS MESSAGE TYPES

In the OSI protocol suite, a Protocol Data Unit (PDU) is a single unit of data, consisting of protocol-specific control information and user data, and is transmitted between peer entities of a communication network. Each layer of the OSI reference model implements protocols that are tailored to a specific mode of information exchange. At the Data Link Layer, the Network Layer, and the Transport Layer, a PDU is referred to as a frame, a packet, and a segment, respectively. In TCP/IP, a Network Layer PDU, that is, a packet is sometimes called an IP datagram.

Figure 2.12 shows three types of OSI PDUs: IS-IS, ES-IS, and CLNP PDUs. IS-IS and ES-IS PDUs can carry multiple fields of variable-length, depending on the function of the PDU. Each field is encoded as a Type, Length, and Value (TLV) field; which stands for type code, total length, and the actual data values.

IS-IS is not designed to run over a Network Layer protocol. Instead, IS-IS messages are encapsulated directly into Data Link frames. IS-IS encodes addressing and adjacency information in IS-IS messages as TLV records, allowing IS-IS to provide excellent flexibility and extensibility. Integrated IS-IS can send LSPs that contain multiple variable-length TLV fields, with some TLVs carrying OSI-specific link-state information, and some containing IP-specific link-state information.

The different IS-IS PDUs (or packets) described in this chapter exclude the Data Link header (with OSI family 0xFEFE). The first byte of the IS-IS header (0x83) is referred to as the Intradomain Routing Protocol Discriminator (IRPD).

IS-IS uses the following four general types of packets (or PDUs) which all share a common header. Each packet type can be a Level 1 or Level 2 packet:

- **Hello PDUs**: These are used to establish and maintain IS-IS adjacencies.
  o Level-1 LAN IS-IS Hello PDU (PDU Type = 15)

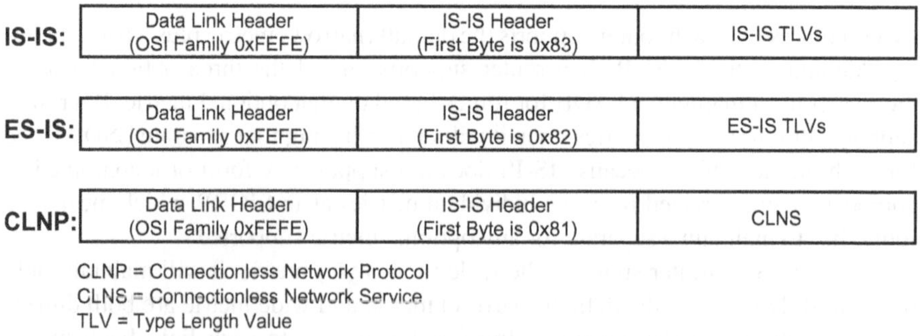

IS-IS:	Data Link Header (OSI Family 0xFEFE)	IS-IS Header (First Byte is 0x83)	IS-IS TLVs

ES-IS:	Data Link Header (OSI Family 0xFEFE)	IS-IS Header (First Byte is 0x82)	ES-IS TLVs

CLNP:	Data Link Header (OSI Family 0xFEFE)	IS-IS Header (First Byte is 0x81)	CLNS

CLNP = Connectionless Network Protocol
CLNS = Connectionless Network Service
TLV = Type Length Value

FIGURE 2.12   OSI PDUs

o Level-2 LAN IS-IS Hello PDU (PDU Type = 16)
o Point-to-Point IS-IS Hello PDU (PDU Type = 17)
- **Link-State Packets (LSPs)**: These are used to distribute link-state information. These are also referred to as LSPs.
  o Level-1 LSP (PDU Type = 18)
  o Level-2 LSP (PDU Type = 20)
- **Complete Sequence Number PDUs (CSNPs)**: These are used to distribute summaries of the LSPs in a router's entire LSDB.
  o Level-1 Complete Sequence Numbers PDU (PDU Type = 24)
  o Level-2 Complete Sequence Numbers PDU (PDU Type = 25)
- **Partial Sequence Number PDUs (PSNPs)**: These are used to acknowledge and request link-state information.
  o Level-1 Partial Sequence Numbers PDU (PDU Type = 26)
  o Level-2 Partial Sequence Numbers PDU (PDU Type = 27)

Each PDU type listed above has a number of subtypes, resulting in nine types of IS-IS PDUs. Each PDU type is identified by a 5-bit type code. Each IS-IS PDU type consists of a header and a number of optional fields of variable-length, encoded in TLV format. A TLV is a tuple that contains specific routing-related information. Each TLV field has a 1-byte *Type* label that describes the information carried in the TLV. The *Value* field of the TLV is the specific information carried in the TLV. Typically, the *Value* in the TLV consists of repeated blocks of similar routing-related information. The length of the *Value* is specified by a 1-byte *Length* field. Each PDU type supports only specific optional TLVs.

The different IS-IS PDU types have slightly different header field formats, but the first eight fields, each 1-byte long, are carried in all PDUs. Each PDU type has its own set of additional header fields, followed by a number of optional TLVs. The additional header fields are PDU type-specific and vary in length, composition, and order of information placement. The optional TLV fields (if any) are appended to the PDU header of the specified PDU type to make up the entire IS-IS PDU.

When Integrated IS-IS is used for IP routing, the LSPs describe IP routing information in a similar manner to how IS-IS describes ESs; IP subnets attached to an IS-IS router are described as if they are ESs in Integrated IS-IS. The LSPs use specific TLV types for IP routing information. Even for IP-only routing, Integrated IS-IS still uses OSI protocols for forming neighbor relationship between routers. IS-IS routers still establish ES and IS adjacencies using IS-IS Hello PDUs. OSI NET addresses are used to identify the routers in Integrated IS-IS, and are also required for SPF calculations, as well as, for Layer 2 forwarding.

To understand the behaviors of IS-IS, it is important to understand the two main types of physical networks defined in IS-IS:

- **Broadcast Media**: These are multiaccess media such as those based on Ethernet technologies that have inherent broadcast and multicast capabilities. These media support the addressing and transmission of information to groups of systems attached to a network segment simultaneously.

- **Non-Broadcast Media**: These are media that do not have inherent broadcast and multicast such as those based on ATM technologies. In such media, systems must be addressed individually and transmissions are directed to the individual systems. These media include point-to-point links, multipoint links, and links that are dynamically established as in ATM networks.

Recognizing these media types, IS-IS supports different packet types and behaviors tailored for only two media types: Broadcast media (for multiaccess broadcast LANs), and point-to-point links (for all other media types). IS-IS recognizes only these two media types. IS-IS has no concept of a Non-Broadcast Multiple Access (NBMA) network as in OSPF, and recommends that all other networks that are not broadcast be treated as point-to-point links. Recommended practice dictates that networks such as hub-and-spoke ATM networks be configured using point-to-point links, and IS-IS run over these point-to-point links.

IS-IS uses various timers to control events and to ensure routing stability. The timers, when properly set, enable IS-IS to converge faster when significant changes in the network occur. A number of timers are used to ensure the integrity of routing information such as enforcing periodic refresh of individual LSPs and aging out of stale LSDB information.

The stability and the convergence speed of IS-IS are affected by the values configured for various timers. Setting the timers involves a trade-off between reaction time to external events, and the amount of CPU and bandwidth resources needed for maintaining the information in the LSDBs. This section also discusses the most relevant IS-IS timers.

IS-IS routers connected by point-to-point links send LSPs to a unicast address. Routers connected to multiaccess broadcast media (e.g., Ethernet networks) send all IS-IS packets to a multicast address. All IS-IS PDUs sent on broadcast media are sent as multicasts **[ISO10589:2002]**. Level 1 PDUs are sent to the multicast (or multidestination) destination MAC address 01-80-C2-00-00-14 (i.e., the IS-IS "All L1 Intermediate Systems" (AllL1ISs) address), while Level 2 PDUs are sent to multicast destination MAC address 01-80-C2-00-00-15 (i.e., the IS-IS "All L2 Intermediate Systems" (AllL2ISs) address). All IS-IS routers on the broadcast segment listen to these addresses (at the appropriate routing level) for IS-IS messages.

ISO 9542 **[ISO9542:1988]** specifies the "All Intermediate Systems" (AllIntermediateSystems) as 09-00-2B-00-00-05, and the "All End Systems" (AllEndSystems) as 09-00-2B-00-00-04.

2.8.1   IS-IS Hello PDUs

IS-IS routers send IS-IS Hello PDUs to establish and maintain adjacencies with neighbor IS-IS routers (Figures 2.13 and 2.14). IS-IS routers send these PDUs to discover the identity of IS-IS neighbor routers, and to determine whether the neighbors are Level 1 or Level 2 routers. IS-IS Hello PDUs are of the following types **[ISO10589:2002]**:

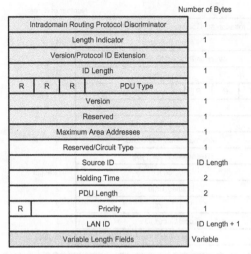

Number of Bytes

	Number of Bytes
Intradomain Routing Protocol Discriminator	1
Length Indicator	1
Version/Protocol ID Extension	1
ID Length	1
R R R PDU Type	1
Version	1
Reserved	1
Maximum Area Addresses	1
Reserved/Circuit Type	1
Source ID	ID Length
Holding Time	2
PDU Length	2
R Priority	1
LAN ID	ID Length + 1
Variable Length Fields	Variable

R = Reserved

Field	Meaning
Intradomain Routing Protocol Discriminator	This is the network layer protocol identifier assigned to IS-IS, as specified by ISO 9577. Its value is 10000011 (binary), 0x83 (hexadecimal), or 131 (decimal).
Length Indicator	This is the length of the fixed header in bytes.
Version/Protocol ID Extension	This is the version of the IS-IS protocol used which currently has value of 1.
ID Length	This indicates the length of the System ID (Source ID) field; must be the same for all nodes in the routing domain. A value of 0 implies a length of 6 bytes; a value of 255 implies 0 length. Other possible values from 1 to 8 indicate the actual length in bytes.
PDU Type	This specifies the type of IS-IS packet. For example, a value of 15 is for Level 1 LAN (Broadcast Link) IS-IS Hello packets and 16 for Level 2 Hello packets.
Version	The value of this field is 1.
Maximum Area Address	This indicates the maximum number of area addresses supported by the local router's IS-IS area. Values between 1 and 254 are for the actual number; value of 0 indicates that a maximum of three area addresses are supported by the IS-IS area.
Reserved/Circuit Type	Most significant 6 bits are reserved and transmitted as zero, they are ignored on receipt. The low order bits (bits 1 and 2) indicate the following: Value = 0 indicates reserved value (and if specified the entire PDU is ignored); Value = 1 indicates Level 1 only; Value = 2 indicates Level 2 only (sender is Level 2 IS and will use this link only for Level 2 traffic); Value = 3 indicates Level 1 and 2 (sender is Level 2 IS, and will use this link both for Level 1 and Level 2 traffic).
Source ID	This indicates the System ID of the router that transmitted the packet.
Holding Time	This is the holding time as configured on the local router and specifies the maximum interval between two consecutive hello packets before the router is considered no longer available. Specifies Holding Timer to be used for the IS. This is similar to the OSPF router dead interval (see RouterDeadInterval RFC 2328)
PDU Length	This specifies the length of the entire PDU, including header and TLVs.
Priority	This 7-bit value designates the priority to be the Designated Intermediate System (DIS) (Level 1 or Level 2) on the LAN. The Priority value is used in DIS election and carried in in LAN Hello PDUs. Bit 8 is reserved and transmitted as zero, ignored on receipt. Bits 1 to 7 are used to indicate the priority of the node for being a LAN Level 1 or 2 DIS. A higher number indicates that the node has a higher priority as a LAN Level 1 or 2 DIS. This value is copied from the IS-IS Hello of the DIS.
LAN ID	This field is composed of the System ID (1 to 8 bytes) of the LAN Level 1 DIS, plus a low order byte assigned by LAN Level 1 DIS (that is, the pseudonode ID (circuit ID) to differentiate LAN IDs on the same DIS). This is copied from the LAN Level 1 DIS's IS-IS Hello PDU.
Variable Length Fields	This is a variable length field that carries a number of TLVs as described in ISO/IEC 10589.

FIGURE 2.13   Level 1 and Level 2 LAN (Broadcast Link) IS-IS Hello PDU Format

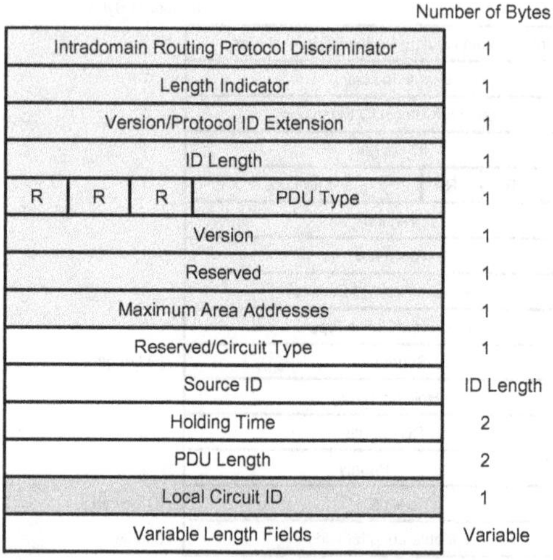

Number of Bytes

Field				Number of Bytes
Intradomain Routing Protocol Discriminator				1
Length Indicator				1
Version/Protocol ID Extension				1
ID Length				1
R	R	R	PDU Type	1
Version				1
Reserved				1
Maximum Area Addresses				1
Reserved/Circuit Type				1
Source ID				ID Length
Holding Time				2
PDU Length				2
Local Circuit ID				1
Variable Length Fields				Variable

Field	Meaning
Local Circuit ID	This is a 1 byte unique link ID assigned to this circuit when it is created by this IS router. This is the circuit ID on the sending interface (and is carried in point-to-point IS-IS Hello PDUs).

FIGURE 2.14   Point-to-Point IS-IS Hello PDU Format

- Level 1 LAN IS-IS Hello PDU (PDU Type 15)
- Level 2 LAN IS-IS Hello PDU (PDU Type 16)
- Point-to-Point IS-IS Hello PDU (PDU Type 17)

Level-1 routers send Level-1 LAN Hello PDUs on broadcast LANs, while Level-2 routers send Level-2 LAN Hello PDUs on broadcast LANs. However, IS-IS routers connected to non-broadcast networks (treated as point-to-point links) send Point-to-Point Hello PDUs. LAN Hello PDUs and Point-to-Point (P2P) PDUs have slightly different formats as shown in Figures 2.13 and 2.14. All Hello PDUs contain information that describe an ISO system, its capabilities, and parameters.

Point-to-Point IS-IS Hello PDUs have a 1-byte Local Circuit ID field in place of the LAN ID field (of length ID Length + 1) in LAN IS-IS Hello PDUs. Also, Point-to-Point IS-IS Hello PDUs do not have the Priority field found in LAN IS-IS Hello PDUs.

An IS-IS router on a multiaccess broadcast network segment (e.g., based on Ethernet technology) uses the 7-bit Priority field in its Level 1 and Level 2 LAN Hello PDUs to advertise its priority to become the Designated IS (DIS) for the network segment. Multiaccess broadcast networks are networks that have inherent broadcast and multicast capabilities (such as Ethernet networks) and support the attachment of more than two routers. IS-IS routers use the advertised priorities to elect a DIS for the network. The DIS acts on behalf of the other routers and is responsible for sending LSPs that describe all the routers attached to the network, plus flooding these advertisements throughout a single IS-IS area. Note that the priority value indicated in the LAN Hello PDU Priority field is meaningful only on a broadcast network and has no meaning on a point-to-point links.

The priority level of router for becoming the DIS (at Level 1 or Level 2 routing) is indicated by an arbitrary number from 0 to 127 (= 2^7), which is configured on the IS-IS router and advertised in the Priority field of the LAN Hello PDU. The router with the highest priority value becomes the DIS for the broadcast segment. If the IS-IS routers on the broadcast network segment have the same priority values, then the router with the highest MAC address is elected as DIS. The default priority setting in Cisco and Juniper routers is 64.

IS-IS Hello PDUs are assigned a predefined length. An IS-IS router does not have to resize any Hello PDU to match the maximum transmission unit (MTU) on its interface. Each router interface must be capable of supporting the maximum IS-IS PDU of 1,492 bytes (the IS-IS parameter *RecieveLSPBufferSize*), and IS-IS PDUs may be padded to meet this maximum value. When an IS-IS router sends a PDU to a neighbor router, the originating and receiving interfaces must support the maximum PDU size.

An IS-IS router pads LAN IS-IS Hello PDUs (with trailing pad TLV option fields containing arbitrary valued bytes) so that the lower layer packet containing the IS-IS Hello PDU has a length of at least *maxsize* – 1 bytes, where *maxsize* for Level 1 IS-IS Hello PDUs is the maximum of the IS-IS parameters *dataLinkBlocksize* and *originatingL1LSPBufferSize* [ISO10589:2002]. The minimum length of the trailing pad that may be added is 2 bytes, which is also the size of the TLV option header. Where possible, the router should pad the PDU to *maxsize*, but if the PDU length is already *maxsize* – 1 bytes, no padding is possible (or required).

The requirement for Level 2 IIH PDUs is *maxsize* must be the maximum of *dataLinkBlocksize* and *originatingL2LSPBufferSize*. This is to ensure that only systems which are capable of exchanging PDUs of length up to *maxsize* bytes can form an adjacency. In the absence of this check, an adjacency can still be formed with a lower maximum block size, but some LSPs and Sequence Number PDUs (SNPs) would not be exchanged (i.e., those PDUs that are larger than the lower maximum block size, but less than *maxsize*). See corresponding requirements for point-to-point links below.

The benefit of padding IS-IS Hello PDUs is to allow early detection of errors, for example, when large PDUs are transmitted, or when MTU mismatches occur on adjacent router interfaces [CISCISISWP02]. The drawbacks of padding IS-IS Hello PDUs are that it could lead to more consumption of router buffers, and link bandwidth wastage. Some IS-IS router implementations allow padding of IS-IS Hello PDUs to be turned off for some or all interfaces on a router [CISCISISWP02].

2.8.1.1   IS-IS Router Manual Area Addresses

It is often convenient, in an OSI domain, to associate more than one Area ID or Address with an IS-IS area. The following are some of reasons why it may be useful to assign more than one Area Address:

- Due to administrative and management reasons or constraints, multiple addressing authority may be involved in the assignment of addresses to a routing domain, and at the same time, it is not possible or even efficient to require a separate area for each addressing authority.

- In some cases, it may be necessary to reconfigure a routing domain by combining a number of existing areas into a single area, or dividing an area into two or more areas. Also, performing these reconfigurations would not be possible during normal operation of the routing domain if only a single Area Address were to be assigned per area.

For these reasons, **[ISO10589:2002]** permits a number of synonymous Area Addresses to be associated with an area.

The IS-IS parameter *manualAreaAddresses* contains a list of all synonymous Area IDs or Addresses associated with a particular IS router in the OSI domain. All of the *manualAreaAddresses* of the IS-IS router, when combined with the System ID of the router, are valid NETs for the router. The management parameter *manualAreaAddresses*, which is set locally for each router by system management, is used to define several synonymous Area IDs for an IS-IS router.

Each Level 1 router announces its *manualAreaAddresses* in the Area Addresses TLV option field of its Level 1 LSP. This allows Level 2 routers in the OSI domain to create a composite list of all Area Addresses in use within a given IS-IS area. The Level 2 routers, in turn, will include the composite list in the Area Addresses TLV option field of their Level 2 LSPs, and advertise them throughout the Level 2 subdomain. This allows all the Area Addresses associated with the entire routing domain to be known throughout the domain.

One key requirement for establishing an adjacency between two Level 1 routers is that there be at least one Area ID in common in their *manualAreaAddresses* lists. The corresponding requirement for establishing an adjacency between a Level 1 router and an ES is that, the Area ID of the ES must match an entry in the router's *manualAreaAddresses* list. Therefore, the system management is responsible for ensuring that each Area ID associated with a router is included. In particular, the system management must ensure that the Area IDs of all Level 1 routers and ESs adjacent to a given Level 1 router are included in that router's *manualAreaAddresses* list.

It is possible that the union of all Area IDs of the routers in an area may exceed the configured maximum number of Area Addresses of one or more routers in the area. To enable all routers to agree on the Area IDs of an area, each router has the IS-IS parameter *maximumAreaAddresses* (see Maximum Area Addresses field in Figure 2.13) which is established by the system management. Each router communicates its *maximumAreaAddresses* value in the Hello PDUs it sends, and this in turn is checked by other routers to ensure that all routers in an area have the same *maximumAreaAddresses* parameter value.

According to **[ISO10589:2002]**, all IS-IS routers must support a *maximumAreaAddresses* value of at least 3, however, the system management may set a lower value if desired. Failure to set the *maximumAreaAddresses* parameter consistently among the routers in an area may cause an area to become partitioned, and/or adjacencies that try to initialize, to fail.

2.8.1.2 Variable-Length Fields in IS-IS Hello PDUs

The variable length TLV fields are used for carrying other IS-IS information. This section describes some of TLVs (and their codes) defined for IS-IS Hello PDUs.

Each TLV has a 1-byte *Type* field containing the TLV Code (or Type), a 1-byte *Length* field specifying the length of the *Value* field, and a *Value* field carrying data of length equal to the value specified in the *Length* field. Typically, the *Value* contains repeated blocks of similar information.

The following TLVs (each having a 1-byte Type field and 1-byte Length field) are defined for Level 1 and Level 2 LAN IS-IS Hello PDUs:

- **Area Addresses TLV (Code 1)**: This TLV specifies the set of *manualAreaAddresses* associated with the local IS-IS router. The Value field contains one or more entries with each entry consisting of the tuple <Address Length, Area Address>. Address Length is a 1-byte field that specifies the length of Area Address in bytes.
- **Intermediate System Neighbors**: This is the set of IS-IS routers on the local broadcast LAN to which adjacencies of *neighborSystemType* "Level 1 Intermediate System" exist in the state "Up" or "Initializing" (i.e., those IS-IS routers from which Level 1 IS-IS Hello PDUs have been received). This field can appear more than once in a PDU. Two types of TLV Codes are defined for this field:
  o **Intermediate System Neighbors with 6-byte MAC Address TLV (Code 6)**: The Value field can contain one or more entries with each entry specifying the 6-byte MAC Address of an IS-IS neighbor. The TLV has 1-byte Type field specifying the TLV Code 6, a 1-byte Length field specifying the total length of the Value field, and a variable-length Value field containing an array of the MAC addresses of the neighbor routers [*MAC_Address_1*, ..., *MAC_Address_Length*], where each element is a 6-byte MAC address.
  o **Intermediate System Neighbors with Variable length SNPA Address TLV (Code 7)**: The Value field contains one or more entries with each entry consisting of the tuple <LAN Address Length, LAN Address>. LAN Address Length is a 1-byte field that specifies the length of SNPA of an IS-IS neighbor. LAN Address is a variable length SNPA of the IS-IS neighbor; this is not to be used for an IS-IS neighbor with SNPA of 6 bytes long.
- **Padding TLV (Code 8)**: This option is used to pad the PDU to at least *maxsize* – 1 and may occur more than once. The router inserts trailing pad option fields to make PDU Length (see Figure 2.13) equal to at least *maxsize* – 1, where *maxsize* is the maximum of *dataLinkBlocksize* and *originatingL1LSPBufferSize* (or *originatingL2LSPBufferSize*) as specified in **[ISO10589:2002]**.
- **Authentication Information TLV (Code 10)**: This TLV carries information for performing authentication of the PDU's originator. The Value field consist of the tuple < Authentication Type, Authentication Value>. Authentication Type is a 1-byte field that identifies the type of authentication to be carried out. The following Authentication Type values are defined in **[ISO10589:2002]**: 0 = RESERVED; 1 = Cleartext Password; 2–254 = RESERVED; 255 = Routing Domain Private Authentication method. Authentication Value is determined by the value of the Authentication Type. If Cleartext Password Authentication (also referred to as Plaintext or Simple Password Authentication) as defined in **[ISO10589:2002]** is used, then the Authentication Value is a byte long string.

Point-to-Point IS-IS Hello PDUs are transmitted by IS-IS routers on non-broadcast circuits. A router inserts trailing pad fields to make the PDU Length (see Figure 2.14) equal to at least *maxsize* − 1, where *maxsize* is the maximum of *dataLinkBlocksize*, *originatingL1LSPBufferSize*, and *originatingL2LSPBufferSize* (as defined in **[ISO10589:2002]**).

The first Point-to-Point IS-IS H PDU a router sends must have this size (i.e., the PDU the router transmitted as a result of receiving a Hello PDU, rather than as a result of timer expiration). This is to ensure that only systems which are capable of exchanging PDUs of length up to *maxsize* bytes can form an adjacency. In the absence of this check, it would be possible for the routers to form an adjacency with a lower maximum block size, but some LSPs and SNPs would not be exchanged (i.e., those larger than the lower maximum block size, but less than *maxsize*) **[ISO10589:2002]**.

The network manager must ensure that the value of *dataLinkBlocksize* on a point-to-point circuit used to form an adjacency between two IS-IS routers, is set to a value greater than or equal to the maximum of the *LSPBufferSize* value. The adjacency will fail to initialize if this is not done. The reason it is not possible to enforce this requirement is that, it is not known until initialization time if the neighbor on the circuit will be an ES or an IS-IS router. An adjacency with an ES may operate with a lower value for *dataLinkBlocksize*.

The following TLVs (each having a 1-byte Type field and 1-byte Length field) are defined for Point-to-Point IS-IS Hello PDUs:

- Area Addresses TLV (Code 1)
- Padding TLV (Code 8)
- Authentication Information TLV (Code 10)

2.8.1.3   IS-IS Hello Messages and Adjacency Formations

When two IS-IS neighbor routers in the same area operate at both Level 1 and Level 2 routing (i.e., both neighbors are Level 1-2 routers), they establish two separate adjacencies, one for each routing level. The routers store the routing information for the Level 1 and Level 2 adjacencies in separate Level 1 and Level 2 LSDBs. On broadcast media like Ethernet networks, the routers establish the two adjacencies with specific Level 1 and Level 2 IS-IS Hello PDUs.

Because all IS-IS routers attached to a common broadcast network segment must agree on certain parameters, they include these parameters in Hello PDUs. Differences in some of these parameters may prevent neighbors from forming adjacencies. Routers on the broadcast network segment establish adjacencies with a virtual node created by a DIS called the *Pseudonode*. All routers on the segment including the DIS itself form adjacencies with the Pseudonode. Each router is able to send LSPs to all routers on the segment via the Pseudonode. In OSPF, routers establish adjacencies only with the Designated Router (DR) and the Backup DR (BDR) (see Chapter 1).

IS-IS routers on point-to-point links use a common IS-IS Hello format (see Figure 2.14), but part of this PDU (the Reserved/Circuit Type field) still indicates whether the Hello relates to a Level 1, Level 2, or Level 1-2 sender. As explained in Figure 2.13, the Reserved/Circuit Type field indicates if a Point-to-Point IS-IS Hello PDU is sent by a Level 1, Level 2, or Level 1-2 router.

The IS-IS Hello PDUs a router sends (both point-to-point and broadcast PDUs) carry the Area Address(es) of the router as one of the attributes in the PDUs' variable-length TLV fields (i.e., the Area Addresses TLV [Code 1]). Level 1 neighbor routers must share the same Area Address (embedded in their NETs) to be able to establish an adjacency; Level 2 routers do not have this restriction.

An adjacency is established based on the Area Address and the routing level of the router as specified in the (the Reserved/Circuit Type field of the) IS-IS Hello PDU. A router encodes its routing level capabilities for an adjacency in the Reserved/Circuit Type field of the Hello PDU:

- Level 1 routers exchange IS-IS Hello PDUs with Level 1-only and Level 1-2 routers in the same area to establish Level 1 adjacencies.
- Level 2 routers exchange IS-IS Hello PDUs with Level 2-only and Level 1-2 routers in the same area or different area to establish Level 2 adjacencies.
- Two Level 1-2 routers in the same area exchange IS-IS Hello PDUs to establish both Level 1 and Level 2 adjacencies.
- Two Level 1-2 routers in different areas exchange IS-IS Hello PDUs to establish only a Level 2 adjacency.
- Two Level 1 routers that happened to be physically connected but are in different areas (including a Level 1-only router connected to a Level 1-2 router in a different Level 1 area) may exchange Level 1 IS-IS Hello PDUs but will ignore them because their Area Addresses do not match. This means, these routers cannot establish any form of adjacency. Also, a Level 1 router cannot establish an adjacency with a Level 2 router.

2.8.1.4   OSI End System and Intermediate System Hello Messages

IS-IS uses ES Hello PDUs to establish adjacencies with ESs as shown in Figure 2.15. This figure shows the three types of Hello PDUs: ESH PDUs, sent by an ES to an IS; ISH PDUs, sent by an IS to an ES; and IS-IS Hello PDUs, used between two ISs (routers). An IS uses IS-IS Hello PDUs to establish and maintain relationships with its IS neighbors. After any two ISs have established an adjacency, they exchange link-state information using LSPs.

Even though an IS-IS router sends IS-IS Hello PDUs on a point-to-point link, it will also send an ISH PDUs. IS-IS requires that routers connected by a point-to-point link, exchange ISH PDUs, as the first step in establishing the adjacency between the two neighbors. *Basically, routers on point-to-point links exchange ISH PDUs as a way to initialize the link, and to allow each router to determine if the other system at the other end of the link is a router, before IS-IS Hello PDUs are exchanged.*

Level 1 routers transmit Level 1 LAN IS-IS Hello PDUs to the multicast address 01-80-C2-00-00-14 (AllL1ISs), and also listen on that address for messages. The routers also listen for ESH PDUs on the multicast address 09-00-2B-00-00-05 (*AllIntermediateSystems*). The list of neighbor IS-IS routers a router maintains contains only Level 1 routers within the same area. (i.e., Adjacency type of *neighbourSystemType* "L1 Intermediate System".)

ISs on a network send out ISH PDUs to ESs, and each ES listen for these PDUs. An ES will randomly choose an IS (the IS that sends the first ISH PDU) to forward

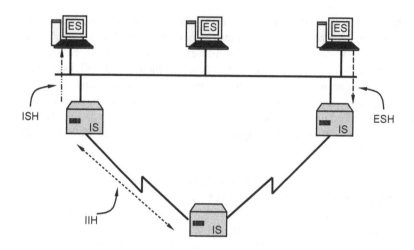

IIH = Intermediate System-to-Intermediate System Hello
ESH = End System Hello
ISH = Intermediate System Hello

FIGURE 2.15   Three Types of IS-IS Hello PDUs

all its packets to. This means, ESs on an OSI network require no special configuration in order to forward packets to destinations in the network or other internetworks. ISs (IS-IS routers) also listen to ESH PDUs to discover all the ESs on a network segment. ISs then include the discovered ES information in the LSPs to be sent to other ISs. For some destinations, an IS might send redirect messages to ESs (similar to ICMP Redirect messages **[RFC792]**) informing them about an optimal route out of the network segment.

2.8.1.5   IS-IS Hello PDUs Related Timers and Parameters
The following timers and parameters are related to the transmission of IS-IS Hello PDUs **[ISO10589:2002]**:

- *dSISHelloTimer*: This specifies the length of time (in seconds) the DIS on a broadcast LAN must wait since the last periodic IS-IS Hello PDU was sent, before sending the next Hello PDU. Basically, this timer specifies the period between Hello PDUs. In Juniper Network Operating System (JUNOS), the default value is 3 seconds for the DIS. The default interval is 3.3 seconds in Cisco Internetwork Operating System (IOS) ($\approx 10/3$).
- *iSISHelloTimer*: This specifies the length of time (in seconds) that a router that is not a DIS on a broadcast LAN must wait since the last periodic IS-IS Hello PDU was sent, before sending the next Hello PDU. In JUNOS, the default value is 9 seconds for non-DIS routers. The default value is 10 seconds in Cisco IOS.

    Routers transmit Hello PDUs periodically to neighbors at the expiration of the *iSISHelloTimer* (or *dSISHelloTimer* for a DIS). However, any change in network conditions that causes changes in the optional TLV information that a router advertises in its most recent Hello (such as election or resignation of a DIS on LAN segment), also triggers immediate transmission of Hello PDUs.

- ***ISISHoldingMultiplier***: This specifies the number an IS-IS router uses to multiply the iSISHelloTimer to obtain the Holding Time for Level 1 and Level 2 IS-IS Hello PDUs (the default value is 10 in **[ISO10589:2002]**). The *iSISHelloTimer* specifies the elapsed time since the last LAN IS-IS Hello PDU transmission by a non-DIS router. The Holding Time for a Hello PDU is set to *ISISHoldingMultiplier × iSISHelloTimer*.

  The Holding Time is the amount of time a router will wait for a Hello PDU from a neighbor before declaring it as unavailable or unreachable ("dead"). It is the maximum amount of time allowed between receipt of two consecutive IS-IS Hello PDUs. For the DIS on a LAN, the Holding Time is *ISISHoldingMultiplier × dSISHelloTimer*. The default is 3 in both JUNOS and Cisco IOS (resulting in a Holding Time of 27 seconds and 30 seconds, respectively, for non-DIS routers).

  If an IS-IS router does not receive a Hello PDU from a neighbor before the Holding Time expires, it will tear down the adjacency. The router resets the Holding Time anytime a Hello PDU is received.

- ***Jitter***: This specifies a small random amount of time that is introduced into the value of a timer to prevent multiple timers in different systems from expiring and starting at the same time, therefore, becoming synchronized. The value specified is a percentage of the timer setting and is applied to the generation of periodic PDUs. The default value is 25% in **[ISO10589:2002]**.

  When the timers of individual systems expire and PDUs are transmitted, there is a possibility that the timers may become synchronized. This can result in the systems sending traffic at the same time thereby generating traffic that will contain peaks. When the PDU generation is synchronized, both the transmission medium and the systems receiving the PDUs can be overloaded. In order to prevent this from occurring, "jitter" can be introduced in the expiration of the periodic timers so that they do not all expire and start at the same time.

2.8.2   IS-IS Link-State PDUs

An IS-IS router uses LSPs to distribute link-state information to neighbor routers. LSPs contain information about the state of adjacencies to IS-IS neighbor routers, or Pseudonodes. IS-IS uses a two-level routing hierarchy (two routing levels), and separate LSPs (namely, Level 1 LSPs and Level 2 LSPs) to distribute link-state information. (see Figure 2.16):

- Level 1 LSP (PDU Type 18)
- Level 2 LSP (PDU Type 20)

Level-1 IS-IS routers send Level-1 LSPs, while Level-2 IS-IS routers send Level-2 LSPs. Level-1-2 IS-IS routers can send both Level-1 and Level-2 LSPs. The LSPs are

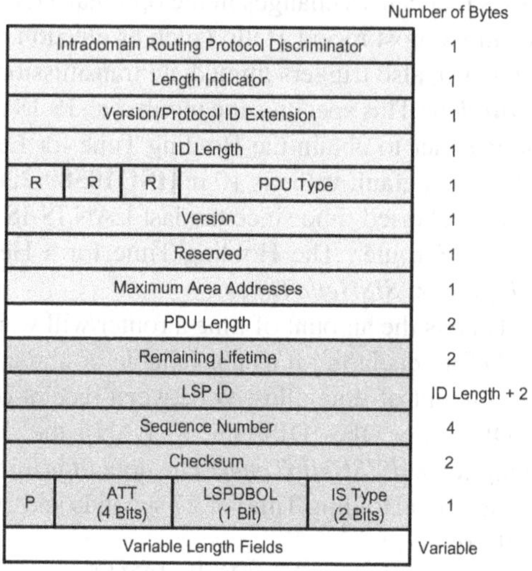

Number of Bytes

Intradomain Routing Protocol Discriminator	1
Length Indicator	1
Version/Protocol ID Extension	1
ID Length	1
R R R PDU Type	1
Version	1
Reserved	1
Maximum Area Addresses	1
PDU Length	2
Remaining Lifetime	2
LSP ID	ID Length + 2
Sequence Number	4
Checksum	2
P ATT (4 Bits) LSPDBOL (1 Bit) IS Type (2 Bits)	1
Variable Length Fields	Variable

Field	Meaning
Remaining Lifetime	This indicates the time (in seconds) before this LSP expires.
LSP ID	This field also includes the System ID of the source of the Link State PDU. The structure of this field consists of three components: System ID (or Source ID), pseudonode ID, and LSP number. The length of this field is the System ID length + 2 bytes.
Sequence Number	This indicates the sequence number of this LSP.
Checksum	The checksum is computed from the Source ID to the end of the LSP. Note that the Source ID is a subfield within the LSP ID
P (Partition) Bit	This is bit 8 of the byte; and when set, means the originator of LSP supports the partition repair optional function.
ATT (Attached) Bits	These are bits 4 to 7 of the byte. When any of these bits is set, it indicates that the originator of the LSP is attached to another area using one of the following metrics: Bit 4 - the Default Metric; Bit 5 - the Delay Metric; Bit 6 - the Expense Metric; Bit 7 - the Error Metric.
LSPDBOL (LSP Database Overload) Bit	This is bit 3 and when set, it indicates the originator's LSP database is overloaded and should not be used in path calculations to other destinations.
IS Type Bits	These are bits 1 and 2 and are used to indicate the type of IS router. When only bit 1 is set, it indicates a Level 1 IS. If both are set, this indicates a Level 2 IS. Other settings are not used.

FIGURE 2.16 Level 1 and Level 2 IS-IS Link-State PDU Format

flooded periodically throughout their appropriate IS-IS areas. Flooding involves propagating an LSP to all neighbor routers except that neighbor from which the LSP was received. Receiving routers detect and drop duplicate LSPs. Four types of LSPs are used on multiaccess broadcast network segments: Level 1 Pseudonode LSPs, Level 1 Non-Pseudonode LSPs, Level 2 Pseudonode LSPs, and Level 2 Non-Pseudonode LSPs. The concept of Pseudonode and Pseudonode LSPs will be discussed in the "Designated Intermediate System (DIS) and Pseudonodes" section below.

Any two IS-IS neighbors that have established an adjacency exchange LSPs that depend on the routing level of the adjacency. An LSP may contain specific information about the systems attached to the issuing router; information that describes IS-IS neighbor routers (used to build the topology map of the network), type of adjacencies, connected network prefixes, neighbor ESs, the IP subnets attached to the router (described as ESs in Integrated IS-IS), path costs, authentication information (used to secure IS-IS routing updates), and Area Addresses. This information is carried in the main body of the LSP in multiple variable-length TLV fields (Figure 2.16). IS-IS routers receive LSPs and use the information they contain to build and maintain their LSDBs.

The size of an LSP is limited to the IS-IS *ReceiveLSPBufferSize* parameter value at the receiving IS-IS router. A router must provide buffers of at least *ReceiveLSPBufferSize* for the reception, storage, and forwarding of received LSPs and Sequence Numbers PDUs. If any of such PDUs is received and is larger than this buffer size, the router treats the PDU as if it had an invalid Checksum (i.e., it is ignored). This means when buffers are limited, it may not be possible for a router to include information about all of its neighbors in a single LSP.

In such cases, the sending router may choose to use multiple LSPs to transport this information to its neighbor. Each LSP in this set will carry the same Source ID field, but the LSP Number field will be set individually. The receiving neighbor will recognize that the multiple LSPs are all originated by a common source because they all carry the same Source ID. The maximum size a Level 1 or Level 2 LSP can take when generated by an IS-IS router is controlled by the management parameters, *originatingL1LSPBufferSize* or *originatingL2LSPBufferSize*, respectively.

When reporting a link to neighbor routers, the router must indicate that the link has a value defined for at least the IS-IS default metric; the default metric is not optional. It is permissible for the two endpoints (routers) of a link to report or advertise different values of the same routing metric for the link. In this case, the routes that are calculated by the two routers may end up being asymmetric.

2.8.2.1 The Remaining Lifetime Field

The Remaining Lifetime field of an LSP indicates the time left (the remaining time) until the expiration of the LSP. The Remaining Lifetime field specifies a time that is used to age out LSPs. Routers use the Remaining Lifetime value in the LSP aging process to ensure that LSPs that are "too old" and may be invalid, are removed from the LSDB after a specified period. A router first sets the Remaining Lifetime of an LSP to a *MaxAge* of 1,200 seconds (20 minutes) and counts down from that value to 0.

Before a router will transmit an LSP to a neighbor, it will decrement the Remaining Lifetime in the LSP by at least one or more seconds if the transit time to that neighbor is estimated to be greater than one second. When the value of Remaining Lifetime field reaches 0, the router must purge that LSP from its LSDB. In order to keep the LSDBs of routers in the network (area) synchronized, the purging of an LSP with an expired Remaining Lifetime is also synchronized by flooding the expired LSP (with Remaining Lifetime set to 0). For all LSDBs in the area to be identical, all routers must purge the expired LSP.

The *MaxAge* specifies the maximum life span of an LSP from the moment it is generated by the source IS-IS router. *MaxAge* a network-wide protocol constant, and must have a consistent value on all IS-IS routers in the network. If a router receives an LSP with a Remaining Lifetime value greater than the configured *MaxAge* value, the LSP is considered corrupted and is, therefore, discarded.

Usually, the source router refreshes an LSP before its Remaining Lifetime reaches zero. In some situations, another IS-IS router in the network may initiate the purge of a corrupted LSP from the network by generating an LSP and setting its Remaining Lifetime to zero and, then reflooding it to neighbors. Routers receiving this LSP will purge it from their LSDBs. The *ZeroAgeLifetime* is the amount of time an LSP is retained in the LSDB after its Remaining Lifetime has expired, before it is deleted.

2.8.2.2 The LSP ID Field

The LSP ID field, with length equal to the ID Length plus 2 bytes, consists of three subfields (see Figure 2.17):

- Source ID (System ID) with length equal to the System ID Length (typically 6-bytes)
- 1-byte Pseudonode ID
- 1-byte LSP Number

An IS-IS router can generate up to 256 LSPs at each routing level. The LSPs generated are identified by LSP (Fragment) Numbers in the range 0–255. LSP Number 0 has special properties, and IS-IS routers treat an LSP with LSP Number 0 in a special way as described in **[ISO10589:2002]**. If a router checks its LSDB and determines that an LSP with LSP Number 0 and corresponding Remaining Lifetime is greater than 0 is not present for a particular IS-IS system in its LSDB, then the router must not process any other LSPs with non-zero LSP Number for that system which may be stored in its LSDB.

The LSP with LSP Number 0 has special significance because an IS-IS router takes the following information only from this LSP. Any of these parameters, if present in other non-zero LSPs for that system, must be ignored by the local router. These fields (see Figure 2.16) are meaningful only when they are present in the LSP with LSP Number 0:

- Setting of the LSP Database Overload (LSPDBOL) bit.
- Value of the 2-bit IS Type field.

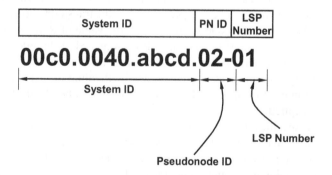

FIGURE 2.17   Link-State Packet ID (LSP ID) Format

- Area Addresses option field (carried in TLV Code 1). This field is present in the LSP with LSP number 0 only for Non-Pseudonode LSPs; it is never present in an LSP with a non-zero LSP Number.
- Setting of any of the ATT (Attached) Flag bit.

These fields have no meaning in LSPs that are numbered 1 through 255. When the values in any of these fields are changed, an IS-IS router has to resend the LSP with LSP Number 0, to inform other routers of the change. LSPs numbered 1 through 255 are not to be reissued for such changes. All routers must treat the LSP with LSP Number 0 in a special way. If a router detects that the LSP with LSP Number 0 and Remaining Lifetime greater than zero is not present for a particular router in its LSDB, then it must not process any LSPs with non-zero LSP Number which may be stored in the LSDB for that router.

2.8.2.3   The Sequence Number Field

The Sequence Numbers in LSPs enable IS-IS routers that receive LSPs to ensure that only the most recent LSPs are used in best route calculations. This also ensures that no duplicate LSPs are installed in the LSDBs. Each router also refreshes LSPs periodically (on broadcast LANs) using Complete Sequence Number PDUs (CSNPs), and new LSPs are acknowledged or requested via Partial Sequence Number PDUs (PSNPs). Any router that receives newer LSP information in a CSNP will purge the out-of-date entries in its LSDB, and then will update the LSDB accordingly.

The Sequence Number is a 4-byte unsigned value (see Figure 2.16). Sequence Numbers must increase sequentially from 0 to ($SequenceModulus - 1$). However, when a router initializes, it must start with Sequence Number of 1 for its own locally generated LSPs. The router starts with a Sequence Number of 1 instead of 0 so that the value 0 can be reserved (as a marker). The reserved Sequence Number of 0 is guaranteed to be less than the Sequence Number of any LSP actually generated by the router. Also, the Sequence Numbers that an IS-IS router generates for its LSPs having different LSP Number values are independent.

If a Router R somewhere in the IS-IS domain receives an LSP or CSNP/PSNP that shows that the current Sequence Number for an LSP for source Router S in its LSDB is greater than that held by Router S, Router R will return a LSP to Router S with Router R's understanding of Router S's current Sequence Number. When Router S receives this LSP, it will change the Sequence Number of its next LSP to be the next number greater than the new one received. Router S will then generate an LSP with this updated Sequence Number.

If an IS-IS router needs to increment the Sequence Number of an LSP, but the Sequence Number is already equal to $SequenceModulus - 1$, the Sequence Number generation process will be disabled for a period of at least $MaxAge + ZeroAgeLifetime$, to ensure that any versions of the LSP with the high Sequence Number will expire (see timer definitions below). When the process is re-enabled, the router will restart with Sequence Number of 1.

It is possible that an LSP generated by an IS-IS router in a previous incarnation will still be alive in the IS-IS domain and have the same Sequence Number as that of a current LSP. To ensure consistency among the LSDBs of the IS-IS routers, it is essential for the routers to be able to distinguish between such PDUs. A router can do this efficiently by comparing the Checksum in a received LSP or an LSP entry in a received CSNP/PSNP with the Checksum of the LSP stored in its LSDB.

If the Sequence Numbers of the received and stored LSP match, but the Checksums do not, and the LSP is not part of the current set of LSPs generated by the router, then the received LSP is treated as if its Remaining Lifetime had expired. The router will store a copy of the LSP with its Remaining Lifetime set to 0, and then flood it. The router floods this "bad" LSP to purge it from the network. The router that originated this LSP will regenerate a new copy of the LSP, and reflood it to ensure all routers on the network have the current good copy in their LSDBs. If the LSP is in the current set of LSPs generated by the router, then the router will change the Sequence Number of the LSP to be one plus that of the received LSP, and will regenerate the LSP for flooding.

If a router detects that the Remaining Lifetime in a received LSP, or in an LSP entry in an CSNP or PSNP is 0, it will retain only the header of the LSP, and will record the time the LSP's Remaining Lifetime became 0. The router will wait until *ZeroAgeLifetime* has elapsed since the LSP's Remaining Lifetime became 0, and then purged it from the LSDB.

If the Remaining Lifetime in the received LSP, or in an LSP entry in a received CSNP or PSNP is non-zero, but the router has an LSP in its LSDB with the same Sequence Number and zero Remaining Lifetime, the router considers the LSP in the LSDB the most recent. Otherwise, router considers the LSP with the larger Sequence Number the most recent. Note that when a router initiates a purge of an LSP with non-zero Remaining Lifetime, it retains the LSP header for *MaxAge* before deleting it.

If an IS-IS router receives an LSP from a particular source with zero Remaining Lifetime, it will perform the following actions:

- If there is no LSP from the source in the LSDB, then the router will send an acknowledgement for the LSP, but will not retain the LSP after the acknowledgement has been sent.
- If the LSP is newer than the one in its LSDB (i.e., received LSP has a higher Sequence Number, or the same Sequence Number and the LSDB's LSP has a non-zero Remaining Lifetime), the router will store the new LSP in the LSDB, overwriting the existing LSDB's LSP for that source (if any).
- If the router finds that it is the source of the LSP (or the source is a Pseudonode), and there is an LSP from this source that has not expired (i.e., its own LSP) in the LSDB, then the router will not overwrite it with the received LSP, but will change the Sequence Number of the unexpired LSP, and generate a new LSP.

2.8.2.4   The LSP Checksum Field

An IS-IS router must compute an LSP Checksum for each LSP it generates (see Checksum field in Figure 2.16). The Checksum must never be modified by any other system. The Checksum allows systems to detect corrupted LSPs, thus, preventing incorrect routing information to be used and propagated further to other systems. The Checksum is calculated over all fields that appear after the Remaining Lifetime field.

All fields up to the Remaining Lifetime field are excluded from the Checksum computation so that IS-IS routers can age the LSP without requiring recomputation.

An IS-IS router receiving an LSP carrying an incorrect LSP Checksum, or having an invalid PDU syntax, will discard the PDU. Also, an LSP received with a zero Checksum is treated as if the Remaining Lifetime of the LSP is 0. The Remaining Lifetime, if not zero, must be overwritten with zero.

2.8.2.5 The LSP Database Overload (LSPDBOL) Bit

It is possible that a router, due to network misconfiguration, or certain transitory network conditions, may have insufficient memory and processing resources to handle a received LSP. When this happens and if its LSDB becomes inconsistent with those of other routers, the router needs to take certain steps to ensure that these other routers do not rely on forwarding paths that go through it (the overloaded router).

When the overloaded router detects that it cannot store an LSP, it will ignore the LSP and enter the *Waiting State*. The router will start a timer for an interval equal to *waitingTime*, and will generate and flood its own LSP with zero LSP Number and the LSPDBOL bit set. This prevents other routers from using the overloaded router as a forwarding path.

An IS-IS implementation may partition the available memory resources between its Level 1 and Level 2 LSDBs. This allows an overload condition to exist independently for Level 1 or Level 2 (or both). The IS-IS uses the status attributes *l1State* and *l2State* to indicate the condition for the Level 1 and Level 2 LSDBs, respectively. A router upon entering Level 1 "Waiting State" will generate the *lSPL1DatabaseOverload* event, and will generate the *lSPL2DatabaseOverload* event upon entering Level 2 "Waiting State".

An IS-IS router must not utilize a link to a neighbor router which has set the LSPDBOL bit in its LSPs. Such paths, when used, may introduce routing and forwarding loops (black holes) since the overloaded router does not have a complete LSDB. A router receiving the LSPDBOL bit, however, can utilize the link to reach ES neighbors on the overloaded router since these paths are guaranteed to be loop-free. The router can still send packets to the directly connected networks of the overloaded router, but will not use this router for transit traffic. A router must not utilize a link to an IS-IS neighbor router unless both routers report the link as "up". This check is not applicable to links to an ES.

2.8.2.6 Variable-Length Fields in IS-IS Link-State PDUs

The contents of an LSP are encoded as variable-length TLVs.

2.8.2.6.1 Level 1 LAN LSP TLVs

The following TLVs (each having a 1-byte Type field and 1-byte Length field) are defined for Level 1 LAN LSPs [ISO10589:2002]:

- **Area Addresses TLV (Code 1)**: This TLV specifies the set of *manualAreaAddresses* of the local IS-IS router; the Value field contains one or more entries of <Address Length, Area Address>. For Non-Pseudonode LSPs, this TLV option is always present in LSP Number 0, and is never to be carried in an LSP with a non-zero LSP Number. It must appear before any Intermediate System Neighbors options (TLV Code 2), or End System Neighbors options (TLV Code 3). This option must never be present in Pseudonode LSPs.

The Area Addresses option must appear only in an LSP with LSP Number 0. If there are more Area Addresses than will fit in a single instance of the Area Addresses TLV option field, then the router must only enter 12 Area Addresses in each instance of the field except the last.

A Level 1 or Level 2 router must compute the values of *areaAddresses* (the set of Area Addresses for the local Level 1 area), by forming the union of the sets of *manualAreaAddresses* announced in the Area Addresses field of all Level 1 LSPs with LSP Number 0 (received from all IS-IS routers which are reachable via Level 1 routing) in the local router's LSDB. If more than *maximumAreaAddresses* are present, the router must retain only those areas with numerically lowest Area Address.

- *originatingLSPBufferSize* **TLV (Code 14)**: This TLV specifies the local value for *originatingL1LSPBufferSize* (value from 512 to 1,492 bytes). This TLV must appear before any Intermediate System Neighbors TLV options or End System Neighbors TLV options. This option may appear only once in an LSP having any LSP Number.

- **Intermediate System Neighbors TLV (Code 2)**: This TLV specifies the IS-IS neighbors and Pseudonode neighbors. It is present in an LSP with any LSP Number and is permitted to appear more than once. However, all the Intermediate System Neighbors TLV options must appear before the End System Neighbors TLV options; meaning they must appear before any End System Neighbors options in the same LSP, and no End System Neighbors options must appear in an LSP with lower LSP Number. The Value field contains one or more entries with each entry consisting of the tuple <Virtual Flag, Default Metric, Delay Metric, Expense Metric, Error Metric, Neighbor ID>. It should be noted that all of these four metrics are narrow metrics (i.e., 6-bit metrics).

  o **Virtual Flag** is a 1-byte field and is a Boolean. If set to 1, it indicates the link is a Level 2 path used to repair an area partition. Level 1 routers would always report this byte as 0 to all neighbors.

  o **Default Metric** is a 1-byte field that specifies the default metric value for the link to the listed IS-IS neighbor. Bit 8 of this field is reserved. Bit 7 (designated the I/E bit) indicates whether the metric type is Internal (value set to 0) or External (value set to 1). Bit 7 must be set to 0, indicating an Internal metric.

  o **Delay Metric** is a 1-byte field hat specifies the delay metric value for the link to the listed IS-IS neighbor. If the local router does not support this metric it must set bit 8 (the "S" bit) to 1 to indicate that the metric is not supported. Bit 7 (designated the I/E bit) indicates the metric type, and must be set to 0, indicating an Internal metric.

  o **Expense Metric** is a 1-byte field that specifies the expense metric value for the link to the listed IS-IS neighbor. If the local router does not support this metric it must set bit 8 (the "S" bit) to 1 to indicate that the metric is not supported. Bit 7 (designated the I/E bit) indicates the metric type, and must be set to 0, indicating an Internal metric.

o **Error Metric** is a 1-byte field that specifies the error metric value for the link to the listed IS-IS neighbor. If the local router does not support this metric it must set bit 8 (the "S" bit) to 1 to indicate that the metric is not supported. Bit 7 (designated the I/E bit) indicates the metric type, and must be set to 0, indicating an Internal metric.

o **Neighbor ID** field (with length equal to ID Length +1 byte) consists of the System ID of the IS-IS Non-Pseudonode neighbor, followed by a byte that is set to the value zero. For a Pseudonode neighbor, the first ID Length bytes is the System ID of the LAN Level 1 DIS, and the last byte is a non-zero quantity defined by the LAN Level 1 DIS.

• **End System Neighbors TLV (Code 3)**: This TLV specifies the ES neighbors and may appear more than once, and in an LSP with any LSP Number. The Value field contains one or more entries with each entry consisting of the tuple <Virtual Flag, Default Metric, Delay Metric, Expense Metric, Error Metric, Neighbor ID> similar to the Intermediate System Neighbors TLV option above. The Neighbor ID field specifies the System ID of the ES neighbor. Only adjacencies with identical routing costs can appear in the same list.

• **Authentication Information TLV (Code 10)**: This TLV specifies information for authenticating the originator of the PDU.

A Level 1 Non-Pseudonode LSP contains the following information in its variable length TLV fields: Area Addresses option, Intermediate System Neighbors option, End System Neighbors option, and Authentication Information field.

A Pseudonode LSP specifies the following information in its variable length TLV fields: Intermediate System Neighbors option, End System Neighbors option, and Authentication Information field. In all cases of a Pseudonode LSP, a value of zero must be used for all supported routing metrics. *The Area Addresses TLV option is not present in a Pseudonode LSP. This information is not required since the set of Area Addresses for the DIS (which creates the Pseudonode LSP) will already have been included the DIS's own Non-Pseudonode LSP.*

2.8.2.6.2   *Level 2 LAN LSP TLVs*

The following TLVs (each having a 1-byte Type field and 1-byte Length field) are defined for Level 2 LAN LSPs **[ISO10589:2002]**:

• **Area Addresses TLV (Code 1)**: This TLV specifies the set of *partitionArea-Addresses* of the local IS-IS router, if the router supports partition repair; otherwise the field specifies the set of *areaAddresses* of the router. The Value field contains one or more entries with each entry consisting of the tuple <Address Length, Area Address>. This option must never be present in Pseudonode LSPs. For Non-Pseudonode LSPs, this TLV option must always be present in the LSP with LSP Number 0, and must never be present in an LSP with a non-zero LSP Number. This option must appear before any Intermediate System Neighbors TLV options or Prefix Neighbors TLV options.

- **Partition Designated Level 2 Intermediate System TLV (Code 4)**: This TLV specifies the System ID of the Level 2 DIS for the partition (Partition Designated Level 2 Intermediate System). This option must never be present in Pseudonode LSPs. For Non-Pseudonode LSPs sent by IS-IS routers which support the partition repair optional function, and which are currently attached (see IS-IS attach conditions in Section 7.2.9.2 of **[ISO10589:2002]**), this TLV option must always be present in the LSP with LSP Number 0, and must never be present in an LSP with non-zero LSP Number. The TLV must appear before any Intermediate System Neighbors TLV options or Prefix Neighbors TLV options.
- *originatingLSPBufferSize* **TLV (Code 14)**: This TLV specifies the local value for *originatingL2LSPBufferSize* (in the range 512 to 1,492). This option must appear only once in an LSP with any LSP Number, and it must appear before any Intermediate System Neighbors TLV options or Prefix Neighbors TLV options.
- **Intermediate System Neighbors TLV (Code 2)**: This TLV specifies IS-IS routers and Pseudonode neighbors. The Value field contains one or more entries with each entry consisting of the tuple <Virtual Flag, Default Metric, Delay Metric, Expense Metric, Error Metric, Neighbor ID> similar to the Level 1 LSP TLV (with narrow metrics). This TLV is permitted to appear more than once, and in an LSP with any LSP Number. However, all the Intermediate System Neighbors TLV options must appear before the Prefix Neighbors TLV options, that is, they must appear before any Prefix Neighbor TLV options in the same LSP, and no Prefix Neighbor TLV options must appear in an LSP with lower LSP Number.
- **Prefix Neighbors TLV (Code 5)**: This TLV specifies the address prefix neighbors that are reachable. The Value field contains one or more entries with each entry consisting of the tuple <Default Metric, Delay Metric, Expense Metric, Error Metric, Address Prefix Length, Address Prefix>. This option may appear more than once, and in an LSP with any LSP Number. Only adjacencies with identical routing costs can appear in the same list.
  - o **Address Prefix Length** is a 1-byte field that specifies the length in semibytes of the address prefix following. A length of zero indicates a prefix that matches all NSAPs.
  - o **Address Prefix** is a field (of length Address Prefix Length divided by 2 and rounded up) that specifies a reachable address prefix (encoded as described in Section 7.1.6 of **[ISO10589:2002]**). If the length in semibytes is odd, the prefix is padded out with a trailing zero semibyte to an integral number of bytes.
  - o Note that the Area Addresses listed in the Area Addresses TLV option field (TLV Code 1) of Level 2 LSP with LSP Number 0, are understood to be reachable address neighbors with routing cost of zero. These Area Addresses are not listed separately in the Prefix Neighbors TVL options (TLV Code 5).
- **Authentication Information TLV (Code 10)**: This TLV specifies information for authenticating the originator of the PDU.

2.8.2.6.3   Level 1 Area Partition Repair

It is possible that the failure of one or more links in a Level 1 area may result in the partitioning of the area. However, if each of the resulting partitions has a connection to the IS-IS Level 2 backbone, it is possible to repair the partitioned Level 1 area via the Level 2 backbone, provided that the Level 2 backbone itself is not partitioned (see details in Section 7.2.10 of **[ISO10589:2002]**).

Election of a Partition Designated Level 2 router provides a mechanism for repairing a Level 1 area in an IS-IS routing domain that has been partitioned. Partition Designated Level 2 Routers are elected from among the Level 2 routers in the partitions, and then an IS-IS Virtual Link is established over the Level 2 routing subdomain between these Designated Level 2 Routers. The Virtual Link interconnects the Level 1 area partitions, and intra-area traffic between the partitions is forwarded over the Virtual Link.

If a Level 1 area becomes partitioned, IS-IS allows the partition to be repaired as long as there are Level 2 routes linking the partitions. However, if the Level 2 backbone becomes partitioned, IS-IS has no provision for using Level 1 links to repair the Level 2 partition. IS-IS requires that the Level 2 routers forming the backbone be contiguously connected.

A single Level 2 capable router (i.e., Level 1-2 router) may lose connectivity to the Level 2 routing subdomain. When this happens, the Level 1-2 router will indicate in its Level 1 LSPs (by setting the ATT bit) that it is no more "attached", thereby, allowing the Level 1 routers in the local area to route traffic destined for outside the area to a different Level 1-2 router. This means Level 1 routers in an area will route traffic to destinations outside of the area only to the Level 1-2 routers that indicate in their Level 1 LSPs that they are "attached" to the IS-IS backbone.

2.8.2.6.4   Level 2 Pseudonode and Non-Pseudonode LSPs

A Level 2 Non-Pseudonode LSP contains the following information in its variable length TLV fields: Area Addresses option, Intermediate System Neighbors option, Prefix Neighbors option, and Authentication Information field. A Level 2 Pseudonode LSP contains the following information in its variable length TLV fields: Intermediate System Neighbors option, and Authentication Information field. In all cases of a Pseudonode LSP, a value of zero must be used for all supported routing metrics. *The Area Addresses and Prefix Neighbors TLV options are not present in a Pseudonode LSP.*

2.8.2.7   LSP Related Timers and Parameters

The transmission of LSPs has a number of related timers and parameters **[ISO10589:2002]**:

- *MaxAge*: This specifies the maximum length of time since a stored LSP was originated by its source before it is considered expired and invalid. *MaxAge* is an architectural constant and the default value is 1,200 seconds (20 minutes). An LSP that has expired can be deleted from the LSDB after a further *ZeroAgeLifetime* has expired. The setting of *MaxAge* is larger than *maximumLSPGenerationInterval*, so that a system is not purged merely because it lacks events for reporting LSPs.

- **ZeroAgeLifetime**: This specifies the amount of time (in seconds) that an LSP with zero Remaining Lifetime should be retained in the LSDB before it is purged even after it has expired in the network (default value is 60 seconds). The *ZeroAgeLifetime* is the minimum amount of time an IS-IS should retain the header of an expired LSP after it has been flooded with zero Remaining Lifetime. A recommended value is $2 \times MaxAge$. The use of the *ZeroAgeLifetime* is to ensure that the LSP header is retained until the zero Remaining Lifetime of the LSP has been safely propagated to all the IS-IS neighbors.

- **maximumLSPGenerationInterval**: This specifies the maximum amount of time a source IS-IS router must wait since the last LSP was generated before generating the next LSP. The *maximumLSPGenerationInterval* must be less than *MaxAge* to ensure that a router periodically generates a newer copy of its LSP into the network before the LSP expires, even when there are no network changes to report. This process helps to ensure that the integrity of the LSP is always maintained throughout the network. Setting this parameter too small adds overhead to the LSP generation process (resulting in a lot of LSPs being sent). Setting this parameter too large (while not violating constraints) causes the LSP generation process to wait a long time to recover in the event that incorrect link-state information sent by the source exists somewhere in the domain. A recommended setting for *maximumLSPGenerationInterval* is 900 seconds (15 minutes). The *maximumLSPGenerationInterval* is also known as the LSP refresh interval.

- **minimumLSPGenerationInterval**: This specifies the minimum amount of time between successive generation of LSPs by an IS-IS router. A source must wait at least this length of time before regenerating one of its own LSPs. The *minimumLSPGenerationInterval* serves as a hold down timer on the generation of each individual LSP. Setting the *minimumLSPGenerationInterval* too large causes a delay in reporting new link-state information. Setting the interval too small can result in too much routing overhead. A recommended setting is 30 seconds.

- **minimumLSPTransmissionInterval**: This specifies the minimum amount of time an IS-IS router must wait before propagating/advertising another LSP received from the same source IS-IS router (the same LSP ID); it is minimum interval between retransmissions of an LSP. Setting the *minimumLSPTransmissionInterval* too large causes a delay in the propagation of routing information and the stabilization of the routing algorithm. Setting the interval too small may cause the IS-IS routers to use more CPU and bandwidth resources. Setting *minimumLSPTransmissionInterval* smaller than *minimumLSPGenerationInterval* is desirable to allow IS-IS routers to recover from lost LSPs. The recommended value of *minimumLSPTransmissionInterval* is 5 seconds.

- **minimumBroadcastLSPTransmissionInterval**: This specifies the minimum interval between transmission of LSPs on a broadcast LAN. The *minimumBroadcastLSPTransmissionInterval* attribute indicates the minimum interval between PDU that a router can send and can be processed by the slowest IS-IS router on the LAN.

2.8.3 IS-IS COMPLETE SEQUENCE NUMBERS PDUs

IS-IS routers send SNPs to describe all or some of the LSPs in their LSDBs. The receiving routers use the SNPs to update and synchronize their LSDBs. The two types of SNPs defined in **[ISO10589:2002]** are CSNPs and PSNPs. A CSNP contains summaries of all LSPs in the LSDB of the originating router, and these are exchanged between neighbor routers to ensure that their LSDBs are synchronized.

An IS-IS router uses CSNPs to inform other routers on a network segment (broadcast and point-to-point) about the LSPs in its local LSDB so that they can determine which LSPs may be missing or outdated from their own LSDBs. The CSNPs are sent to ensure that all routers have the same routing information in their LSDBs and are synchronized. CSNPs are similar to OSPF Database Description Packets as described in Chapter 1. CSNPs are of the following two types (Figure 2.18):

- Level 1 Complete Sequence Numbers PDU (PDU Type 24)
- Level 2 Complete Sequence Numbers PDU (PDU Type 25)

On multiaccess broadcast network segments, instead of each IS-IS router sending explicit acknowledgements for each received LSP, the Designated Intermediate System (DIS) will multicast periodically a CSNP to all routers on the network segment. The CSNP contains a list of all LSPs in its LSDB, and includes enough information so that other routers receiving the CSNP can compare with their LSDBs to determine whether they and the DIS have synchronized LSDBs.

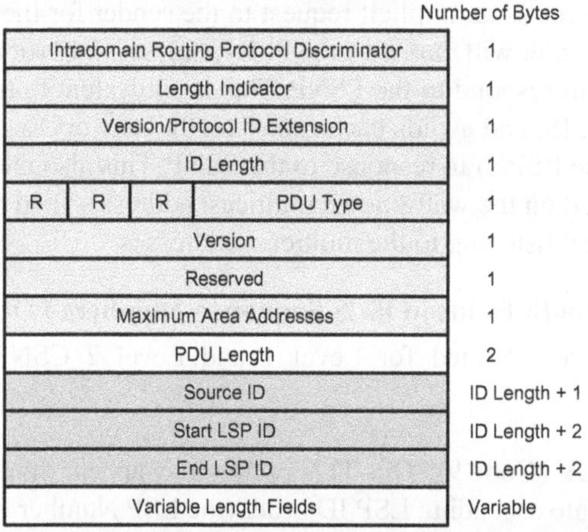

	Number of Bytes
Intradomain Routing Protocol Discriminator	1
Length Indicator	1
Version/Protocol ID Extension	1
ID Length	1
R R R PDU Type	1
Version	1
Reserved	1
Maximum Area Addresses	1
PDU Length	2
Source ID	ID Length + 1
Start LSP ID	ID Length + 2
End LSP ID	ID Length + 2
Variable Length Fields	Variable

Field	Meaning
Source ID	This is the system ID of IS router (with zero Circuit ID) that generated this Sequence Numbers PDU.
Start LSP ID	This is the LSP ID of the first LSP in the range of LSPs covered by this Complete Sequence Numbers PDU.
End LSP ID	This is the LSP ID of the last LSP in the range of LSPs covered by this Complete Sequence Numbers PDU.

FIGURE 2.18 Level 1 and Level 2 IS-IS Complete Sequence Number PDU Format

The DIS on a broadcast network segment sends CSNPs at specified intervals, with the default interval being 10 seconds. The DIS sends periodic CSNPs summarizing the following information: Remaining Lifetime, LSP ID, LSP Sequence Number, and Checksum. Adjacent IS-IS routers on a point-to-point link send CSNPs only when the neighbor relationship is established for the first time. If a router's LSDB is large, it sends multiple CSNPs. When a point-to-point circuit starts or restarts, each router sends a CSNP on that circuit.

In practice, the information a router may want to transmit may be greater than what will fit in a single CSNP. Therefore, each CSNP sent contains an inclusive range of LSP IDs marking the LSP information carried. The router sends the complete set of LSP information by transmitting a series of individual CSNPs, each carrying a subset of the complete range. The sender must ensure that the ranges of the complete set of CSNPs is contiguous (although they may not necessarily be transmitted in order), and must cover the entire range of possible LSP IDs.

The DIS generates CSNP in order for all routers connected to a broadcast media to synchronize their LSDBs. The other IS-IS routers on the network segment use CSNPs to keep their LSDBs up to date. If a router, upon receipt of a CSNP, detects that the sender has out of date LSP information, it will multicast the missing information. A Level 1 DIS periodically multicasts complete sets of Level 1 CSNPs to the well-known multicast address 01-80-C2-00-00-14 (AllL1ISs). A Level 2 DIS periodically multicasts complete sets of Level 2 CSNPs to the multicast address 01-80-C2-00-00-15 (AllL2ISs).

If a router detects that the sender of an LSP has more up to date information, it will multicast a PSNP, containing information about the older LSPs it wants to update. This PSNP serves as an implicit request to the sender for the missing information. Although the router will multicast the PSNP, only the DIS (of the appropriate IS-IS routing level) will respond to the PSNP. This is equivalent to transmitting the PSNP directly to the DIS, and avoids each router on the network segment unnecessarily sending the same LSP(s) in response to the PSNP. This also allows all routing messages to be received on the well-known multicast addresses, and hence by a network interface (adapter) listening to the multicast addresses.

2.8.3.1   Variable-Length Fields in IS-IS Sequence Numbers PDUs

The following TLVs are defined for Level 1 and Level 2 CSNPs and PSNPs **[ISO10589:2002]**:

- **LSP Entries TLV (Code 9)**: This TLV option may appear more than once but must be sorted into ascending LSP ID order (the LSP Number byte of the LSP ID is the least significant byte). The Value field contains one or more entries with each entry consisting of the tuple <Remaining Lifetime, LSP ID, LSP Sequence Number, Checksum>. Remaining Lifetime is a 2-byte field that specifies the remaining time before LSP expires. LSP ID is a field of length ID Length plus 2, and specifies the System ID of the source of the LSP. LSP Sequence Number is a 4-byte field that specifies the Sequence Number of the LSP. Checksum is a 2-byte field that carries the computed Checksum of the LSP.

- **Authentication Information TLV (Code 10)**: This TLV specifies information for authenticating the originator of the PDU.

2.8.3.2 CSNP Related Timers and Parameters

The *CompleteSNPInterval* is a CSNP related timer which specifies the amount of time the DIS on a broadcast LAN must wait before sending the next periodic complete set of CSNPs **[ISO10589:2002]**. Setting the *CompleteSNPInterval* too large delays the convergence of the routing algorithm when LSPs are lost. Setting the interval too small results in extra routing information traffic overhead. The lower the *CompleteSNPInterval* value, the faster the speed of LSDB synchronization. However, a *CompleteSNPInterval* that is too low will trigger intensive PSNP transmissions. A recommended value of *CompleteSNPInterval* is 10 seconds **[ISO10589:2002]**.

A complete set of CSNPs is a set that has the Start LSP ID and End LSP ID ranges covering the complete possible range of LSP IDs, that is, there is no LSP ID value which does not appear within the range of one of the CSNPs in the set. When a DIS transmits more than one CSNP on a broadcast medium, it must separate them by an interval of at least *minimumBroadcastLSPTransmissionInterval* **[ISO10589:2002]**.

2.8.4 IS-IS PARTIAL SEQUENCE NUMBERS PDUs

IS-IS routers use PSNP to acknowledge receipt of one or more LSPs, or request newer versions of one or more complete LSP. An PSNP contains a summary of only a subset of LSPs known by the originating router. PSNPs consist of the following two types (Figure 2.19):

- Level 1 Partial Sequence Numbers PDU (PDU Type 26)
- Level 2 Partial Sequence Numbers PDU (PDU Type 27)

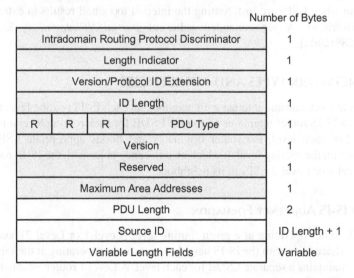

				Number of Bytes
Intradomain Routing Protocol Discriminator				1
Length Indicator				1
Version/Protocol ID Extension				1
ID Length				1
R	R	R	PDU Type	1
Version				1
Reserved				1
Maximum Area Addresses				1
PDU Length				2
Source ID				ID Length + 1
Variable Length Fields				Variable

FIGURE 2.19 Level 1 and Level 2 IS-IS Partial Sequence Number PDU Format

A router sends (multicasts) a PSNP when it detects that it is missing an LSP or when an LSP in its LSDB is out of date. When an IS-IS router receives a CSNP, it checks the link-state entries in the PDU against its own local LSDB. If it detects missing information, the router sends a request using a PSNP. The router sends the PSNP to the router that originated the CSNP, effectively requesting that the missing LSP be retransmitted. That originating router, in turn, forwards the missing LSP to the requesting router. The requesting router receives the LSP and stores it in the local LSDB, and then sends an acknowledgment back to the originating router. PSNPs are similar in function to the OSPF Link-State Request Packets and Link-State Acknowledgment Packets (see Chapter 1).

A PSNP that a neighbor router sends will list only the Sequence Numbers of recently received LSPs. A router can send a PSNP to acknowledge the receipt of multiple LSPs at a time. If a router wants to update its LSDB, it can send PSNPs to request a new LSP from a neighbor. All routers that have LSDBs that are not synchronized with the DIS and need additional LSPs in their databases, will send PSNPs.

On point-to-point links, an IS-IS router sends an explicit acknowledgement encoded as a PSNP containing the following information: Source ID, PDU Type (Level 1 or 2), LSP Sequence Number, Remaining Lifetime, and Checksum. The router does this for all received LSPs which are newer than those in its LSDB, or for duplicates of the LSPs in its LSDB. If the router receives an LSP which is older than a corresponding copy in the LSDB, the router will instead send a newer LSP.

2.8.4.1 PSNP-Related Timers and Parameters

The *partialSNPInterval* is an attribute related to the transmission of PSNPs. This specifies the amount of time between periodic transmission of PSNPs [ISO10589:2002]. The *partialSNPInterval* must be set less than *minimumLSPTransmissionInterval*. Setting the interval too large delays the convergence of the routing algorithm when LSPs are lost. Setting the interval too small results in extra control traffic overhead. A recommended value of *partialSNPInterval* is 2 seconds [ISO10589:2002].

2.9   NETWORK TYPES AND ADJACENCIES

IS-IS routers operate and interact with neighbors at each IS-IS routing level independently. An IS-IS router maintains a separate LSDB for each routing level it is operating on. For each level, the router originates and floods appropriate LSPs. Also, depending on the routing level at which an adjacency is formed, the IS-IS router will send Level 1 or Level 2 LSPs to its neighbors.

2.9.1   IS-IS ADJACENCY FORMATION

An IS-IS router operating at a given routing level (Level 1 or Level 2), establishes separate adjacencies with the IS-IS neighbor routers also operating at the same level, and also maintains a separate LSDB for each level. A Level 1 router establishes adjacencies only with IS-IS neighbors that are also configured as Level 1 routers.

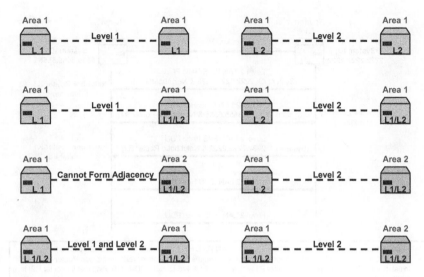

FIGURE 2.20 Rules for Establishing IS-IS Adjacencies

Similarly, a Level 2 router establishes adjacencies only with neighbors also configured as Level 2 routers.

The rules for establishing IS-IS adjacencies are as follows (see also Figure 2.20):

- *Level 1 routers* form *Level 1 adjacencies* with *Level 1 and Level 1-2 routers* only in the same area.
- *Level 2 routers* form *Level 2-only adjacencies* with *Level 2 and Level 1-2 routers* in the same area or another area.
- *Level 1-2 routers* form *Level 1 and Level 2 adjacencies* with other *Level 1-2 routers* in the same area.
- *Level 1-2 routers* form *Level 2-only adjacencies* with other *Level 1-2 routers* in another area.
- A *Level 1 router does not* form an adjacency with a *Level 2 router*, regardless of the areas to which they belong.
- The System ID of each router in the IS-IS area must be unique.
- The network types of the two IS-IS router interfaces wishing to form an adjacency must be consistent (broadcast LAN or point-to-point). Note that two Ethernet interfaces attached to a common network segment can be configured to establish a point-to-point adjacency.
- The Hello intervals (specified by the IS-IS *iSISHelloTimer* attribute) and Holding Times of IS-IS neighbors do not necessarily have to match for them to form an adjacency.
  - o The Hello interval (in seconds) defines the frequency at which a router sends IS-IS Hello packets out of an interface.
  - o The Holding Time specifies how long (in seconds) a router should wait before declaring a neighbor as unavailable or unreachable when an IS-IS Hello packet is not received from that neighbor.

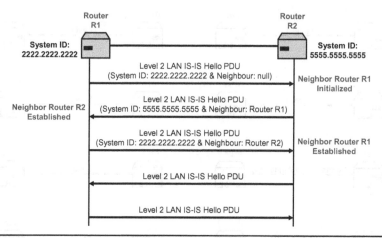

1. Router R1 broadcasts out of its interfaces a Level-2 LAN IS-IS Hello PDU with no neighbor ID specified.
2. Router R2 receives this broadcast message and sets the status of the neighbor relationship with Router R1 to Initial. Router R2 then responds to Router R1 with a Level-2 LAN IS-IS Hello PDU, indicating that it (Router R2) is a neighbor of Router R1.
3. Router R1 receives this message and sets the status of the neighbor relationship with Router R2 to Up. Router R1 then sends a Level-2 LAN IS-IS Hello PDU to Router R2 indicating that it (Router 2) is a neighbor of Router R1.
4. Router R2 receives this message and sets the status of the neighbor relationship with Router R1 to Up. Router R1 and Router R2 then establish a neighbor relationship successfully.

FIGURE 2.21   Process of Establishing a Level 2 Neighbor Relationship on a Broadcast Network

As an example, Figure 2.21 describes process of establishing a Level 2 neighbor relationship on a broadcast network.

Two IS-IS neighbor Level 1-2 routers will establish two independent adjacencies, one for each routing level (see Figure 2.20). In addition, since each IS-IS router must belong to a single area (and only a single NET [which carries the Area identifier] is usually configured on a router), Level 1 adjacencies are created only between routers with the same Area ID. Level 2 adjacencies must also be contiguous as shown in Figure 2.22.

Two IS-IS routers are considered neighbors only when each router announces that the other is directly reachable over one of its SNPAs. On a point-to-point, the two routers assume that a neighbor relationship has been established when IS-IS Hello PDUs are exchanged. However, a router that is malfunctioning might report another router as a neighbor when in fact it is not. To detect this type of failures, a router checks that a link to a neighbor that is reported as "Up" in a received LSP, is reported by both routers to be indeed in that state. A link that is considered as "Down" must not be advertised in LSPs.

On broadcast network segments, this type of failure is detected by the DIS, which is responsible for establishing the set of routers that can all communicate on the network segment. The DIS includes only the routers reported as "Up" in the Pseudonode LSPs it generates for the Pseudonode representing the broadcast network segment (see "Designated Intermediate System (DIS) and Pseudonodes" section below).

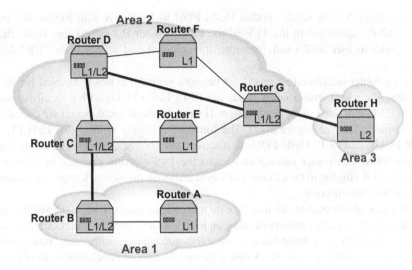

FIGURE 2.22 Level 2 Adjacencies Must be Contiguous

2.9.2 THREE-WAY HANDSHAKE FOR FORMING IS-IS LAN ADJACENCIES

A router attached to a multiaccess broadcast network segment sends IS-IS Hello PDUs on its interfaces as soon as the interface is enabled. A three-way handshake is defined in the ISO standard **[ISO10589:2002]** for initiating adjacencies on broadcast LANs. The following describe the three-way handshake process when an adjacency is being formed between two IS-IS routers, Router A and Router B, on a broadcast LAN:

- Starting the adjacency in the "Down" state, IS-IS Router A sends out a LAN IS-IS Hello PDU on its OSPF-enabled interface.
  - o Note that the Data Link (or Layer 2) Layer of a Hello PDU carries the source Layer 2 address of the sending interface, that is, the SNPA that identifies it at Layer 2.
  - o The destination Layer 2 address is a multicast address: Level 1 PDUs are sent to the multicast destination MAC address 01-80-C2-00-00-14 (AllL1ISs address), while Level 2 PDUs are sent to the multicast destination MAC address 01-80-C2-00-00-15 (AllL2ISs address).
- If Router B on the LAN receives the LAN IS-IS Hello PDU, it places the adjacency in the "Initializing" or "Init" state. Router B then sends out an IS-IS Hello PDU to the initiating neighbor, Router A, including the SNPA of Router A (as indicated by the Layer 2 address associated with Router A's Hello PDU) in a TLV of the Hello PDU (i.e., the Intermediate System Neighbors TLV Code 6 or 7).
- The initiating IS-IS Router A receives the IS-IS Hello PDU from the neighbor Router B with Router A's own SNPA identified in the PDU. On receipt of this Hello PDU, the initiating IS-IS Router A is now aware that the new IS-IS neighbor Router B knows of its presence.

- Router A then sends another Hello PDU to Router B with Router B's own SNPA identified in the TLV of the PDU. Router B receives this Hello PDU with its own SNPA and, therefore, transitions the adjacency to the "Up" state.

When a router interface to a broadcast network segment is newly enabled for IS-IS routing, the router immediately sends out IS-IS Hello PDUs with a locally defined LAN ID that consist of its own System ID and a unique local circuit ID called the Pseudonode ID (see Figure 2.17). The router also begins to listen for ESH PDUs, ISH PDUs, and IS-IS Hello PDUs to discover any connected neighbors and to form adjacencies. The router subsequently runs the DIS election process, to determine whether it is eligible to be a Level 1 or Level 2 DIS on the network segment, depending on its configuration.

If a new router enables an interface on a broadcast network segment and the segment already has other routers running including a DIS, the new router interface will follow the three-way handshake to establish adjacencies with every other active router on the network segment. A new DIS may be elected depending on the Priority field values specified in the IS-IS LAN Hello PDUs (Figure 2.13) sent by the routers (see "Election of the DIS" section below). When the router interface transitions to the "Up" state, each router on the network segment will see the SNPAs of all routers listed in the TLV 6 or 7 of the Hello PDU they transmit to each other. When this happens, each router has bidirectional visibility with every other router on the network segment.

The way in which the router processes received IS-IS Hello PDUs depends on its configuration (i.e., Circuit Type and IS Type). The Circuit Type and IS Type are described in Figures 2.13 and 2.16, respectively. The router checks all received IS-IS Hello PDUs for configuration conformity and authentication. The router must make sure that the ID Length and Maximum Area Addresses fields in the received IS-IS Hello PDUs (Figure 2.13) match local values, and authentication is successful before the adjacency is further processed. Some of additional information carried in the IS-IS Hello PDUs are the neighbor's System ID, Holding Time, Level 1 or Level 2 priority to be DIS, and the configured Area Addresses.

When the IS-IS router receives an IS-IS Hello PDU, it checks if an adjacency with the sending neighbor exists. If an adjacency exists, the router resets the Holding Time to the value in the received Hello PDU. If the neighbor is not known, the router creates an adjacency, indicating the type of adjacency (Level 1 or Level 2), and sets the adjacency state to "Initializing" until it receives subsequent Hello PDUs confirming two-way communication.

IS-IS routers include the SNPA (MAC addresses) of all neighbors on the broadcast network segment that they have received Hello PDUs from, providing a simple mechanism to confirm two-way communication. As described above, a router confirms two-way communication when subsequent Hello PDUs received from the neighbor contain the router's own SNPA in the Intermediate System Neighbors TLV Code 6 or 7 field. Otherwise, a router deems the communication with a neighbor as one-way, and the adjacency stays at the "Initializing" state. An adjacency must be in the "Up" state before a router can begin to send or process LSPs.

2.9.3 THREE-WAY HANDSHAKE FOR FORMING IS-IS POINT-TO-POINT ADJACENCIES

An IS-IS router upon receiving IS-IS Hello PDUs performs checks to confirm various parameters in the Hello PDU's header, such as, System ID length, Maximum Area Addresses, and so on. Routers advertise system capabilities in the appended TLVs.

2.9.3.1 Original IS-IS Point-to-Point Adjacency Formation Process

The ISO standard does not specify a corresponding three-way handshake process for point-to-point adjacencies. The absence of information about IS-IS neighbors in Point-to-Point IS-IS Hello PDUs (as specified in the original IS-IS Hello PDU format specified in [ISO10589:2002]) can cause reliability issues when forming point-to-point adjacencies.

When an IS-IS router receives an ISH PDU on a newly enabled point-to-point link, it verifies whether an adjacency already exists with the sending neighbor by comparing the source System ID in the Hello PDU against its Adjacency Database. The router ignores the Hello PDU if an adjacency already exists. If no adjacency exists, the router creates a new adjacency and sets its adjacency state to "Initializing" and the system type of the neighbor to "Unknown".

The router then sends an IS-IS Hello PDU to the new neighbor in response. Upon receiving a subsequent IS-IS Hello PDU from the new neighbor, the router then transitions the adjacency to the "Up" state, and changes the system type of the neighbor to "IS" (i.e., IS-IS router). All along this process, the local router is unable to determine whether the Hello PDUs it has sent are being received by the remote end. In the original IS-IS specification in [ISO10589:2002], a router assumes that an adjacency is "Up" as soon as a Point-to-Point IS-IS Hello PDU is received from the neighbor.

As specified in [ISO10589:2002], unlike LAN IS-IS Hello PDUs, Point-to-Point IS-IS Hello PDUs do not include the Intermediate System Neighbors TLV Code 6 or 7. Therefore, it is not possible to use a three-way handshake mechanism on point-to-point links to confirm whether IS-IS Hello PDUs generated locally by a router are reaching its neighbors. This lack of further neighbor information on point-to-point links can lead to situations where one end of an adjacency will be up and the other end will not. Reference [ISO10589:2002] has no mechanism for verifying bidirectional visibility between two routers on a point-to-point link. This native approach leads to problems if the point-to-point link, or the router interfaces are not able to successfully transfer PDUs in both directions between the two routers.

Reference [RFC5303] proposes a more reliable mechanism (using a three-way handshake process) for forming point-to-point IS-IS adjacencies, which is also backward-compatible to the procedure in [ISO10589:2002]. As described above, routers attached to a broadcast network segment are able to confirm two-way communication with adjacent nodes using a three-way handshake procedure made possible by the presence of the Intermediate System Neighbors TLV (Code 6 or 7) field in Level 1 or Level 2 LAN IS-IS Hello PDUs. On the other hand, reliable point-to-point adjacency formation is made possible by the presence of the TLV 240 in Point-to-Point

IS-IS Hello PDUs and the process is similar to the broadcast network segment process.

2.9.3.2   Extensions for Reliable IS-IS Point-to-Point Adjacency Formation

A Point-to-Point Adjacency State TLV (TLV 240) was defined for Point-to-Point IS-IS Hello PDUs **[CISCTEAPAQ00] [RFC5303]** to allow a three-way handshake to be used on IS-IS point-to-point links. Similar to the LAN adjacency process, an IS-IS router checks for its own SNPA in the Point-to-Point Hello PDU sent by a neighbor before placing the adjacency in the "Up" state. Using this extension, if it happens that an IS-IS router (configured for point-to-point communications) receives a LAN Hello PDU, it will recognize the mismatch, and will ignore the neighbor.

The three-way handshake process for point-to-point adjacencies uses the new TLV with Code 240 **[RFC5303]**. To allow backward compatibility, IS-IS implementations that receive an IS-IS Hello PDU with this new TLV and do not support it, will simply ignore it, and follow the procedures described in **[ISO10589:2002]** for forming the adjacency. This is necessary to allow newer systems that conform to **[RFC5303]** and those that do not, to coexist in the same network.

The three-way handshake requires a TLV 240-compliant system to transition the state of the point-to-point adjacency to "Up" only after confirming that its IS-IS Hello PDUs are reaching the remote router, and that, there is bidirectional communication with the remote end. A TLV 240-compliant router includes the System IDs of neighbors from which they have received IS-IS Hello PDUs in TLV 240. A router knows that the IS-IS Hello PDUs it sends are reaching a neighbor over the point-to-point link when it receives an IS-IS Hello PDU from that neighbor listing its parameters in the TLV 240.

The TLV Code 240 consists of the following information **[RFC5303]**:

- **1-byte Type Field**: This carries the TLV Code (decimal 240, hexadecimal 0xF0)
- **1-byte Length Field**: This specifies the total length of the Value field (5–17 bytes)
- **Value field consists of the following subfields**:
  - **1-byte Adjacency Three-Way State field**: This carries the three-way adjacency state of the point-to-point link; "Up" = 0, "Initializing" = 1, or "Down" = 2. This is the state of adjacency as seen by the local router.
  - **4-byte Extended Local Circuit ID field**: This specifies the Unique ID assigned to the circuit when it was created by the local IS-IS router. This is the Local Circuit ID of the local router's interface.
  - **Neighbor System ID field (Equal to ID Length)**: This specifies the System ID of the IS-IS neighbor if known. The length of this field is equal to "ID Length" described in the Point-to-point IS-IS Hello PDU. The local router sets this value to the System ID of the neighbor router whose IS-IS Hello PDUs have been successfully received.
  - **4-byte Neighbor Extended Local Circuit ID field**: This specifies the Extended Local Circuit ID of the other end of the point-to-point adjacency

if known. The local router sets this value to the Extended Local Circuit ID field value carried in the neighbor's IS-IS Hello PDUs.

To describe the three-way point-to-point adjacency formation process, let us consider two routers, Router A and Router B attached to a point-to-point link. Let us assume that neither router has received an IS-IS Hello PDU from the other yet, and both are treating the point-to-point adjacency as being in the Down state:

- Router A receives an IS-IS Hello PDU from Router B with the three-way adjacency state set to "Down"; Router A has communication visibility of Router B. However, it is not certain if Router A is visible to Router B. Router A starts sending its IS-IS Hello PDUs with the three-way adjacency state set to "Initializing" to inform Router B that it can receive its PDUs.
  - o TLV 240 to Router B carries {Adjacency State A = "Initializing"; Local Circuit ID = A_CirID; Neighbor System ID = B_SysID; Neighbor Extended Local Circuit ID = B_ExtID}.
- When Router B receives an IS-IS Hello PDU from Router A with the three-way adjacency state set to "Initializing", it knows that these IS-IS Hello PDUs are essentially sent in response to its own Hello PDUs, and that Router A is in fact indicating to Router B it can receive its PDUs. Router B now sees that bidirectional communication is possible with Router A. Therefore, Router B starts sending its IS-IS Hello PDUs with the three-way adjacency state set to "Up".
  - o TLV 240 to Router A carries {Adjacency State B = "Up"; Local Circuit ID = B_CirID; Neighbor System ID = A_SysID; Neighbor Extended Local Circuit ID = A_ExtID}.
- When Router A receives an IS-IS Hello PDU from Router B with the three-way adjacency state set to "Up", it knows that Router B can receive its PDUs. Router A is now also certain that bidirectional communication is possible with Router B and starts sending its IS-IS Hello PDUs with the three-way adjacency state set to "Up", concluding the three-way handshake process.
  - o TLV 240 to Router B carries {Adjacency State A = "Up"; Local Circuit ID = A_CirID; Neighbor System ID = B_SysID; Neighbor Extended Local Circuit ID = B_ExtID}.

The "Down" state is the initial state of the three-way point-to-point adjacency process. In this state the IS-IS router has not received any IS-IS Hello PDU containing the "Point-to-Point Three-Way Adjacency" TLV option (TLV 240) on the point-to-point link. In the "Initializing" state, the router has received an IS-IS Hello PDU containing the TLV 240 option from a neighbor but does not know whether its Hello PDU are being received by the neighbor. In the "Up" state, the router knows that its IS-IS Hello PDUs are being received by the neighbor.

An IS-IS router uses the 8-bit Local Circuit ID field in the Hello PDU header (see Figure 2.14) to assign a locally unique link identifier to each point-to-point link. Local Circuit IDs on point-to-point links are only carried in IS-IS Hello PDUs and, routers on the link use these only for the detection of a change in identity at the other end of the link. On broadcast network segments, an equivalent of the Local Circuit ID, the Pseudonode ID, is used when a router is elected the DIS on that interface. In this case, the 8-bit Pseudonode ID (see Figure 2.17) associated with the DIS is required to be unique only for interfaces to the broadcast network segment on which the router is a DIS. The non-zero Pseudonode ID is assigned by the DIS when elected. DIS advertises the Pseudonode ID in the Pseudonode LSPs it sends.

The TLV 240 enhances the reliability of IS-IS point-to-point adjacency formation in two ways. First, an IS-IS router can confirm that two-way communication has been established with its neighbor by checking for the presence of its System ID in the TLV 240 carried in IS-IS Hello PDUs received from the neighbor. The router can then adjust its local adjacency state based on the existing local state, and the state specified in the Value field of the TLV 240 in the received Hello PDU.

Second, IS-IS routers can leverage the Extended Local Circuit ID field in the TLV 240 to provide unique link IDs for point-to-point links beyond the 256-limit specified in [ISO10589:2002]. The 8-bit Local Circuit ID field in Point-to-Point IS-IS Hello PDUs (Figure 2.14) allows up to only 256 unique point-to-point links to be defined in an IS-IS router. The 32-bit Extended Local Circuit ID field used in the three-way point-to-point handshake mechanism was defined to remove the 256 point-to-point interfaces limit imposed by IS-IS specification [ISO10589:2002].

The three-way point-to-point adjacency formation process is similar conceptually to the adjacency process on broadcast network segments. An IS-IS router on a broadcast network segment will lists the SNPAs of all neighbor routers from whom it has received IS-IS Hello PDUs on its broadcast interface. If the router receives an IS-IS Hello PDU from a neighbor and finds its own SNPA is indicated in that PDU, it knows that the neighbor router can receive its own PDUs, and will move the adjacency to the "Up" state. If IS-IS Hello PDUs from a neighbor do not contain the local router's SNPA, the adjacency is kept in the "Initializing" state.

2.9.4   ES-IS ADJACENCIES

The ES-IS protocol is developed for ES-to-IS (i.e., host-to-router) communication in a pure ISO environment. In an IP environment, the ES-IS protocol only facilitates IS-to-IS (i.e., router-to-router) adjacency formation. IP hosts in a network running IS-IS do not participate in the ES-IS protocol exchange. Instead, IP hosts rely on ARP (Address Resolution Protocol) for IP address-to-MAC address (i.e., Layer 3 address-to-Layer 2 address) resolution. IP hosts use ARP to determine the MAC addresses of other hosts connected to a common network segment and that of the default IP gateway when the IP address is known.

In CLNP environments, ESs and ISs (routers) via the ES-IS protocol, send ESH PDUs and ISH PDUs to well-known multicast Layer 2 (MAC) addresses to discover each other. ESs send ESH PDUs targeted at the IS-IS routers, to the multicast address 09-00-2B-00-00-05 (AllIntermediateSystems). IS-IS routers send the ISH PDUs directed to ESs to the multicast address 09-00-2B-00-00-04 (AllEndSystems). The ES-IS protocol allows ESs to locate the closest IS-IS router that can be used to access other networks that are not directly connected.

IS-IS routers, in turn, use the adjacency information provided by the ES-IS protocol to learn about the location of ESs. IS-IS routers also use Level 1 IS-IS LSPs to distribute the System ID of known ESs to other routers within the same IS-IS area. Other ES-IS protocol functions include sending redirect messages to ESs informing them about more suitable IS-IS router through which they can forward packets (PDUs).

2.10 DESIGNATED INTERMEDIATE SYSTEM (DIS) AND PSEUDONODES

This section describes the use of IS-IS in broadcast multiaccess environments and the concept of establishing adjacencies. With OSPF, one router on a broadcast multiaccess network segment (the DR) is assigned the responsibility of sending (flooding) link-state advertisements (LSAs) for LSDB synchronization on behalf of the other OSPF routers on the network segment. In IS-IS, this router, as illustrated in Figure 2.23, is referred to as the DIS. The DIS creates a *Pseudonode*, that represents the other routers on the network, and sends separate Level 1 and Level 2 LSPs on the network segment behalf of the *Pseudonode* **[ISO10589:2002]**.

2.10.1 IS-IS PSEUDONODE CONCEPT

The SPF algorithm used in OSPF and IS-IS requires a single source to construct a weighted directed graph of the shortest paths to all other vertices (i.e., known network destinations) from the source vertex. For broadcast media in IS-IS, the single source vertex is referred to as a *Virtual Router* (or *Pseudonode*). The DIS is elected and creates the *Pseudonode* representing the center point (a Virtual Router) of a starshape topology connecting all other routers attached to the broadcast media (see Figure 2.23).

IS-IS treats a broadcast network segment as a Pseudonode (with links to each attached system), instead of treating it as a fully connected topology (Figure 2.23). Without a Pseudonode, the five routers on the broadcast network segment would be represented in the LSDB as a fully meshed topology. Each of the five routers on the network segment would need to create its LSP with five entries, pointing toward each other router. Each router would have to exchange LSPs, CSNPs and PSNPs with each other router on a one-to-one basis in order to synchronize the LSDBs. A single broadcast network segment with N routers would have $N(N-1)$ links, the total number of links would grow quadratically with the number of routers. For N equal to 5, the total number of links is 20. A large N would cause increased memory requirements and possibly higher SPF tree computations.

With a Pseudonode, the broadcast network itself is represented as a virtual node. IS-IS views the entire broadcast network itself as a Pseudonode whose role is played by the elected DIS. Each IS-IS router on the broadcast network segment reports that it is only connected to the Pseudonode (rather than reporting a connection to every other router on the broadcast network segment), and the Pseudonode in turn, reports what resembles a set of point-to-point links to each of IS-IS routers connected to the

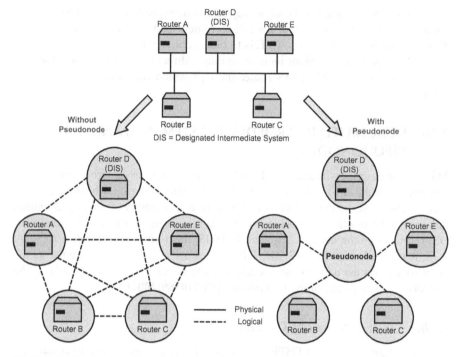

FIGURE 2.23   The Pseudonode Concept

network segment. This model requires only $N \times 2$ links (where each router advertises a link to the Pseudonode, and the Pseudonode advertises a link to each router). For N equal to 5, only 10 virtual (or logical) links are required.

The routers attached to the broadcast network segment will form adjacencies with each other (as described above) and will flood LSPs on the segment, but will deal only with the DIS when exchanging CSNPs and PSNPs. The DIS constructs special LSPs on behalf of the Pseudonode (called Pseudonode LSPs) reporting the links to all the routers on the broadcast segment, but with each supported routing metric in these special LSPs set to zero. The metric values are set to zero since the links reported to the Pseudonode (by the other nodes in the LSPs they send) had already been assigned metrics. Otherwise, if the Pseudonode assigns non-zero metrics in the Pseudonode LSPs, this would result in doubling the actual metric values.

The Pseudonode is identified by the Source ID of the DIS, followed by a non-zero Pseudonode ID assigned by the DIS (see Figure 2.17). The Pseudonode ID is locally unique to the DIS. A DIS is determined separately for each routing level (Level 1 and Level 2); LAN Level 1 DIS and LAN Level 2 DIS.

2.10.2   FUNCTIONS OF THE DIS

On multiaccess broadcast network segments, the election of a DIS to provide the Pseudonode functionality is an effective way of managing the potentially complex LSDB synchronization process between the many possible adjacent routers. The DIS performs two main tasks: responsible for creating and updating a special LSP (the Pseudonode LSP) describing the broadcast LAN topology, and performing LSP flooding (including its own LSPs), as well as, sending CSNPs and responding to PSNPs sent by other routers:

- **Generating and Sending Pseudonode LSPs**: The DIS sends out Pseudonode LSPs on behalf of the Pseudonode, reporting point-to-point links to all of the routers on the broadcast network segment which reduces the potential $N(N-1)$ links to N links. The DIS reports all the IS-IS neighbor routers (including itself) in the Pseudonode LSPs. In addition, only the Pseudonode LSP includes the list of ESs on the broadcast network segment, thereby eliminating the potential duplication of link-state information.
- **Exchanging CSNPs and Responding to PSNPs**: The DIS is also responsible for sending periodic CSNPs (that provide a complete summary of the current contents of its LSDB) over the broadcast LAN to the other member routers. All IS-IS routers on the broadcast LAN establish adjacencies with the DIS and with all other routers on the broadcast LAN segment, and send their Non-Pseudonode LSPs, essentially, to the Pseudonode, although these LSPs are flooded on the network segment. All the IS-IS neighbor routers communicate with the DIS (which creates the Pseudonode) via their Non-Pseudonode LSPs.

Recall that when a new IS-IS router joins the broadcast network segment, it will follow the three-way handshake procedure to establish two-way communication and adjacencies with all the existing routers on the network segment. The new router may even assume the role of DIS if its LAN IS-IS Hello PDU Priority field value is higher than that of the current DIS (see "Election of the DIS" section below).

A Pseudonode LSP, when created by the DIS, is uniquely identified by the following parameters:

- 6-byte System ID of the DIS that generated the LSP
- 1-byte Non-zero Pseudonode ID; equivalent to a "Local Circuit ID" for the broadcast network segment
- 1-byte LSP Number (in the range 0–255)
- 32-bit Sequence Number

The DIS uses the non-zero Pseudonode ID to differentiate a Pseudonode LSP from a Non-Pseudonode LSP. It is chosen by the DIS to be unique among any other broadcast network segments for which it is also the DIS at the specified IS-IS routing level. Each of other non-DIS routers on the broadcast network segment performs the following activities:

- Floods Non-Pseudonode LSPs that are absent from or are newer than the LSPs that are listed in the CSNPs sent by the DIS.
- Send a PSNP to request LSPs that are described in the CSNPs sent by the DIS but are missing from the local LSDB, or request up-to-date LSPs described in the CSNPs to replace older LSPs in its local LSDB.

These activities allow all routers on the network segment to efficiently and reliably synchronized their LSDBs. For Non-Pseudonode LSPs, the System ID carries the System ID of the originating router, but the Pseudonode ID is always set to 0.

When an IS-IS router on the broadcast segment receives a new LSP, rather than sending this new LSP only to the DIS, it will flood the LSP to all routers on the entire network segment. *On the specific issue of flooding, the only thing the DIS does is to send out periodic CSNPs describing its LSDB and responding to PSNPs.* The DIS multicasts CSNPs on the broadcast network segment in place of sending explicit acknowledgments for each received LSP. The CSNP the DIS sends on the broadcast segment serves as an implicit acknowledgment of any flooded LSP. If an IS-IS router fails to receive a particular LSP that was flooded by another router on the network segment, it will notice the missing LSP in its LSDB (by examining the CSNPs), and will request it from the DIS using a PSNP.

Note that the DIS will also flood all new LSPs that it receives from outside to all routers on the broadcast network segment. These new LSPs are sent as Non-Pseudonode LSPs. The DIS will also originate and flood its own Non-Pseudonode LSPs when necessary. This is done in addition to the Pseudonode LSPs, CSNPs, and requested LSPs (via PSNPs) it sends.

The DIS is responsible for creating and flooding a new Pseudonode LSP for each IS-IS routing level (Level 1 or Level 2) in which it is participating, and for each broadcast LAN to which it is connected. A DIS may be responsible for all broadcast LANs it is connected to, or a subset of the connected LANs, depending on the IS-IS Priority field value in the LAN IS-IS Hello PDU or the SNPA (or MAC address). The DIS also generates and floods a new Pseudonode LSP on the broadcast medium when a new neighbor adjacency is established, terminated, or the LSP refresh interval timer (the *maximumLSPGenerationInterval*) expires. The DIS provides a mechanism for reducing the amount of flooding on the broadcast network.

After the DIS generates a Pseudonode for the broadcast network, it sends IS-IS Hello PDUs and CSNPs for each routing level it is participating in (Level 1 and 2) at specified intervals (specified by the IS-IS *dSISHelloTimer* parameter). The IS-IS Hello PDUs indicate to the other routers on the broadcast network that it is the DIS for that routing level. The CSNPs describe a summary of all the LSPs in the DIS's LSDB and include the Remaining Lifetime, LSP ID, LSP Sequence Number, and Checksum. The DIS floods the LSPs to a well-known multicast address (MAC address 01-80-C2-00-00-14 for Level 1 messages, and 01-80-C2-00-00-15 for Level 2 messages). The exchange of CSNPs and PSNPs allow receiving routers to correct for any missing PDUs in their LSDBs. A neighbor router can send a request to the DIS for a missing LSP using a PSNP or, in turn, send a new LSP to the DIS (and the rest of the routers).

2.10.3 ELECTION OF THE DIS

In a broadcast network such as Ethernet, one of the IS-IS routers elects itself the DIS, based on the 7-bit IS-IS Priority field value carried in LAN IS-IS Hello PDU (see Figure 2.13). The priority value is a number that ranges from 0 to 127. If all IS-IS routers on the network have the same priority, the router with the highest (SNPA is elected. IS-IS has no way of making a router ineligible from being the DIS; IS-IS has no equivalent to the priority 0 option in OSPF; 0 is also a valid priority in IS-IS. The priority values in IS-IS does not exclude a router from participating in DIS election.

Every IS-IS router is assigned both a Level 1 priority and a Level 2 priority which are advertised in Level 1 and Level 2 LAN IS-IS Hello PDUs, respectively (Figure 2.13). Because Level 1 or Level 2 routing may have different priority values, there can be two different routers in a broadcast network segment that can be designated as DIS. In this case, one DIS be responsible for all the Level 1 routers, and the other DIS supporting all the Level 2 routers on that segment. No DIS is elected on point-to-point links, and only an adjacency must be formed with the neighbor router.

2.10.3.1 IS-IS DIS Election

The DIS election process is based on selecting the router with the highest configured priority, and if there is a tie, the router with the highest MAC address is selected. Unlike in OSPF, which uses a backup BDR (BDR), no backup DIS exists in IS-IS. This means if a DIS fails, a new DIS election takes place, and that router takes over immediately with little or no impact on the topology of the network.

2.10.3.1.1 No Requirement for Backup DIS

In IS-IS, a backup DIS is redundant because each IS-IS router (including the DIS) on the network segment forms an adjacency with every other router (as discussed in the "Three-way Handshake for Forming IS-IS LAN Adjacencies" section), which allows all routers to flood LSPs and synchronized their LSDBs with each other (via the DIS). Also, the shorter Hello interval used by the DIS allows for failures to be detected quickly and for faster replacement of the DIS. The DIS transmits IS-IS Hello PDUs three times faster (*dSISHelloTimer*) than the interval for other routers (*iSISHelloTimer*) on the broadcast network segment. The default Cisco IOS Hello PDU interval for the DIS, *dSISHelloTimer,* is 3.3 seconds while that for the other routers (*iSISHelloTimer*) is 10 seconds. The faster setting of *dSISHelloTimer* allows for quick detection of DIS failure and its immediate replacement.

2.10.3.1.2 Determinism in DIS Election

In IS-IS, DIS election is deterministic from the set of operational routers attached to the network segment. This makes it possible to predict from the set of active routers, the router that will be elected as DIS. Unlike OSPF, the election of the DIS is also preemptive. If the DIS fails, there will not be much disruption of routing information exchange. If one IS-IS router on the segment loses an LSP when a new DIS is elected, the CSNPs the new DIS sends will allow the loss to be detected. For this reason, the DIS in IS-IS can be preemptively replaced, while the OSPF DR cannot be.

2.10.3.1.3 DIS Preemption

Periodic LSDB synchronization on the broadcast network segment allows the existing DIS to be preempted without significant disruption of IS-IS operation on the network. This implies it is difficult to guarantee that an elected DIS will remain the DIS if a new router with a higher priority powers up on the broadcast network segment. Any eligible router that newly connects to the network segment immediately takes over the DIS role, and assumes responsibility for the Pseudonode functionality. The simplest mechanism for making a particular router eligible to be the DIS, is by configuring the router with the highest priority value relative to the priorities of the other router on the network segment.

Thus, if a new IS-IS router powers up on the broadcast network with a higher priority value, it preempts the existing DIS and becomes the DIS. The new DIS then purges the old Pseudonode LSPs in the network, and generates and floods a new set of Pseudonode LSPs. When a router declares itself as DIS and is in possession of any LSP of the same routing level issued by the previous DIS for that network segment (if any), it must initiate a network-wide purge of that (or those) LSPs as describe above.

2.10.3.1.4   DIS Resignation

An IS-IS router may resign as DIS on a broadcast network segment either because it (or its SNPA on the network segment) is being shut down, or because some other IS-IS router having a higher priority has taken over that function. When a router resigns as DIS, it must initiate a network-wide purge of its Pseudonode LSPs by setting their Remaining Lifetime to zero in order to purge all of its existing LSPs from the network. A LAN Level 1 DIS will purge Level 1 LSPs, and a LAN Level 2 DIS will purge Level 2 LSPs.

A router which has resigned as both Level 1 and Level 2 DIS will purge both sets of LSPs. This means any newly elected DIS or a resigning DIS is also responsible for purging the old Pseudonode LSPs from the broadcast network segment. Because of the DIS's critical role, detection of DIS resignation or failure is expedited using a shorter Hello PDU interval (*dSISHelloTimer*), with default value of 3.3 seconds, rather than the 10 seconds used for non-DIS routers.

2.10.3.2   OSPF Designated Router (DR) and Backup DR (BDR) Election

OSPF has a different behavior for multiaccess broadcast networks. In OSPF, once the DR and a BDR are elected, the other routers on the broadcast network segment establish adjacencies only with the DR and the BDR (see Chapter 1). The BDR is also an elected entity, and when the DR fails, the BDR is then promoted to become the DR. Also, in OSPF, the election DR is sticky. After a DR has been elected, no other router can take over that role unless the existing DR goes down. When a router powers up on the broadcast network segment, it will accept the existing DR regardless of its own priority.

The OSPF DR election process makes it harder to predict which router can become the DR, but has the advantage of ensuring that the DR changes less frequently. As a DR maintains adjacencies only with the BDR and the other routers on the segment, and is required to keep track of which routers have acknowledged LSAs that have been sent, having a preemptive DR process would require a lot of time and protocol messaging for another router to take over in a preemptive manner. In IS-IS, all routers including the DIS establish adjacencies with every other router on the network segment. However, in OSPF, the other routers and the BDR are only required to establish adjacencies with the DR. This makes a sticky DR in OSPF more sensible and a preemptive DR process more disruptive.

The presence of BDR in OSPF allows for a seamless takeover of the DR role in case the DR fails. By allowing all OSPF routers on the network segment to establish adjacencies and synchronize LSDBs only with the DR and BDR, the BDR is always fully synchronized with the DR. Forming adjacencies with only the DR/BDR allows the complexity of routing information exchange to be reduced, which also minimizes LSA flooding on the network segment. Once the DR and BDR have received (and acknowledged) a new LSA, the DR will reflood the new LSA to all of the other routers on the broadcast segment using a well-known multicast address (see Chapter 1).

2.11 HANDLING IP ROUTING WITH INTEGRATED IS-IS

Integrated IS-IS **[RFC1195]** is a single routing protocol (an IGP) based on the IS-IS routing protocol specified in **[ISO10589:2002]** with IP-specific routing functions added; IP-specific fields are added to existing IS-IS PDUs. By supporting both OSI and IP routing functions, Integrated IS-IS supports routing to OSI End Systems, IP hosts, and dual (OSI and IP) End Systems. Integrated IS-IS supports pure-OSI environments, pure-IP environments, and dual environments. In addition, Integrated IS-IS allows the interconnection of dual (OSI and IP) routing domains with other dual domains, with OSI-only domains, and with IP-only domains.

2.11.1 ROUTING CHARACTERISTICS OF INTEGRATED IS-IS

Augmented with IP-specific functions, Integrated OSI IS-IS provides explicit support for IP subnetting, VLSMs, type-of-service (TOS) based routing, and exchange of external routing information. Integrated OSI IS-IS forwards both OSI and IP packets "as is", that is, IS-IS routers transmit both packet types directly over the underlying Data Link Layer services without the need for mutual encapsulation; no need to encapsulate an IP packet in OSI or vice versa. Similar to OSPF, Integrated IS-IS is based on the SPF (Dijkstra) algorithm.

2.11.1.1 Router Types

Integrated IS-IS allows for mixing of OSI-only, IP-only, and dual (OSI and IP) routers **[RFC1195]**:

- **An OSI-only router**: This a router that uses IS-IS as the routing protocol for OSI traffic as specified in **[ISO10589:2002]**.
- **An IP-only IS-IS router (or "IP-only" router)**: This is a router that uses IS-IS for routing IP traffic, does not support other OSI protocols, and is not able to forward OSI CLNP packets.
- **A dual IS-IS router (or "dual" router)**: This is a router that uses IS-IS as a single integrated protocol for routing both OSI and IP traffic.

IP-only and dual routers as defined in **[RFC1195]**, are still required to conform to the standard requirements of IP routers as defined in **[RFC1812]**, just like routers in a pure IP environment. Integrated IS-IS does not change how IP packets are received, processed and forwarded (relayed). Similar to RIP, EIGRP, and OSPF, Integrated

IS-IS is an IGP that provides routing within a TCP/IP routing domain (i.e., an autonomous system).

Similarly, Integrated IS-IS does not change how OSI packets are handled. It does not change the contents and the handling of ISO 8473 PDUs and Error Reports, nor does it change the way ISO 9542 ESH PDUs and redirect messages are processed. ISO 9542 ISH PDUs transmitted on multiaccess broadcast network segments are similarly not changed. Integrated IS-IS does not change ISO 9542 ISH PDUs transmitted on point-to-point links except adding some IP routing related information. Similarly, it does not change other OSI PDUs (specifically PDUs used in the IS-IS intradomain routing protocol) except for the addition of IP routing related information.

2.11.1.2   Routing Domain Types

Integrated IS-IS specifies the following three types of routing domains [RFC1812]:

- **Pure OSI Routing Domain**: All routers in a pure OSI routing domain must be OSI-capable, and OSI-only routers may be freely mixed with dual routers. However, the OSI-only routers will ignore some of the fields specifically related to IP operation included by dual routers. The OSI-capable routers (both OSI-only and dual) will route only OSI traffic in a pure OSI domain, and may discard any IP traffic.
- **Pure IP Routing Domain**: All routers in a pure IP routing domain must be IP-capable, and the network operator may freely mix IP-only routers with dual routers. However, the IP-only routers will ignore some of the fields specifically related to OSI operation included by the dual routers. The IP-capable routers (both IP-only and dual) will route only IP traffic in a pure IP domain and may discard any OSI traffic (except IS-IS PDUs necessary for the operation of the routing protocol).
- **Dual Routing Domain**: In a dual routing domain, the network operator may mix OSI-only, IP-only, and dual routers on a per-area basis. Each area in a dual domain, in turn, may be defined to be pure OSI, pure IP, or dual. *If both OSI and IP traffic are to be routed between areas in a dual domain, then all the Level 2 routers must be dual.*

2.11.1.3   Area Types in a Dual Routing Domain

Integrated IS-IS specifies the following three types of areas in a *dual routing domain* [RFC1812]:

- **Pure OSI Area within a Dual Domain**: In this area type, OSI-only and dual routers may be freely mixed, but the Level 1 routers will route only OSI traffic within the pure OSI area.
- **Pure IP Area within a Dual Domain**: In this area type, the network operator may freely mix IP-only and dual routers, but the Level 1 routers will route only IP traffic within the pure IP area.
- **Dual Area within a Dual Routing Domain**: In this area type, only dual routers may be used, and both OSI and IP traffic can be routed within the dual area.

Note that in a dual environment, a single routing protocol (Integrated IS-IS) is used for routing both IP and OSI traffic and not separate protocols for independently routing OSI and IP traffic.

It is possible due to network configuration error for a dual router to forward an OSI packet to an IP-only router, or forward an IP packet to an OSI-only router. When this happens, the router receiving such a packet must discard it. The router may transmit an error report, in accordance with IP forwarding rules for IP packets **[RFC1812]**, or ISO 8473 specification for ISO packets. The router would specify the reason for discard in the error report as "destination host unreachable" (for IP packets), or "destination unreachable" (for OSI packets).

Similarly, due to configuration errors, an IP-only router may forward an IP packet to an OSI-only router. Again, the receiving ISO-only router must discard the packet, as described above. Note that this forwarding situation may only occur if IP-only and OSI-only routers are mistaken configured to be in the same area, which is a configuration error.

2.11.2 AREA AND DOMAIN REQUIREMENTS AND RESTRICTIONS

Integrated IS-IS is a single routing protocol that can be used to route both OSI and IP traffic, and still adhere to the same two-level routing hierarchy defined in **[ISO10589:2002]**. Reference **[RFC1195]** specifies the following requirements and restrictions when using Integrated IS-IS:

- **Area Definitions**: Each area in the IS-IS routing must be specified as either OSI-only (only OSI traffic can be routed in that particular area), IP-only (only IP traffic can be routed in that area), or dual (both OSI and IP traffic can be routed in that area).
- **Overlapping of Areas**: OSI and IP areas must not be overlapped. For example, if one area is defined as OSI-only, and another area as IP-only, no router is allowed to belong to both areas. Similarly, both areas must attach to a single backbone in the routing domain; the routing domain cannot have independent OSI and IP backbones.
- **Intra-Area Routing**: Within an IP-only or dual area, routers maintain routing information about specific IP destinations similar to the intra-area routing behaviors specified in **[ISO10589:2002]**. For example, IP-capable Level 1 routers maintain intra-area routing information, and will route directly to IP destinations within the area. IP-capable Level 1 routers will not maintain routing information about destinations that are outside the area. Similar to normal IS-IS routing in **[ISO10589:2002]**, IP-capable Level 1 routers will send traffic to destinations outside of the area to the nearest Level 1-2 router.
- **Handling IP Host Identifiers**: IP-capable Level 1 routers routes IP traffic to IP subnets (using the IP address prefix or IP address plus subnet mask), rather than to specific IP End Systems. IP routers do not need to maintain or distribute lists of IP host identifiers since routes to IP hosts can be announced using an IP address and a subnet mask consisting of all ones.

- **Relationship between IS-IS Areas and IP Addresses**: No specific relationship between the IS-IS area structure and IP subnet addresses is required. The network operator can assign IP addresses completely independently of the IS-IS area structure and OSI addresses. However, to achieve greater efficiency and scalability of the routing algorithm, it is better to provide some correspondence between the assignment of IP addresses and the IS-IS area structure.

2.11.3 LEVEL 1 AND LEVEL 2 IP ROUTING

Level 1 routers within an area exchange LSPs that identify the IP addresses that are reachable by each router. Specifically, each router will include zero or more [*IP address, subnet mask, metric*] combinations in each LSP. The network operator will configure each Level 1 router with the [*IP address, subnet mask, metric*] combinations which are reachable on each of its interfaces. A Level 1 router routes packets as follows:

- If the router determines that a specified destination address matches an [*IP address, subnet mask, metric*] reachable within its local area, it will route the packet via Level 1 routing.
- If the router determines that a specified destination address does not match any [*IP address, subnet mask, metric*] combination listed as reachable within its local area, it will route the packet toward the nearest Level 1-2 router.

An area (or domain) may use different IP address masks (VLSMs) for different IP subnets in the area (or domain). IP routers perform IP Forwarding Table lookups using longest matching prefix (LMP) so that, if a specified destination IP address matches more than one [*IP address, subnet mask*] pair, it will route toward the more specific IP address (i.e., the address with more "1" bits in the subnet mask or the longest prefix).

Level 1-2 routers include a complete list of [*IP address, subnet mask, metric*] specifying all IP addresses reachable in their area, in their Level 2 LSPs. The Level 1-2 routers may obtain this information from a combination of the Level 1 LSPs (received from the Level 1 routers in the same area), and/or via manual configuration. Level 2 routers may also advertise addresses which can be reached via routers in other routing domains (or autonomous systems); these are addresses corresponding to external reachability information.

Routers may announce default routes using a subnet mask containing all zeroes. However, great care must be exercised when using default routes, since they can result in "black holes". Reference **[RFC1195]** does not permit the use of default routes at Level 1. However, it permits default routes only at Level 2 as external routes (i.e., routes that can be included in the "IP External Reachability Information" field, as explained below).

2.11.4 INTEGRATED IS-IS TLV EXTENSIONS

The original IS-IS protocol specified in **[ISO10589:2002]** is very flexible and allows enhancements to be added to the protocol through the introduction of new TLV options. A key strength of IS-IS is that it allows new features and capabilities to be added through the introduction of new TLVs rather than the definition of new PDU types. New TLVs were introduced for Integrated IS-IS specified in **[RFC1195]**, and in other standards documents for recent modifications to IS-IS to support Multiprotocol Label Switching (MPLS) Traffic Engineering.

Reference **[RFC1195]** defines additional IP-specific information that can be added to IS-IS routing PDUs. The additional TLVs are for use by the ISO IS-IS intra-domain routing protocol in pure IP and dual environments. These additional TLVs can be exchanged between IS-IS routers during the normal routing information exchange defined in the OSI IS-IS routing specification **[ISO10589:2002]**. We discuss these new TLVs in this section.

In general, an IS-IS PDU may contain multiple TLVs, some of which may carry OSI-specific information (as specified in **[ISO10589:2002]**), and some of which may contain IP-specific information (as specified **[RFC1195]**). IS-IS requires that any router that receives any TLVs in a PDU that it does not recognize should ignored it and passed it on unchanged. This requirement also applies to all OSI-only, IP-only, and dual routers implementing IS-IS. This allows IP-specific information to be exchanged in a routing domain which OSI-only routers can ignore, and also allows OSI-specific information to be exchanged which IP-only routers can ignore.

2.11.4.1 IP-Specific TLVs in LAN and Point-to-Point IS-IS Hello PDUs

Reference **[RFC1195]** defines the following IP-specific TLVs for Level 1 and Level 2 Hello PDUs:

- **Protocols Supported TLV (Code 129)**: This TLV specifies the set of Network Layer Protocol Identifiers (NLPIDs) of the Network Layer protocols that the originating router is capable of relaying. The TLV has a 1-byte Type field specifying the TLV Code 129, a 1-byte Length field specifying the total length of the Value field, and a variable-length Value field containing an array of the NLPIDs supported by the originating router, [*NLPID_1*,..., *NLPID_Length*], where each element is a 1-byte NLPID.

  This TLV allows all IP-capable (both IP-only and dual) routers to know the Network Layer protocols supported by other routers in their local area. A router includes the TLV 129 specifying information about the Network Layer protocols it supports when sending IS-IS Hello PDUs and LSPs. An NLPID is a 1-byte value assigned by ISO to identify Network Level protocols, for example, IP and ISO 8473. TLV 129 must be included in all IS-IS Hello PDUs sent by IP-capable routers. If TLV 129 is not included in an IS-IS Hello PDU, it may be assumed that the Hello PDU was sent by an OSI-only router.
- **IP Interface Address TLV (Code 132)**: This TLV specifies one or more IP addresses of the router interface corresponding to the SNPA (or MAC address) over which the TLV's PDU is to be transmitted. The TLV has a 1-byte Type field specifying the TLV Code 132, a 1-byte Length field specifying the total length of the Value field in bytes, and a variable-length Value field containing an array of the IP addresses supported by the originating router's transmitting

interface, [*IP_Address_1,..., IP_Address_Length*], where each element is a 4-byte IP address. Up to a maximum of 63 IP addresses can be supported on each interface.

TLV 132 is included in all IS-IS Hello PDUs sent by IP-only and dual routers and can only appear once in each PDU. If a router transmits an IS-IS Hello over an interface which is not assigned an IP address, then the router may omit this TLV, or may include it but with the address entries set to all zeros **[RFC1195]**.

This TLV allows IP-capable routers (both IP-only and dual) to learn the IP address of the adjacent interface of neighbor routers in their local area. Routers in the area require the IP addresses so that they can send ICMP redirect messages. An IP-capable router must include the IP address of the appropriate interface of the correct next-hop router when sending an ICMP redirect message to a host. A router must also include the IP interface address(es) of the interface over which the IS-IS Hello PDUs are transmitted. IS-IS allows each physical interface to be assigned multiple IP addresses.

- **Authentication Information TLV (Code 133)**: This TLV contains information that the receiving router can use to authenticate the originator and/or contents of the received PDU. The TLV has a 1-byte Type field specifying the TLV Code 133, a 1-byte Length field specifying the total length of the Value field, and a variable-length Value field containing the authentication information. Reference **[RFC1195]** proposes only Plaintext Authentication where the Authentication Value field carries passwords that are transmitted in the clear without encryption. The use of TLV 133 is optional and routers are not required to be able to interpret the information carried in this TLV.

2.11.4.2   IP-Specific TLVs in ISO 9542 IS Hello PDUs

Routers on a point-to-point link exchange ISH PDUs **[ISO9542:1988]** to initialize the link, and to allow each router to check if the entity at the other end of the link is a router, before IS-IS Hello PDUs are exchanged. All routers implementing IS-IS (whether OSI-only, IP-only, or dual), and have any interfaces on a point-to-point link, must therefore be able to send ISH PDUs on their point-to-point links.

The IP-Specific TLVs carried in ISO 9542 ISH PDUs as specified in **[ISO9542:1988]**, are the Protocols Supported TLV (Code 129), and the Authentication Information TLV (Code 133). TLV 129 must be included in ISH PDUs transmitted by IP-capable routers over point-to-point links to other IS-IS routers.

2.11.4.3   IP-Specific TLVs in LSPs

This section describes the IP-specific TLVs defined for Level 1 and Level 2 LSPs.

2.11.4.3.1   IP-Specific TLVs in Level 1 LSPs

In addition to the Protocols Supported TLV (Code 129), IP Interface Address TLV (Code 132), and Authentication Information TLV (Code 133), Level 1 LSPs support another IP-specific TLV, the IP Internal Reachability Information TLV (Code 128). *TLV 129 can only appear once in LSP Number 0.* TLV 129 must be included in all LSPs with LSP Number 0 sent by IP-capable routers. If TLV 129 is not included in an LSP with LSP Number 0, it may be assumed that the LSP was sent by an OSI-only router.

- **IP Interface Address TLV (Code 132)**: This TLV contains the IP addresses of one or more interfaces corresponding to the SNPAs (MAC addresses) enabled on the originating router (i.e., one or more IP addresses of the originating router). This TLV can be carried in an LSP with any LSP Number, and can appear multiple times. TLV 132 must be included in all LSPs sent by IP-only and dual routers.

  In some cases, an IP-capable router may need to know the IP addresses of all other routers at its routing level (i.e., for Level 1 routers the IP addresses of all other routers in their local area; for Level 2 routers, the IP addresses of all other Level 2 routers in the routing domain). A router may need to send an IP packet to another router for encapsulation, or send network control packets for network management purposes. Routers include the TLV 132 carrying IP address information in the LSPs they send. Each IS-IS LSP carries the TLV 132 with one or more IP addresses of the router originating the LSP. Each IP-capable router is required to include at least one of its IP addresses, possibly, all of its IP addresses, in its LSPs. If a single IP-capable router operates at both Level 1 and a Level 2 (i.e., it is a Level 1-2 router), it is required to include the same IP address(es) in its Level 1 and Level 2 LSPs.

- **IP Internal Reachability Information TLV (Code 128)**: This TLV contains the IP addresses within the routing domain that are reachable directly via one or more interfaces on the originating router. This TLV can be present in an LSP with any LSP Number and can appear multiple times. However, this TLV MUST NOT be present in Pseudonode LSPs. The TLV has a 1-byte Type field, a 1-byte Length field, and a variable-length Value field. The Value field contains one or more entries with each entry consisting of the tuple <Default Metric, Delay Metric, Expense Metric, Error Metric, IP Address, Subnet Mask>. It should be noted that all of these four metrics are narrow metrics (i.e., 6-bit metrics). A 1-byte field is assigned to each of the narrow metrics (see format in [**RFC1195**]), and 4 bytes each to the IP Address and Subnet Mask fields.

TLV 128 identifies zero or more [*IP address*, *subnet mask*, *metrics*] entries that can be directly reached via the router which originates the LSP with TLV 128. Each entry of the TLV must contain a default metric, and may contain a delay metric, expense metric, and error metric. If no IP addresses can directly be reached via the announcing IP-capable router, then this TLV may be omitted, or may include the [*IP address*, *subnet mask*, *metrics*] entries with all zero values.

This TLV allows IP-capable routers to learn IP destination addresses so that they can correctly route packets to their destinations. Level 1 routers need to know the IP addresses that are reachable from each Level 1 router in the local area. Level 1 routers also need to know the Level 1-2 routers in the area to which traffic to IP addresses outside of their area can be forwarded. Level 1-2 routers need to know the IP addresses that are reachable internally in the local area (either directly, or via Level 1 routing). The Level 1-2 routers also need to learn from other Level 2 routers, the IP addresses that are reachable externally from those Level 2 routers. Routers include the IP addresses that are reachable through them in their LSPs.

When TLV 128 is included in Level 2 LSPs, it includes only IP address entries that can be reached through the router which originates the LSP, either through one of its interfaces, or indirectly through Level 1 routing.

2.11.4.3.2   IP-Specific TLVs in Level 2 LSPs

Level 2 LSPs support all the TLVs defined for Level 1 LSPs in addition to two other IP-specific TLVs, the IP External Reachability Information TLV (Code 130), and the Inter-Domain Routing Protocol Information TLV (Code 131). *TLV 129 can only appear once in Level 1 and Level 2 LSPs with LSP Number 0.*

- **IP External Reachability Information TLV (Code 130)**: This TLV contains the IP addresses that are outside the routing domain and are reachable via the interfaces of the originating Level 2 IP-capable router. TLV 130 includes only IP address entries that are reachable through the router which originates the LSP, via a direct link to an external router in another routing domain. These are routes external to the IS-IS routing domain, such as those imported from other routing sources or protocols via redistribution. TLV 130 can carried in a Level 2 LSP with any LSP Number and can appear multiple times. *However, TLV 130 must not be present in Pseudonode LSPs.*

  Internal reachability information (for routing within the routing domain) is advertised in TLV 128 in Level 2 LSPs. External reachability information (for addresses outside the routing domain) is announced separately in TLV 130 in Level 2 LSPs. The Value field of TLV 130 also contains one or more entries with each entry consisting of the tuple <Default Metric, Delay Metric, Expense Metric, Error Metric, IP Address, Subnet Mask> similar to what is defined for TLV 128.

  TLV 130 contains reachable IP addresses that include a default metric, and may include the other routing metrics. In general, the metrics of external routes may be of type "internal" (meaning they are directly comparable with internal metrics), or of type "external" (meaning they are not comparable with internal metrics). A route using internal metrics (i.e., either announced via TLV 128 (IP Internal Reachability Information), or announced via TLV 130 (IP External Reachability Information) with an internal metric) is always preferred to a route using external metrics (i.e., announced via TLV 130 with an external metric).

  TLV 130 identifies zero or more [*IP address, subnet mask, metrics*] entries that can be reached through the router which originates the Level 2 LSP with TLV 130. Each entry of the TLV must contain a default metric, and may include a delay metric, expense metric, and error metric. Each address entry may specify metrics of type "internal", or of type "external" as discussed above. If a Level 2 router does not know any external routes (via neighbor routers in other routing domains), then it may omit this TLV, or may include the TLV with zero entries.

- **Inter-Domain Routing Protocol Information TLV (Code 131)**: This TLV contains IDRP information that can be carried transparently through the Level 2 routing subdomain to be used by any IDRP that may be running in the boundary routers of the IS-IS routing domain. TLV 131 is transmitted for the convenience of any IDRP, and is not used by IS-IS. For example, the information carried in TLV 131 may be used by border routers to discover each other. Note that, as discussed above, internal and external reachability information is advertised separately in Level 2 LSPs. TLV 131 may be present in Level 2 LSPs sent by Level 2 IP-capable routers. This TLV can be carried in an LSP with any LSP Number and can appear multiple times.

  The Value field of this TLV contains one or more entries of the tuple <Inter-Domain Information Type, External Information>. The 1-byte interdomain Information Type field indicates the type of the external information carried in the Value field of the TLV. The variable-length External Information field contains the actual interdomain routing information, which is passed transparently by the IS-IS protocol.

2.11.4.4   IP-Specific TLVs in Sequence Number PDUs

According to **[RFC1195]**, CSNPs and PSNPs support only the Authentication Information TLV (Code 133) as IP-specific information. In some cases, LSPs and CSNP may be too large to fit into one PDU. Similar to ISO IS-IS, Integrated IS-IS allows these PDUs to be split into multiple smaller PDUs. Also, IP-specific fields in IS-IS PDUs may be split across multiple PDUs. CSNPs may be split into multiple smaller PDUs, with the range to which each PDU applies ("Start LSP ID" and "End LSP ID") explicitly reported in the PDU. PSNPs can easily be split into multiple PDUs if necessary, since they are inherently partial.

2.11.5   Addressing Routers in IS-IS PDUs

IS-IS PDUs carry the System IDs (contained in the PDU header) and the Area Addresses (carried in the Area Addresses TLV [Code 1]) of routers. The Area Addresses are used to determine area membership of routers. Therefore, all routers using the IS-IS protocol must have NSAP addresses assigned to them. For IP-only routers, NSAP addresses are used only in the operation of the IS-IS protocol, and not for any other purpose (such as the operation of ARP, ICMP, or other TCP/IP protocols).

For OSI-only and dual routers, the assignment of NSAP addresses uses mechanisms that are specified by standards bodies which also allow globally unique OSI NSAP addresses to be used. For IP-only routers, NSAP addresses may be obtained for use with the IS-IS protocol via a number of ways:

- NSAP addresses can be assigned in the normal manner to IP-only routers to be used (only) in the operation of the IS-IS protocol. This approach for address assignment is recommended even for pure IP routing environments, since it simplifies future migration from IP-only to dual operation.

- In some cases, routers may already have IP addresses, and it may be involving to have to go through the normal process of obtaining NSAP addresses. Instead, the network operator may decide to use an alternate mechanism (which may include algorithmically techniques) for generating valid NSAP addresses from existing IP addresses.

The network operator must assign the Area Addresses such that each area in the routing domain has a unique area value. The System ID must be assigned such that every router in the routing domain has a unique value. For example, since Ethernet MAC addresses are assigned to be globally unique, basing the System IDs on MAC addresses clearly assures uniqueness in the area.

2.11.6   ROUTING BASED ON IS-IS ROUTING METRIC TYPE

Integrated IS-IS uses the same routing metric feature of the IS-IS protocol in **[ISO10589:2002]** which allows routing based on the default metric (typically based on throughput), delay, expense, or residual error probability. A router can route any particular packet based on any one of these four (narrow) metrics. Both IS-IS and Integrated IS-IS do not support routing on the basis of any general combination of these routing metrics. Recall that support for delay, expense, or residual error probability as a routing metric is optional. Note that when an IS-IS router supports routing based on any of the optional (in addition to the default metric which is mandatory), it must also support a LSDB for that particular optional routing metric. If all four narrow metrics are supported, then the router must support four LSDBs (one for each metric).

   If a particular packet calls for a specific routing metric, and the router performing the forwarding determines that the correct best path to the destination consists of routers that all support that particular routing metric, then the router will route the packet via that optimal path. However, if the router determines that there is no path to the destination with routers supporting that particular routing metric, then the router will route the packet using the default metric instead. This allows for routing to be based on a particular routing metric in environments where it is needed and possible, while still providing routing in the case where an unsupported routing metric is requested.

2.11.7   ROUTE PREFERENCE ORDER AND THE SPF ALGORITHM COMPUTATION

In Integrated IS-IS, the combination of the [*IP address*, *subnet mask*] is referred to as the "IP reachability entry". Using the SPF algorithm, a router calculates routes to each distinct IP reachability entry. To use the SPF algorithm, each IP reachability entry is treated as if it is an OSI ES. However, each IP reachability entry is treated as distinct from any OSI ESs which may also be reachable in the same area or routing domain. The actual calculation of routes to IP reachability entries is therefore similar to the calculation of routes to OSI ESs.

   The use of the SPF algorithm does not take into consideration whether a router is OSI-only, IP-only, or dual. The topological restrictions of pure OSI, pure IP, or dual is only to ensure that OSI packets will only be sent via OSI-capable routers, and IP packets only via IP-capable routers.

With Integrated IS-IS, routers prefer routes within the local area (i.e., via Level 1 routing) whenever possible. If Level 2 routes must be used, then routers prefer routes within the local routing domain (specifically, routes using internal routing metrics) to routes that are outside of the local routing domain (using external routing metrics).

As with standard IP forwarding, Integrated IS-IS uses longest prefix matching (LPM) when looking up the destination IP address of a packet in the IP Forwarding Table. An IP packet whose destination is not directly attached to the forwarding router, is forwarded first to the appropriate IP subnet, and then forwarded (by the next-hop router) on that subnet to another next-hop router or to the destination host. Specifically, the IP Forwarding Table does not contain explicit routes to the individual "IP interface addresses" each router will list in the IP Interface Address TLV (Code 132) and also send in its LSPs. However, Integrated IS-IS may include host routes (i.e., IP address with a subnet mask of all ones) in the IP reachability entries, and these will be handled in the same way as other IP reachability entries.

In order to ensure that different Integrated IS-IS router implementations can interoperate correctly, reference **[RFC1195]** specifies the order of preference of possible routes. With IS-IS, when a router advertises a route to a given destination, or advertises a link to a neighbor router, it may assign routing metric values based on some or all of the specified routing metric types to that destination or link. However, the router must always assign a default metric value to a link. This ensures that if a route using any of the optional routing metrics is available to other routers, then a route using the default metric will also be available (except when the route using the default metric has a total cost greater than *MaxPathMetric* within the area, or within the Level 2 backbone). When determining the route to a particular destination for any specified routing metric, a router considers only routes using either the requested routing metric, or the default routing metric.

2.11.7.1 Level 1 Routing Order of Route Preference

If an Integrated IS-IS router determines that a given destination is reachable within an area via a route using either the requested specific routing metric or the default metric, the router will always use a path within that area (via Level 1 routing), irrespective of the existence of an alternate path outside of the area (via Level 2 routing). The router will always prefer an intra-area route over an inter-area route (going over the IS-IS backbone) to a destination. In this case, the router selects routes within the area as follows:

1. If the specified destination IP address matches more than one [*IP address, subnet mask*] pair among the routes in the area, then the route with the longest prefix is preferred (i.e., the more specific address with more "1" bits in the subnet mask).
2. When multiple routes exist in the area with equal prefix lengths, routes on which the requested routing metric is supported (if any) are always preferred to routes on which the requested metric is not supported.

3. When multiple routes exist in the area using the same routing metric and have equal prefix lengths, the shortest routes are preferred. When determining the shortest path, if a route is available on which the specified routing metric is supported, then that route is used, otherwise the route using the default metric is used. If multiple routes of equal cost exist, the router may perform load sharing on these routes.

If a Level 1 only capable router (i.e., a router which is not performing Level 2 routing, or a Level 1-2 router which has not set the ATT ("Attached") bit in its LSPs), determines that a given destination is not reachable within its area, it will always use Level 1 routing to route traffic to a Level 1-2 router as follows:

1. If multiple routes exist in the area to attached Level 1-2 routers, routes on which the requested routing metric is supported (if any) are always preferred to routes on which the requested routing metric is not supported.
2. If multiple routes exist in the area with the same routing metric to attached Level 1-2 routers, the shortest routes are preferred. When determining the shortest path, if a route is available on which the specified routing metric is supported, then that route is used, otherwise the route using the default metric is used. If multiple routes of equal cost exist, the router may perform load sharing on these routes.

2.11.7.2 Level 2 Routing Order of Route Preference

For Level 1-2 routers, routes learned via Level 1 routing, using either the requested specific routing metric or the default metric, are always preferred to routes learned through Level 2 routing. For destinations that are not reachable via Level 1 routing, or for strictly Level 2 only routers, Level 2 routes are selected as follows:

1. The router always prefers routes using internal routing metrics only to routes using external metrics.
2. If the router determines that a route using internal metrics only is available, it performs the following:
   a. If the specified destination address matches more than one [*IP address, subnet mask*] pair, then the route with the longest prefix is preferred.
   b. If multiple routes with equal prefix length exist, the routes on which the requested routing metric is supported (if any) are always preferred to routes on which the requested routing metric is not supported.
   c. If multiple routes with the same routing metric and with equal prefix length exist, the shortest path is preferred. When determining the shortest path, if a route is available on which the specified routing metric is supported, then that route is used, otherwise the route with the default metric is used. If multiple routes of equal cost exist, the router may perform load sharing on these routes.
   NOTE: Internal routes (i.e., routes to destinations reported in the IP Internal Reachability Information TLV [Code 128]), and external routes using internal

routing metrics (i.e., routes to destinations reported in the IP External Reachability Information TLV (Code 130) plus indicating that they are using an routing metric type "internal") are treated identically from the perspective of the order of route preference, and the SPF algorithm calculation.

3. If the router determines that a route using internal routing metrics only is not available, but a route using external routing metrics is available, it will perform the following:

   a. If the specified destination address matches more than one [*IP address, subnet mask*] pair, then the route with the longest prefix is used.

   b. If multiple routes with equal prefix length exist, then routes on which the requested routing metric is supported (if any) are always preferred to routes on which the requested routing metric is not supported. NOTE: An external route is considered to support the requested routing metric, only if the internal route to the border router reporting the route, supports the requested routing metric, and the external route reported by the border router also supports the requested routing metric.

   c. If multiple routes with the same routing metric and with an equal prefix length exist, the shortest path is preferred. When determining the shortest path, the router performs the following:

      i. Routes having a smaller reported external routing metric are always preferred.

      ii. If multiple routes with an equal external routing metrics exist, then routes with a smaller internal routing metric are preferred. If multiple routes of equal cost exist, the router may perform load sharing on these routes.

In some cases when Level 2 routers announce manually configured summary addresses in their Level 2 LSPs, some IP addresses may exist which fall within the range of addresses summarized by the manually configured summary addresses, but are not addresses that are actually reachable via Level 1 routing in the area. Generally, the router handles packets to such addresses according to the following rules:

1. If the router determines that the specified destination is reachable via Level 1 routing, then it will deliver the packet via Level 1 routing according to the order of preference of routes specified above.

2. If the router determines that the specified destination is not reachable via Level 1 routing, but is reachable via Level 2 routing, and there are other Level 2 routers that provide better routes according to the rules specified above (e.g., a route with the best prefix match, or a route with the longest prefix match which also supports the requested routing metric), then the router will forward the packet via Level 2 routing according to the more desirable route.

3. If the router (i.e., Level 1-2 router) determines that the specified destination is not reachable via Level 1 routing, and the manually configured summary address that the router itself advertises represents the most desirable route, then the destination is unreachable and the packet must be discarded.

2.11.8 MULTIACCESS BROADCAST LANs, DESIGNATED INTERMEDIATE SYSTEM, AND PSEUDONODE

In an environment where OSI-only and dual routers, or IP-only and dual routers, are mixed on the same broadcast network segment (LAN) in a pure OSI area, or a pure IP area, respectively, any router on the LAN may be elected DIS. However, reference **[RFC1195]** highlights a fundamental difference in the way that OSI and TCP/IP work on LANs, and other broadcast network segments. In an OSI environment, ESs and routers use the ES-IS protocol **[ISO9542:1988]** to automatically establish connectivity, and also allows all ESs on the LAN to potentially route via any of the routers on the LAN.

In contract, in a TCP/IP environment, each broadcast LAN is explicitly assigned an IP subnet identifier (IP address and subnet mask). In some cases, a single physical broadcast LAN could be logically partitioned (into IP-based virtual LANs [VLANs]), and with each logical partition/VLAN assigned an IP subnet identifier; implying a physical LAN can have multiple subnet identifiers assigned to it. Each logical subnet is connected to the single physical router interface via a subinterface (or virtual interface). In this architecture, IP hosts (end systems) which have an IP address on one logical subnet can only send IP packets directly to a router that recognizes the logical subnet to which the host belongs.

Each router in this environment will be manually configured to recognize which logical subnets it can reach on each interface. In the case where there are multiple logical subnets (VLANs) on the same broadcast LAN, each router ensures that IP packets to a particular logical subnet are only sent to those end systems which are on that logical subnet. An important implication of this real-world architecture is that, it is not sufficient for the Pseudonode LSPs sent by the elected DIS to simply announce all subnets on the broadcast LAN (i.e., list all [*IP address, subnet mask*] pairs that are reachable on the LAN).

This makes it necessary for each router to report in its LSPs the specific (logical) IP subnets which are reachable on each of its interfaces, including interfaces attached to broadcast LANs. Recall that, as discussed above, the Pseudonode LSP created by the DIS does not specify the IP addresses that can be reached on the broadcast LAN (i.e., the Pseudonode LSP does not contain the IP Internal Reachability Information TLV [Code 128]) **[RFC1195]**. This means each router including the DIS must report in its LSPs, the subnets (including IP hosts) that are reachable on each interface. A host identifier is simply an IP address with a subnet mask composed of all ones.

2.11.9 MAINTAINING ROUTER ADJACENCIES IN INTEGRATED IS-IS

Two IS-IS routers that are connected and meet the IS-IS adjacency requirements can establish an adjacency between themselves independent of their IP interface addresses. Multiple logical subnets may exist on the same physical LAN to which some routers may be physically attached. Any two routers that share this physical LAN can still potentially establish an adjacency between themselves even if they do not have a logical IP subnet in common (as long as the conditions for IS-IS adjacency formation are met as specified in **[ISO10589:2002]**). This implies IS-IS routers that are IP-capable must be able to forward IP packets over existing IS-IS adjacencies to routers that share a common physical network segment, even if the IP address of the adjacent interface of the neighbor router is on a different logical IP subnet.

IS-IS requires that any two routers connected by a point-to-point link, exchange ISO 9542 ISH PDUs **[ISO9542:1988]**, as the first step in establishing the adjacency on the link. All IS-IS routers interconnected by point-to-point links are therefore required to transmit and receive ISO 9542 ISH PDUs.

The Protocols Supported TLV (Code 129) must be present in all IS-IS Hello PDUs sent by IP-only and dual routers. If this TLV is missing in a Hello PDU, then it is assumed that the Hello PDU was transmitted by an OSI-only router. Similarly, ISO 9542 ISH PDUs that are sent over a point-to-point link (where there could be another IS-IS router at the other end of the link), must also contains TLV 129. If this TLV is mistakenly sent in an ISO 9542 ISH PDU where the system at the other end of the link is an ordinary OSI-only ES, then in accordance to **[ISO9542:1988]**, the ES is required to ignore the TLV and interpret the ISH PDU accordingly. This requirement makes it safe and not harmful even if routers decide to always include this TLV in ISH PDUs they send over point-to-point links.

Dual routers in a routing domain running IS-IS must operate in a dual fashion on every link. Thus, the contents of TLV 129 must be identical on every link. That is, all the Hello PDUs and LSPs transmitted by any dual router running IS-IS, must contain the same Protocols Supported TLV (Code 129) values.

2.11.10 ROUTE SUMMARIZATION IN INTEGRATED IS-IS

Integrated IS-IS route summarization allows multiple groups of IPv4 addresses to be aggregated or summarized into a single IPv4 addresses for a given IS-IS routing level. It is also possible to summarize routes that have been redistributed from other routing protocols.

IS-IS route summarization helps to reduce the size of the IS-IS LSDBs and the IS-IS Routing Database. It also helps to reduce the propagation of route flapping from one IS-IS area into other areas. Route flapping occurs when a router alternately advertises a destination network as reachable on one route and then reachable on another route in quick succession (or a route is advertised as unavailable then available again).

Level 1-2 routers include a list of all [*IP address, subnet mask, metric*] combinations that are reachable in their local area in the Level 2 LSPs they send. A Level 1-2 router may determine this information from the Level 1 LSPs received from all routers in the area. A Level 1-2 router may simply duplicate all the [*IP address, subnet mask, metric*] entries received from all the Level 1 routers in its local area and include them in its Level 2 LSP. However, in order to allow the hierarchical routing scheme in the OSI routing domain to scale higher, it is highly desirable to summarize the reachable IP addresses received from the Level 1 routers and advertise that information.

Address summarization is accomplished via manual configuration where each Level 1-2 router is configured with one or more [*IP address, subnet mask, metric*] entries for advertisement in its Level 2 LSPs. The Level 1-2 router will compare the set of IP addresses that are reachable in Level 1 LSPs with the configured reachable summary addresses. The router does not include redundant information obtained from the Level 1 LSPs in the Level 2 LSPs. The configured summary addresses must contain at least one more-specific Level 1 IP address to be announced in the Level 2 LSPs. The Level 1-2 router will include the manually configured summary addresses in Level 2 LSPs only if they correspond to at least one IP address that is reachable in the area. Each configured summary address/subnet mask pair hierarchically supersedes the multiple more-specific address entries in Level 1 LSPs.

The routing metric values associated with the manually configured summary IP addresses announced in the Level 2 LSPs are also manually configured. The configured summary IP addresses will supersede the reachable IP address entries in Level 1 LSPs based only on the IP address and subnet mask. The routing metric values of the Level 1 IP addresses obtained from a Level 1 LSP are not considered when determining if a particular configured summary address supersedes an address.

If the Level 1-2 router obtains an IP address from a Level 1 LSP that is not superseded by any manually configured summary addresses, this more-specific address will be included directly ("as is") in the Level 2 LSPs. In this case, the routing metric value advertised for this more specific address in the Level 2 LSPs is calculated from the sum of the metric value advertised in the corresponding Level 1 LSP (for the address), plus the metric from the Level 1-2 router to the appropriate advertising Level 1 router.

Note that if this total metric value is greater than 63, assuming a narrow metric is used, then the value 63 must be used. Recall that the maximum narrow metric value that can be reported in Level 2 LSPs is 63. The Level 1-2 router will include delay, expense, and error metrics (which are optional metrics other than the default metric) in the Level 2 LSPs only if:

    i. the Level 1-2 router supports the specific narrow metric;
   ii. the path from the Level 1-2 router to the appropriate advertising Level 1 router consists of links which support the specific narrow metric; and
  iii. the Level 1 router which can reach the reported IP address directly, also supports the specific narrow metric for this route, as indicated in its Level 1 LSP.

It may happen that multiple Level 1 routers in the same area may announce the same [*IP address*, *subnet mask*] pair in the Level 1 LSPs they send. In this case, if the reported IP address is not superseded by a manually configured summary address, the Level 1-2 router will include only one such entry in the Level 2 LSPs. The total routing metric value(s) advertised in the Level 2 LSPs will correspond to the minimum of the total routing metric value(s) calculated for each of the Level 1 LSP entries.

A Level 1-2 router may have IP addresses that are directly reachable via its own interfaces. If some or all of these "directly reachable" addresses can be summarized, the Level 1-2 router will treat and include these IP reachable address information in Level 2 LSPs, similar to addresses received in Level 1 LSPs.

A manually configured summary IP address at a Level 1-2 router may hierarchically supersede multiple reachable IP address entries announced in Level 1 LSPs. However, it can happen that some IP addresses may be covered (summarized) by the manually configured summary address, but will not be reachable via Level 1 routing in the local area. If a Level 1-2 router receives an IP packet whose destination IP address matches a manually configured summary address that is included in the router's Level 2 LSP, but which is not reachable via Level 1 routing in the area, then the router will discard the packet. In this case, the router may return an error report specifying destination unreachable.

2.11.11 ROUTE REDISTRIBUTION IN INTEGRATED IS-IS

Route redistribution is the process of taking routes learned by one routing protocol and passing them to another routing protocol. Routes can be exchanged between IS-IS and other routing protocols like RIP, EIGRP, OSPF, and BGP. An IS-IS router may also support the redistribution of static routes into IS-IS, and the redistribution of routes between IS-IS routing levels. In the latter case, routes can be redistributed between Level 1 and Level 2, at an IS-IS Level 1-2 router.

In IS-IS, communications with routers outside of a routing domain (external routes) is done only by Level 2 routers. IS-IS ISO as defined in **[ISO10589:2002]**, allows external OSI routes to be reported using the Prefix Neighbors TLV option (Code 5) in Level 2 LSPs as "reachable address prefixes". Note that the Default Metric, Delay Metric, Expense Metric, and Error Metric fields in TLV 5 can be defined as internal or external by setting their I/E bits to 0 or 1, respectively.

Integrated IS-IS also allows IP addresses that are reachable via interdomain routing (i.e., external IP reachable addresses) to be advertised via the IP External Reachability Information TLV (Code 130) in Level 2 LSPs. In Integrated IS-IS, the Default Metric field in TLV 130 can be defined as internal or external by setting the I/E bit to 0 or 1, respectively. Level 2 routers handle external OSI and external IP routes independently. The routes that are announced in the TLV 130 entries include all routes leading to addresses outside the local routing domain. This includes routes learned from RIP, EIGRP, OSPF, or BGP.

External routes, when imported into Integrated IS-IS, may use "internal" or "external" routing metrics. Internal routing metrics are comparable with the routing metrics used within the local routing domain for internal routes. Thus, when a router is choosing between an internal route, and an external route that is assigned internal metrics, the router can directly compare the two metric values. In contrast, internal routing metrics cannot be directly compared with an external route having external metrics. Routers within the routing domain always prefer any route having internal routing metrics to any route having external routing metrics. When a router must use an external route having external routing metrics, it will prefer the lowest external routing metric value regardless of the internal cost to reach the appropriate exit point out of the routing domain.

The Inter-Domain Routing Protocol Information TLV (Code 131) provides a mechanism for border routers of a routing domain to discover each other, and to exchange external information in a form understood only by them. Border routers are routers in the same routing domain that have the ability to route traffic to destinations external to the local domain. Note that TLV 131 is not included in Pseudonode LSPs. To make TLV 131 general enough, an Interdomain Information Type field is included to indicate the type of IDRP information enclosed if there happens to be multiple types of external IDRP information exchanged between the border routers. The information contained in the External Information field of TLV 131 is carried in Level 2 LSPs, and will be stored by all Level 2 routers in the routing domain. However, this information will be used only by the Level 2 routers that are directly involved in external routing.

2.12 EXTENSIONS FOR DOMAIN-WIDE IP PREFIX DISTRIBUTION WITH INTEGRATED IS-IS

Reference **[RFC5302]** describes extensions to the Integrated IS-IS protocol as defined in **[RFC1195]** to support optimal routing within a two-level routing domain. Also, reference **[RFC5302]** extends the semantics presented in **[RFC1195]** to allow a routing domain running both Level 1 and Level 2 routers to distribute IP prefixes between Level 1 and Level 2, and vice versa. To accomplish this, **[RFC5302]** discusses certain restrictions that are required during IP prefix distribution to ensure that persistent routing loops are not form. The goal of the domain-wide IP prefix distribution proposed in **[RFC5302]** for a two-level routing domain is to increase the granularity of the routing information that can be exchanged within the domain beyond what is proposed in **[RFC1195]**.

As discussed in **[RFC1195]**, a Level 1-2 router can be manually configured with a set of IP prefixes that summarizes the IP prefixes that are reachable in its local Level 1 area. The Level 1-2 router can then inject these summary addresses into the Level 2 routing subdomain. Other than this capability and related issues as discussed above, **[RFC1195]** does not specify any further interactions between Level 1 and Level 2 for IPv4 prefixes.

As per **[RFC1195]**, Level 1-2 routers can advertise IP routes that were learned via Level 1 routing into Level 2. These advertised routes can be viewed as inter-area routes. Reference **[RFC1195]** further specifies that these Level 1-to-Level 2 inter-area routes must be advertised in the "IP Internal Reachability Information" TLV (TLV 128) in Level 2 LSPs. However, it should be noted that intra-area Level 2 routes are also advertised in Level 2 LSPs in TLV 128. This makes Level 1-to-Level 2 inter-area routes essentially indistinguishable from Level 2 intra-area routes.

Particularly, **[RFC1195]** does not define Level 2-to-Level 1 inter-area routes. Thus, **[RFC5302]** provides a simple extension that allows a Level 1-2 router to advertise routes learned via Level 2 routing in its Level 1 LSP. However, to prevent routing loops, Level 1-2 routers MUST NOT advertise Level 2-to-Level 1 inter-area routes that they have learned via Level 1 routing back into Level 2. Therefore, **[RFC5302]** provides a simple way to distinguish Level 2-to-Level 1 inter-area routes from Level 1 intra-area routes.

Reference **[RFC5305]** defines the "up/down bit" in the "Extended IP Reachability" TLV (TLV 135) to be discussed below for this purpose. TLVs 128 and 130 defined in **[RFC1195]** for carrying IP routes have a Metric field that consists of four narrow metrics. The first metric, the default metric which is mandatory, has bit 8 (the high-order bit) reserved. In **[RFC1195]**, routers must set this bit to 0 on transmitting a TLV 128 or 130, and ignore it on receipt. Recall that TLV 128 is used to carry IP prefixes that are directly connected to IS-IS routers, while the "IP External Reachability Information" TLV with TLV Code 130 (carried only in Level 2 LSPs) is used to carry IP routes learned from outside a particular IS-IS domain.

Reference **[RFC5302]** redefined this high-order bit (bit 8) in the default metric field in TLVs 128 and 130 as the up/down bit. Level 1-2 routers must set this bit to 1 for IP prefixes that are derived from Level 2 routing and are advertised in Level 1 LSPs within the local IS-IS area. The up/down bit must be set to 0 for all other IP prefixes in Level 1 or Level 2 LSPs. Level 1-2 routers must not advertise IP prefixes with the up/down bit set that were learned via Level 1 routing back into Level 2.

Other issues covered by reference **[RFC5302]** include the following:

- **Clarification of External Route Type and External Metric Type**: [RFC5302] does not change the preference rule for IP routes as specified in **[RFC1195]** but provides some clarification on the route type "external" and metric type "external". According to **[RFC1195]**, the internal metric type should be used when the metric of the IP prefix is comparable to the metrics used to rank/cost links within the IS-IS domain. The external metric type (which can only be used for external IP prefixes) should be used if the metric of the IP prefix cannot be directly compared to internal metrics. The route type can be derived from the type of TLV in which the IP prefix is advertised. Note that internal routes are advertised in TLV 128, and external routes in TLV 130. The metric type (internal or external) is derived from the I/E metric type bit in the Metric field (bit 7) (see discussion under "Intermediate System Neighbors TLV (Code 2)"). Recall that reference **[RFC1195]** states that, internal routes (i.e., routes to destinations reported in the TLV 128) and external routes using internal metrics (i.e., routes to destinations reported in TLV 130 and with a metric of type "internal") are treated identically for the purpose of the order of preference of routes, and the SPF calculation. However, it also notes that less preference must be given to IP routes advertised in TLV 130 with the external metric type than the same IP route advertised with the internal metric type, regardless of the value of the metrics.
- **Inclusion of External IP Prefixes in Level 1 LSPs**: As noted above, reference **[RFC1195]** does not define TLV 130 for Level 1 LSPs. **[RFC1195]** defines that IP routes learned via Level 1 routing must always be advertised in Level 2 LSPs in an a TLV 128. However, **[RFC5302]** loosens the restrictions in **[RFC1195]** to allow for the inclusion of TLV 130 in Level 1 LSPs, plus allow for the advertisement of routes learned via Level 2 routing into Level 1. The rules defined in **[RFC1195]** are extended in **[RFC5302]** to allow this to be possible as follows. When a Level 1-2 router advertises a Level 1 route into Level 2, where that Level 1 route was learned via an IP prefix advertised in a TLV 130, the Level 1-2 router should advertise that IP prefix in its Level 2 LSP in TLV 130. Level 1 routes learned via a TLV 128 should still be advertised in TLV 128. [RFC5302] states that these rules should also be applied when a Level 1-2 router advertises IP routes derived from Level 2 routing into Level 1. In this case, the up/down bit must also be set.

- **Overview of All Types of IP Prefixes in IS-IS LSPs**: Reference [RFC5302] describes extensively the types of IP prefix advertisement in IS-IS and how they are encoded (e.g., Level 1 intra-area routes, Level 1 external routes, Level 2 intra-area routes, Level 2 external routes, and Level 1-to-Level 2 inter-area routes).
- **Order of IP Route Preference in IS-IS**: IS-IS cannot depend on routing metrics alone for route selection. Some types of routes must always be preferred over others, regardless of the metrics that were computed in the SPF calculation. So, [RFC5302] defines the route preference rules as they apply to only TLVs 128 and 130.

Reference [RFC5302] also addresses inter-operability issues when these new functionalities are used with older implementations following [RFC1195].

2.13   NEWER IS-IS TLV EXTENSIONS

Other than the TLVs defined in [ISO10589:2002] and [RFC1195], a number of TLVs have been defined for IS-IS to be used for various purposes. This section describes these newer TLVs. The TLVs described here all use IS-IS wide metric, unlike the narrow metrics used in [ISO10589:2002] and [RFC1195].

2.13.1   IS-IS TRAFFIC ENGINEERING AND MPLS TLVs

Reference [RFC5305] defines TLV extensions to the IS-IS protocol for supporting network traffic engineering (TE). These new TLVs can be carried in LSPs to describe the characteristics of a particular link. The information contained in the TLVs may be used for traffic engineering computations.

Reference [RFC5305] describes a new way of encoding routing information in IS-IS by introducing a new object called a *Sub-TLV*. Sub-TLVs use the same concepts as regular TLVs and are similar to regular TLVs. However, whereas regular TLVs are contained within IS-IS PDUs, Sub-TLVs are carried inside regular TLVs. TLVs provide a way of adding extra information to IS-IS PDUs, while Sub-TLVs provide a mechanism for adding extra information to a particular TLV.

Each Sub-TLV consists of, a 1-byte Sub-TLV Type field, a 1-byte Sub-TLV Length field, and a Sub-TLV Value field of zero or more bytes. The Sub-TLV Type field specifies the type of items carried in the Sub-TLV Value field. The Sub-TLV Length field specifies the length of the Sub-TLV Value field in bytes. Each Sub-TLV can contain multiple items in its Sub-TLV Value field and the number of items can be calculated from the length of the complete Sub-TLV, if the length of each item is known. Routers that receive unknown Sub-TLVs can ignore and skip them. The following TLVs have been defined in [RFC5305].

2.13.1.1   The Extended IS Reachability TLV (Code 22)

The Intermediate System Neighbors TLV (Code 2) defined in [ISO10589:2002] contains information about the IS-IS neighbors of a router. For each neighbor, the TLV entry contains the default metric, delay metric, expense metric, error metric, and the 7-byte Neighbor ID of the adjacent neighbor. Out of this information, the default metric is the routing metric commonly used. The default metric (a narrow metric) in TLV 2 occupies a 1-byte space, with one bit used to indicate whether the default metric is "internal" or "external", and one other bit later defined as the up/down bit in **[RFC5302]**. The remaining 6 bits of the 1-byte space contains the actual metric, resulting in the narrow IS-IS metric range of 0–63. Reference **[RFC5305]** uses wide metrics to remove the limitation of the narrow metrics.

The remaining three metrics of TLV 2 (delay, expense, and error) are typically not used, and therefore consume unnecessary overhead in the TLV in actual IS-IS implementations. The Neighbor ID field of TLV 2 is, typically 6 bytes, plus one byte indicating the Pseudonode ID. This means TLV 2 consumes 11 bytes per neighbor, consisting of 4 bytes used for metrics and 7 bytes for Neighbor ID. When multiple adjacencies are described in TLV 2, then the 11 bytes are repeated within the TLV. However, since the TLV can only contain up to 255 bytes, a single TLV can describe up to 23 neighbors. Note that TLV 2 can be repeated within LSP fragments to describe more neighbors. Therefore, TLV 22 improves the encoding in TLV 2 by reusing the spaces that are commonly left unused **[RFC5305]**.

TLV 22 was defined to replace TLV 2 with the primary objective of providing support for wide metric values, and also to carry additional information (via Sub-TLVs) that can be used for IS-IS-dependent MPLS traffic engineering. Recall that TLV 2 announces the IS-IS neighbors, including Pseudonode neighbors, of the originating router. TLV 22 extends the capabilities of TLV 2 and consists of the following **[RFC5305]**:

- 1-byte (regular) IS-IS Type field set to the value 22
- 1-byte (regular) IS-IS Length field
- Variable-length (regular) IS-IS Value field consisting of TLV 22-specific information:
  - o 7-byte System ID and Pseudonode ID field
  - o 3-byte Wide (Default) Metric field
  - o 1-byte Sub-TLVs Total Length field
  - o 0–244 bytes for Sub-TLVs, where each Sub-TLV consists of the following sequence:
    - 1-byte Optional Sub-TLV Type field
    - 1-byte Optional Sub-TLV Length field
    - 0- to 242-byte Optional Sub-TLV Value field

If no Sub-TLVs are used in TLV 22, then only 11 bytes are required, and TLV 22 can contain up to 23 neighbors. If any particular link is advertised with the maximum link metric, $(=2^{24} - 1)$, the normal SPF computation must not consider this link. However, this link will be advertised for purposes other than building the normal Shortest-Path Tree, for example, the link will be available for traffic engineering, but not for hop-by-hop routing.

The IETF manages the Sub-TLV type space. The following optional Sub-TLVs have been defined in **[RFC5305]** (see specification for descriptions):

- Administrative Group (color, resource class): Sub-TLV 3
- IPv4 Interface Address: Sub-TLV 6
- IPv4 Neighbor Address: Sub-TLV 8
- Maximum Link Bandwidth Sub-TLV 9
- Maximum Reservable Link Bandwidth Sub-TLV 10
- Unreserved Bandwidth: Sub-TLV 11
- Traffic Engineering Default Metric Sub-TLV 18

TLV 22 can be present in a Level 1 or Level LSP with any LSP Number and can appear multiple times similar TLV 2.

2.13.1.2   The Extended IP Reachability TLV (Code 135)

The IP Internal Reachability Information TLV (Code 128) and IP External Reachability Information TLV (Code 130) defined in **[RFC1195]** carry IP prefixes with four metrics similar to the Intermediate System Neighbors TLV (Code 2) in **[ISO10589:2002]**. Out of the four metrics, only the default metric is commonly used. Furthermore, route leaking (or route redistribution) as defined in **[RFC1195]** allows a router to advertise prefixes upward (from Level 1 to Level 2) in the routing hierarchy. However, **[RFC1195]** does not define mechanisms in TLVs 128 and 130 for advertising prefixes downward in the routing hierarchy (i.e., from Level 2 to Level 1).

To address these two issues, TLV 135 which is defined in **[RFC5305]**, contains a 32-bit routing metric, and also adds a bit that can be set to indicate that an IP prefix has been redistributed "down" (i.e., from Level 2 to Level 1) in the routing hierarchy. TLV 135 was designed to replace TLV 128 with the primary goal to support wide metric values for IP prefixes, and to carry Sub-TLVs for distributing MPLS traffic engineering resource information. Recall that TLV 128 announces the IP prefixes within a routing domain that are reachable directly via one or more interfaces of the originating router. TLV 135 consist of the following information **[RFC5305]**:

- 1-byte (regular) IS-IS Type field set to the value 135
- 1-byte (regular) IS-IS Length field
- Variable-length (regular) IS-IS Value field consisting of TLV 135-specific information:
  - o 4-byte Metric field
  - o 1-byte of control information, consisting of:
    - 1 "up/down" bit
    - 1 bit indicating the presence of Sub-TLVs
    - 6-bit Prefix Length field
  - o 0- to 4-byte IPv4 Prefix field
  - o 0–250 optional bytes of Sub-TLVs, if present consisting of:
    - 1-byte Sub-TLVs Total Length field
    - 0–249 bytes of Sub-TLVs, where each Sub-TLV consists of the following sequence:
      - 1-byte Optional Sub-TLV Type field

- 1-byte Optional Sub-TLV Length field
- 0- to 247-byte Optional Sub-TLV Value field

TLV 135 can be present in a Level 1 LSP with any LSP Number and can appear more than once similar TLV 128. The TLV 135-specific information can appear multiple times within the TLV, as long as the maximum length of the TLV is not exceeded. The up/down bit is set to 0 when an IP prefix is first introduced into IS-IS. If a router advertises an IP prefix from a higher routing level to a lower level (e.g., from Level 2 to Level 1), it must set the bit to 1, indicating that the IP prefix has been passed down the routing hierarchy.

An IP prefix that has its up/down bit set to 1 may only be advertised down the routing hierarchy (to lower levels). This mechanism ensures that prefixes will travel in the IS-IS routing hierarchy only in one direction, thereby ensuring that there are no persistent routing loops. When a router learns an IP prefix via Level 2 routing and advertises that prefix down into a Level 1 area, the prefix will not be advertised back up into the Level 2 routing subdomain.

Reference [RFC5305] introduced TLV 135 but the route preference rules defined in [RFC5302] were not explicitly adapt for this new TLV. Thus, in addition to other extensions, reference [RFC7775] defines explicit route preference rules that are applicable TLV 135. Recall that [RFC5302] defined only the route preference rules for TLVs 128 and 130. It should be noted that [RFC7775] provides only clarifications and not corrections of [RFC5302] and [RFC5305]. [RFC7775] also describes extensively the types of route supported in IS-IS for IPv4 when using TLV 135.

TLV 135 can contain Sub-TLVs that apply to a particular IP prefix. If TLV 135 does not contain Sub-TLVs associated with a prefix, the bit indicating the presence of Sub-TLVs must be set to 0. If this bit is set to 1, the first byte after the IPv4 Prefix field is interpreted as the length of all Sub-TLVs associated with the IPv4 prefix.

2.13.1.3 The Traffic Engineering Router ID TLV (Code 134)

The Traffic Engineering Router ID TLV (defined in [RFC5305]) contains a 4-byte field that holds the Router ID of the router originating the LSP. The Router ID in TLV 134 describes a single stable address that can always be used to reference a particular router. TLV 134 is useful for traffic engineering purposes because it guarantees that a single stable address will always be available, regardless of the state of the router's interfaces, that can be referenced in a path that is reachable from multiple hops away.

TLV 134 should not appear more than once in an LSP. A router may add or omit the TLV 134 if it does not implement traffic engineering. However, a router must include this TLV in its LSP if it implements traffic engineering. If a router (i.e., a BGP router) advertises TLV 134 in its LSP, and if it advertises prefixes via BGP with the BGP Next-Hop attribute (NEXT_HOP) set to the BGP Router ID, the 4-byte field in TLV 134 should be the same as the BGP Router ID.

2.13.1.4 Shared Risk Link Group TLV (Code 138)

A Shared Risk Link Group (SRLG) in MPLS traffic engineering, is a set of links (or circuits) that share a common resource that is exposed to the same set of risks. All links (circuits) in the set are affected if the common resource fails. These links/

circuits are considered to belong to the same SRLG because they share the same risk of failure. For example, multiple wavelengths sharing the same fiber, or fiber links sharing the same conduit are said to be in the same SRLG. The wavelengths, for instance, are said to be in the same SRLG because if the single fiber experiences a fault, this can cause all links in the group to fail.

SRLG is a mechanism which allows a network operator to establish a backup secondary Label Switched Path or a Fast-Reroute (FRR) Label Switched Path which is disjoint from the path of the primary Label Switched Path. A typical application of SRLG is to allow the automatic placement of secondary or backup Label Switched Paths (or FRR bypass/detour Label Switched Paths) that minimizes the probability of sharing the same risks as the primary Label Switched Path.

An SRLG is represented by a 4-byte number that is unique within an IS-IS routing domain. The SRLG TLV is defined in **[RFC5307]** and contains a data structure consisting of:

- 1-byte (regular) IS-IS Type field set to the value 138
- 1-byte (regular) IS-IS Length field
- Variable-length (regular) IS-IS Value field consisting of TLV 138-specific information:
  - o 6-byte System ID field
  - o 1-byte Pseudonode ID field
  - o 1-byte Flag field
  - o 4-byte IPv4 Interface Address or Link Local Identifier field
  - o 4-byte IPv4 Neighbor Address or Link Remote Identifier field
  - o Variable-length SRLG Values field, where each element in the list has 4 bytes.

The neighbor is identified by its 6-byte System ID, plus one byte to indicate the Pseudonode ID if the neighbor is on a multiaccess broadcast LAN interface. The least significant bit of the 1-byte Flag indicates whether the interface is a numbered interface (bit set to 1), or an unnumbered interface (bit set to 0). All other bits of the Flag field are reserved and must be set to 0. The length of TLV 138 is equal to [16 + 4 × (number of SRLG Values)]. TLV 138 may occur more than once within an LSP. Routers use the SRLG information in TLV 138 to determine which links belong to the same SRLG. In OSPF-TE, SRLG information is advertised in the SRLG Sub-TLV (Type 16) in OSPF-TE LSAs **[RFC4203]**.

A network link may belong to one or more SRLGs. The SRLG of a particular path (segment) in a Label Switched Path is the set of SRLGs for all the links in the path. When the secondary path for a Label Switched Path is being computed, finding a path such that the secondary and primary paths do not have any links in common, is preferable. That is, finding the situation or solution where the SRLGs for the primary and secondary paths are disjoint. This ensures that the occurrence of a single failure on a particular link in the path does not interrupt the operation of both the primary and secondary paths in the Label Switched Path.

When SRLG is configured on a network node, it may use the Constrained Shortest-Path First (CSPF) algorithm to determine paths with the goal of keeping the network links used for the primary and secondary paths mutually exclusive **[JUNISRLGTLV20]**. If the primary path fails, the CSPF algorithm computes the secondary path by looking (as much as possible) for network links that do not share any SRLG with the primary path. In addition, when the CSPF computes the path for a bypass Label Switched Path, it tries to look for links that do not share any SRLG with the protected links.

When SRLG is not configured on a network node, the CSPF algorithm only considers the costs of the links when computing the secondary path **[JUNISRLGTLV20]**. When configured, any change in the SRLG information of a link triggers IS-IS to send LSP with TLV 138 updates for the new link's SRLG information and, the CSPF algorithm will recompute the paths during the next round of reoptimization.

2.13.2 IS-IS DYNAMIC HOSTNAME TLV

IS-IS identifies each node in an area using a variable 1- to 8-byte System ID (which is typically set to 6 bytes). For management and operation purposes, a network operator may need to check the status of IS-IS adjacencies, the content of the IS-IS LSDB, and entries in the IS-IS Forwarding Database. It is obvious that, using hexadecimal representations of System IDs and LSP IDs for looking at diagnostics information, is very cumbersome and tedious than using symbolic names.

Reference **[RFC5301]** defines the optional Dynamic Hostname TLV (Code 137) that can be used to advertise symbolic names in IS-IS LSPs. This allows IS-IS routers to use the transport mechanism of the IS-IS routing protocol itself to build dynamic mappings between symbolic names and System IDs. TLV 137 allows the IS-IS routers to include the name-to-System ID mapping information in their LSPs. This allows a simple and yet reliable way of transporting name mapping information across the IS-IS routing domain. The mapping can be bidirectionally, symbolic names to system IDs, and System IDs to symbolic names.

This method is more efficient than using static configurations where a name-to-System ID mapping is manually maintained on each router. In the static configuration case, each router has to maintain a statically configured database with mappings between router symbolic names and System IDs of all routers in the network. This database also must be modified each time a System ID is added or deleted, or a name change occurs, which creates scalability and maintainability issues as the network grows. Also, using DNS may not be suitable because (i) the response time of DNS services might not be satisfactory when the network has problems, (ii) the DNS services might not even be available, or (iii) some DNS implementations may not support A and PTR Records for CLNS and NSAPs.

TLV 137 has a 1-byte Type field (set to 137), a 1-byte Length field specifying the total length of a Value field which can contain a string of 1–255 bytes. The Value field specifies the symbolic name of the router originating the LSP. The Value field of TLV 137 is encoded in 7-bit ASCII. The symbolic name may be the Fully Qualified Domain Name (FQDN) for the router (e.g., FinanceAreaRouter.example.com), a subset of the FQDN, or any suitable string chosen for the router. An FQDN consists of two parts, the hostname and the domain name, and is the complete domain name for a specific host on the Internet.

The System ID of the router can be derived from the LSP ID field in the LSP (see Figure 2.16). TLV 137 may appear in any fragment of a Non-Pseudonode LSP. If TLV 137 is included in a Pseudonode LSP, then it is not to be interpreted as the DNS hostname of the originating router. When a router originates an LSP, it may include TLV 137 in its LSP. Upon receipt of an LSP with TLV 137, a router may choose to ignore it, or install the symbolic name and System ID specified in the LSP in its local hostname mapping table for the IS-IS network. A DIS may also optionally insert TLV 137 in its Pseudonode LSP to associate a symbolic name to a local broadcast LAN. If a router receives an LSP containing a name-to-System ID mapping that is different from the mapping in its local table (cache), the router will enter the received LSP information replacing the existing mapping in the local table.

2.13.3   IPv6-Specific IS-IS TLVs

Reference [**RFC5308**] specifies two TLVs, the IPv6 Reachability TLV (Code 236), and the IPv6 Interface Address TLV (Code 232), that can be included in IS-IS LSPs for exchanging IPv6 routing information in a routing domain. This allows a single intradomain routing protocol (IS-IS) to route IPv6 traffic along with IPv4 and OSI traffic.

2.13.3.1   IPv6 Reachability TLV (Code 236)

TLV 236 describes the IPv6 addresses that are reachable within an IS-IS routing domain (i.e., internal reachability information) via one or more interfaces on a router, or IPv6 addresses outside the routing domain that are reachable via some interfaces on the router (i.e., external reachability information). To do this, TLV 236 contains the following information:

- 1-byte (regular) IS-IS Type field set to the value 236
- 1-byte (regular) IS-IS Length field
- Variable-length (regular) IS-IS Value field consisting of TLV 236-specific information:
  - o 4-byte Metric field
  - o Up/down (U) bit
  - o External original (X) bit
  - o Sub-TLV present (S) bit
  - o 5 Reserve bits
  - o 1-byte IPv6 Prefix Length field
  - o Variable-length IPv6 Prefix field

The U bit is used to indicate if the IPv6 prefix is being advertised down from a higher routing level, X bit to indicate if the IPv6 prefix is being distributed from another routing protocol, and the optional S bit to indicate the existence of Sub-TLVs in TLV 236. Using the X bit, TLV 236 combines the functions of the IP Internal Reachability Information TLV (Code 128) and the IP External Reachability Information TLV (Code 130) into a single TLV.

TLV 236 may appear any number of times within an LSP. Link-local prefixes must not be advertised in TLV 236. If an IPv6 prefix is distributed from another routing protocol into IS-IS, the X bit must be set to 1. This prevents IPv6 prefixes from being redistributed back from IS-IS to the source routing protocols. If the S bit is set to 0, then no Sub-TLVs are present in TLV 236. Otherwise, the S bit is 1, and the first byte following the IPv6 prefix contains the length of the Sub-TLV portion of TLV 236.

If an IP prefix is advertised with a metric larger than maximum path metric, MAX_V6_PATH_METRIC, (0xFE000000), the normal SPF computation will not consider this prefix. However, the prefix will be advertised for purposes other than building the normal IPv6 Routing Table. If, during the SPF computation, a path metric exceeds MAX_V6_PATH_METRIC, it will be considered to be equal to MAX_V6_PATH_METRIC.

Reference **[RFC5308]** defined TLV 236 and included explicit statements regarding route preference. However, the statements based on the use of the up/down bit in advertisements that are carried in Level 2 LSPs, appear to be inconsistent with the route preference rules defined in **[RFC5302]** and **[RFC5305]**. So, in addition to defining explicit route preference rules for TLV 135, **[RFC7775]** revises the route preference rules for TLV 236 as described in **[RFC5308]**, and provides clarification on the use of the up/down bit when it appears in TLVs in Level 2 LSPs. **[RFC7775]** corrects the IPv6 route preference rules defined in **[RFC5308]** to be consistent with the route preference rules for IPv4.

[RFC7775] also explicitly states that the same route preference rules apply to the MT TLVs 235 and 237 as discussed below (see also **[RFC5120]**). The content of TLV 235 and TLV 237 (after a few TLV-specific fields) has the same format as TLV 135 and TLV 236, respectively. **[RFC7775]** also describes the types of routes supported in IS-IS for IPv6 when using TLVs 236 and 237.

2.13.3.2 IPv6 Interface Address TLV (Code 232)

TLV 232 defined in **[RFC5308]** maps directly to the IP Interface Address TLV (TLV 132) defined in **[RFC1195]** for IPv4. The 0–63 4-byte IPv4 interface addresses contents of TLV 132 are modified to obtain the TLV 232 contents of 0–15 16-byte IPv6 interface addresses. TLV 232 contains the following information **[RFC5308]**:

- 1-byte (regular) IS-IS Type field set to the value 232
- 1-byte (regular) IS-IS Length field
- Variable-length (regular) IS-IS Value field consisting of TLV 232-specific information:
  o 0–15 16-byte IPv6 interface addresses

For Hello PDUs, TLV 232 must contain only the link-local IPv6 addresses assigned to the router interface transmitting the Hello PDU. For LSPs, TLV 232 must contain only the non-link-local IPv6 addresses assigned to the router.

2.13.3.3 IPv6 Network Layer Protocol ID (NLPID)

Reference **[RFC5308]** also specifies a new IPv6 protocol identifier. The value of the NLPID assigned to IPv6 is 142 (0x8E). Similar to **[RFC1195]** and IPv4, if the IS-IS router supports IPv6 routing using IS-IS, it must advertise the Protocols Supported TLV (TLV Code 129) by adding the IPv6 NLPID of 142. This means when an IS-IS supports IPv6 and sends out a Protocols Supported TLV (TLV 129) specifying the set of NLPIDs of the Network Layer protocols that it is capable of relaying, it must include the NLPID of 142.

The following are some important NLPIDs of interest:

- **ISO CLNP (Connectionless Network Protocol)**: 129 (0x81)
- **ISO ES-IS**: 130 (0x82)
- **IS-IS**: 131 (0x83)
- **IPv4**: 204 (0xCC)

2.13.4   MULTI-TOPOLOGY IS-IS TLVs

In addition to OSI and IPv4, IS-IS for IPv6 extends the address families supported by the IS-IS routing protocol to include IPv6. IPv6 enhancements to IS-IS allow IS-IS to advertise OSI and IPv4 routes in addition to IPv6 prefixes. IS-IS with IPv6 routing is similar to IS-IS with IPv4 routing and offers many of the same benefits. IS-IS with IPv6 routing can be run either in a single-topology mode or multiple topology mode.

2.13.4.1   Single-Topology IS-IS Support for IPv6

The IS-IS discussions up to this point deal with essentially a single topology for an entire IS-IS area or domain with a single address family (i.e., IPv4 unicast). However, Single-Topology support for IPv6 allows IS-IS router interfaces to be configured with CLNS and IPv4 along with IPv6. In this case, all routers in an IS-IS Level 1 area or the Level 2 routing subdomain must support identical sets of network layer address families on all interfaces.

Either old-style (narrow metric) or new-style (wide metric) TLVs may be used when single-topology support for IPv6 is used. However, it should be noted that the TLVs used in LSPs to advertise reachability to IPv6 prefixes (as described above) use wide or extended metrics. In single-topology IS-IS for IPv6, the configured metric is always the same for both IPv4 and IPv6 prefixes. In this topology, IPv6 and IPv4 use the same underlying topology and this can be accomplished as specified in **[RFC5308]** which introduces the IPv6 Reachability TLV (TLV 236).

A single SPF algorithm per routing level is used to compute OSI, IPv4, and IPv6 routes. The same IS-IS adjacency is used for all supported address families (e.g., IPv4 and IPv6) and with a single SPF algorithm applied to both IPv4 and IPv6 reachability information. The use of a single SPF algorithm means that both IS-IS for IPv4 and IS-IS for IPv6 routing protocols must operate on the same single network topology. The use of IS-IS for both IPv4 and IPv6 routing, requires that any router interface configured for IS-IS IPv4 routing must also be configured for IS-IS IPv6 routing, and vice versa. All routers within an IS-IS Level 1 area or Level 2 routing subdomain must also support the same set of network layer address families: IPv4 only, IPv6 only, or both IPv4 and IPv6. Any discrepancy in the configuration with respect to the address families that are supported on router interfaces can create routing black holes where packets will be discarded and not get to their destinations.

If single-topology IS-IS is being used to support IPv4 only, IPv6 only, or both IPv4 and IPv6, the network operator may choose to configure both IPv4 and IPv6 on all IS-IS interfaces for Level 1, Level 2, or both Level 1 and Level 2 routing. However, if both IPv4 and IPv6 are configured on the same router interface, both protocols must be running at the same IS-IS routing level. That is, IPv6 cannot be configured to run at IS-IS Level 1 only on a specified router interface while IPv4 is configured to run at IS-IS Level 2 only on the same interface.

2.13.4.2   Multi-Topology IS-IS Support for IPv6

Reference [RFC5120] describes how a set of independent IP topologies, called MTs, can be run within a single IS-IS domain. Network operators can use MTs for a variety of purposes, such as maintaining separate routing domains for unicast and multicast, rolling out of IPv6 without impacting existing IPv4 operations, creating IPv4 and IPv6 topologies which are not identical within the backbone, overlay an in-band management network over the original IGP topology, or causing a subset of an address space to travel over a different topology.

Multi-Topology IS-IS (MT-ISIS) support for IPv6 allows a single IS-IS area or domain to maintain a set of independent topologies [RFC5120]. MT-ISIS removes the restriction that all IS-IS router interfaces must support identical sets of network address families. It also removes the restriction that all routers in an IS-IS Level 1 area or in the Level 2 routing subdomain must support identical sets of network layer address families.

MT-ISIS for IPv6 supports the use of multiple SPF algorithms to compute routes and removes the restriction (imposed by single-topology support for IPv6) that all router interfaces must support all configured network layer address families, and that all routers in an IS-IS Level 1 area or Level 2 subdomain must support the same set of network layer address families. Multiple SPF computations are performed, one for each configured topology in the MTs. In this case, it is sufficient that connectivity exists among a subset of the routers in the Level 1 area or the Level 2 routing subdomain for a given network address family to be routable. For MT-ISIS support for IPv6, only wide or extended metrics are used in the TLVs used to advertise IPv6 reachability information in LSPs.

A router running IS-IS for IPv6 maintains a local Routing Information Base (RIB) in which it maintains all routes to destinations it has learned from its neighbors. At the end of each SPF algorithm computations, IS-IS enters the least-cost or best routes present in the local RIB to each destination in a global IPv6 Routing Table.

2.13.4.3   TLVs for Multi-Topology IS-IS

Reference [RFC5120] defines the following four new TLVs for supporting MT-ISIS.

2.13.4.3.1   Multi-Topology TLV (Code 229)

The Multi-Topology TLV (TLV 229) contains one or more MTs the originating router is participating in. It is advertised in IS-IS Hello PDUs and LSP (fragment) Number 0. TLV 229 can advertise up to 127 MTs. TLV 229 has the following structure [RFC5120]:

- 1-byte (regular) IS-IS Type field set to the value 229
- 1-byte (regular) IS-IS Length field
- Variable-length (regular) IS-IS Value field consisting of one or more entries of the following 2-byte TLV 229-specific information:
  - o O Bit represents the Overload bit for the MT. This bit is only valid in LSP Number 0 for all MT IDs except MT ID #0, otherwise it should be set to 0 on transmission and ignored on receipt.
  - o A Bit represents the ATT (Attached) bit for the MT. This bit is only valid in LSP Number 0 for all MT IDs except MT ID #0, otherwise it should be set to 0 on transmission and ignored on receipt.
  - o Two R Bits are reserved bits, and should be set to 0 on transmission and ignored on receipt.
  - o 12-bit MT ID field containing the ID of the topology being announced.

TLV 229 can appear multiple times in a PDU, and the resulting MT set is obtained by the union of all the MT TLV occurrences in the PDU. Any other occurrence of this TLV in IS-IS PDUs other than Hello PDUs and LSPs with LSP Number 0 must be ignored. If TLV 229 is not present in Hello PDUs and LSPs with LSP Number 0, this must be interpreted as the advertising interface or router is participating in MT ID #0 only. If a router advertises TLV 229, it has to advertise all the MTs it is participating in, including topology MT ID #0 specifically.

Reference **[RFC5120]** defines the following MT IDs:

- **MT ID #0**: Equivalent to the "standard" topology.
- **MT ID #1**: IPv4 in-band management purposes.
- **MT ID #2**: IPv6 routing topology.
- **MT ID #3**: IPv4 multicast routing topology.
- **MT ID #4**: IPv6 multicast routing topology.
- **MT ID #5**: IPv6 in-band management purposes.
- **MT ID #6 to #3995**: Reserved for IETF consensus.
- **MT ID #3996 to #4095**: Reserved for development, experimental and proprietary features

2.13.4.3.2   MT Intermediate Systems TLV (Code 222)

The MT Intermediate Systems TLV (TLV 222) is aligned with the Extended IS Reachability TLV (TLV 22) other than an additional two bytes of MT membership information placed at the beginning this newer TLV. Recall that TLV 22 describes IS-IS neighbors (and possibly, Pseudonode neighbors) of the originating router using wide metric values, and also carries additional information (via Sub-TLVs) that can be used for IS-IS-dependent MPLS traffic engineering. TLV 2, TLV 22, and TLV 222 each contains information about the IS-IS neighbors of a router.

TLV 222 has the following structure **[RFC5120]**:

- 1-byte (regular) IS-IS Type field set to the value 222
- 1-byte (regular) IS-IS Length field

- Variable-length (regular) IS-IS Value field consisting of TLV 222-specific information:
  - o Four R Bits are reserved bits, and should be set to 0 on transmission and ignored on receipt.
  - o 12-bit MT ID field containing the non-zero MT ID of the topology being announced. The TLV must be ignored if the MT ID is zero. This is to ensure a consistent view of the standard unicast topology.
  - o One or more entries of 11–253 bytes of Extended IS Reachability TLV (Code 22) Value field contents (same format as TLV 22)

After the 2-byte MT membership format, the content of TLV 222 is in the same format as TLV 22. TLV 222 can contain up to 23 neighbors of the same MT if Sub-TLVs are not used. This TLV can appear multiple times in a Level 1 or Level 2 LSP with any LSP Number similar to TLV 22.

2.13.4.3.3   Multi-Topology Reachable IPv4 Prefixes TLV (Code 235)

The Multi-Topology Reachable IPv4 Prefixes TLV (TLV 235) is aligned with the Extended IP Reachability TLV (TLV 135) other than an additional two bytes of MT membership information placed at the beginning of this newer TLV. Recall that TLV 135 describes the IP prefixes within a routing domain that reachable directly via one or more interfaces on the originating router (using wide metric values for IP prefixes), and also carries Sub-TLVs for distributing MPLS traffic engineering resource information. Each MT in an MT set demands its own address space, so TLV 235 (and TLV 237) are needed for advertising IP address prefixes stored in each MT being announced other than MT ID #0, which also enables SPF computations per MT.

TLV 235 has the following structure **[RFC5120]**:

- 1-byte (regular) IS-IS Type field set to the value 235
- 1-byte (regular) IS-IS Length field
- Variable-length (regular) IS-IS Value field consisting of TLV 235-specific information:
  - o Four R Bits are reserved bits, and should be set to 0 on transmission and ignored on receipt.
  - o 12-bit MT ID field containing the non-zero MT ID of the topology being announced. The TLV must be ignored if the MT ID is zero. This is to ensure a consistent view of the standard unicast topology.
  - o One or more entries of 5–253 bytes of Extended IP Reachability TLV (Code 135) Value field contents (same format as TLV 135)

After the 2-byte MT membership format, the content of TLV 235 is in the same format as TLV 135. This TLV can appear multiple times in a Level 1 LSP with any LSP Number similar to TLV 135.

2.13.4.3.4   Multi-Topology Reachable IPv6 Prefixes TLV (Code 237)

The Multi-Topology Reachable IPv6 Prefixes TLV (TLV 237) is aligned with the IPv6 Reachability TLV (TLV 236) apart from an additional two bytes placed in front of this newer TLV. Recall that TLV 236 describes the IPv6 addresses that are reachable within an IS-IS routing domain via one or more interfaces on the originating router, or IPv6 addresses outside the routing domain that are reachable via interfaces on the router (i.e., external reachability information).

TLV 237 has the following structure [RFC5120]:

- 1-byte (regular) IS-IS Type field set to the value 237
- 1-byte (regular) IS-IS Length field
- Variable-length (regular) IS-IS Value field consisting of TLV 237-specific information:
  - o Four R Bits are reserved bits, and should be set to 0 on transmission and ignored on receipt.
  - o 12-bit MT ID field containing the non-zero MT ID of the topology being announced. The TLV must be ignored if the MT ID is zero. This is to ensure a consistent view of the standard unicast topology.
  - o One or more entries of 6–253 bytes of IPv6 Reachability TLV (Code 236) Value field contents (same format as TLV 236)

After the 2-byte MT membership format, the content of TLV 237 is in the same format as TLV 236. This TLV can appear multiple times in an LSP similar TLV 236.

2.13.4.4 Multi-Topology Adjacencies

To develop MTs for IS-IS to be backwards-compatible with IS-IS as specified in [ISO10589:2002] and [RFC1195], reference [RFC5120] defined several extensions to the encoding of TLVs and additional SPF procedures. Reference [RFC5120] focused on two main issues in MT-IS-IS, (i) formation of adjacencies, and (ii) advertising of prefixes and reachable intermediate systems within each topology. This specification also addressed how an MT capable router will use the SPF algorithm on the additional MT information it has learned from other routers. The "standard" IS-IS topology (i.e., IS-IS single-topology) is defined as MT ID #0 (zero) in [RFC5120].

2.13.4.4.1 *Establishing and Maintaining Multi-Topology Adjacencies*

Each adjacency formed on a router interface is classified as belonging to a set of MTs. A router accomplishes this by adding TLV 229 in IS-IS Hello PDUs to advertise the topologies the interface belongs to. If an interface belongs to only MT #0, the router may optionally advertise this MT in TLV 229. Also, if TLV 229 is not included in the IS-IS Hello PDUs, this implies the interface belongs to MT ID #0 only. By exchanging the MT capabilities of interfaces via TLV 229, each router will be able to determine the common MT set over its adjacencies.

2.13.4.4.2 *Forming Adjacencies on Point-to-Point Interfaces*

IS-IS routers on point-to-point interfaces that do not implement MT extensions [RFC5120] will form adjacencies as normal as described in [ISO10589:2002] and [RFC1195]. If a router is not participating in specific MTs, it will not advertise those MT IDs in its IS-IS Hell PDUs, and therefore, will not include that neighbor in its

LSPs. On the other hand, if a router does not detect an MT ID in the remote neighbor's IS-IS Hello PDUs, the router must not include that neighbor in its LSPs. The router and the remote neighbor should not form an adjacency if they do not have at least one common MT over the point-to-point link.

2.13.4.4.3   Forming Adjacencies on Broadcast Interfaces

All routers on a broadcast LAN that implement the MT extension **[RFC5120]** may advertise their MT capabilities by including TLV 229 in their IS-IS Hello PDUs. If an MT capable router detects that there is at least one adjacency on its LAN interface that belongs to a specific MT, the router MUST include the corresponding MT Intermediate Systems TLV (TLV 222) in its LSP, otherwise it MAY include this TLV 222 in its LSP if the LAN interface participates in the detected MT set.

Any two routers on the broadcast network segment must always establish an adjacency, regardless of whether or not they share a common MT. This is to ensure that all routers on the network segment can correctly elect the same DIS. A router should not include TLV 222 describing an MT in its LSP if none of the adjacencies on the LAN contain that MT. The MT extension **[RFC5120]** does not change the DIS, CSNP, and PSNP functions.

2.13.4.5   Advertising MT Reachable Intermediate Systems in LSPs

A router participating in a topology must include the MT Intermediate Systems TLV (TLV 222) in its LSPs describing the adjacent nodes that are participating in the corresponding topology, and advertise such TLVs. The standard Intermediate System Neighbors TLV (TLV 2) or Extended IS Reachability TLV (TLV 22) apply to MT ID #0, and are equivalent to TLV 222 for MTs. A router must advertise TLV 222 when there is at least one adjacency on the interface that belongs to a specified MT, otherwise it may announce TLV 222 of an adjacency for a given MT if the originating interface participates in the broadcast LAN.

Reference **[RFC5120]** does not provide any mechanism to prevent a router that does not understand MT extensions from being elected DIS, and responsible for creating the Pseudonode for a broadcast LAN. So, no special TLVs are introduced to be carried in Pseudonode LSPs, and there are no provisions to run distinct DIS elections per MT. Therefore, a Pseudonode LSP generated by a DIS must contain in its Reachable IS TLV (TLV 2 or TLV 22) all nodes on the LAN as usual, regardless of their MT capabilities. Thus, **[RFC5120]** does not specify any changes to the construction of Pseudonode LSPs.

2.13.4.6   Multitopologies and Overload, Partition, and Attached Bits

An MT in an MT set a router belongs to could potentially become partitioned, overloaded, or attached, independently. To minimize complexity, MT extensions do not support MT-based partition repair. The Partition, Attached, and Overload bits in the LSP header only reflect the status of MT #0 (the default topology).

The Overload (O) bit and Attached (A) bit are part of the Multi-Topology TLV (TLV 229) being distributed within a router's LSP Number 0. Since each adjacency in a network with MTs can belong to a different MT, some MTs may be Level 2 attached, while others on the same router may not. The overload bit in TLV 229 can be used to signal that the topology is being overloaded. A system in an MT is considered overloaded if the corresponding Overload bit for that MT is set. Route leaking (or redistribution) between the routing levels can only be performed within the same MT.

2.13.4.7   Multi-Topology SPF Computation

Each MT in an MT set must run its own instance of the SPF algorithm. All topologies use the Pseudonode LSPs during SPF computation. Each non-default topology (i.e., MT ID greater than 0) may have its Overload bit and Attached bit set in the Multi-Topology TLV (TLV 229). Before considering a link between any two routers in an MT, a reverse-connectivity (or two-way connectivity) check must be performed within SPF to ensure bi-directional reachability is possible within the same MT.

Each router stores the results of each SPF computation in a separate RIB, otherwise overlapping addresses in different MTs could lead to forwarding loops. The forwarding logic and configuration on each router must ensure that a packet being forwarded must traverse the same MT from the source to the destination. For correct packet forwarding, the next-hops derived from the SPF computation for a particular MT must belong to the adjacencies conforming to the same MT. To prevent undesirable forwarding behaviors, the network operator must ensure consistent configuration of all routers in the routing domain.

2.13.4.8   IP Forwarding Considerations in Multi-Topology IS-IS

Routers using MT extension for IS-IS routing end up maintaining multiple RIBs. Reference **[RFC5120]** discusses some considerations for IP forwarding in various MT scenarios.

2.13.4.8.1   Each MT Belongs to a Distinct Network Address Family

In this scenario, each route related to a given MT (with a distinct address family) is installed in a separate RIB (corresponding to that MT). An IS-IS interface may support multiple topologies but the interface has to detect the address family of each incoming packet and forward it according to the MT it belongs to. For example, an IPv4 MT and an IPv6 MT can share the same interface and packets are processed according to their respective MTs.

2.13.4.8.2   A Number of MTs Belong to the Same Network Address Family

We consider here the case where a number of MTs in the routing domain belong to the same address family rather than to dedicated address families.

- **Each Interface Belongs to One and Only One MT**: In this case, each MT connected to an interface can be used to forward packets from the same address family, even if the MT address spaces overlap, since each MT has its dedicated interface, and the interface/MT can be associated with its own MT RIB.
- **Multiple MTs Share an Interface with Overlapping Address Spaces**: In the case where an interface or adjacency contains multiple MTs, and there exists an overlap in IP address space among the MTs, additional mechanisms must be used to resolve the MT to which incoming IP packets on the interface belong.

Each router needs some additional mechanism to select the correct RIBs for the incoming IP packets on an interface to make forwarding decisions.

For example, if the MTs are partitioned according to QoS, then the Differentiated Services Code Point (DSCP) bits in the IP packet header can be utilized to decide which RIB to use. A router may check some IP header fields, or even packet data information, to select the correct IP Forwarding Table. For example, a router may use the source IP address in the header to determine the desired forwarding behavior. A router may send packets for a preferred customer through a higher quality Layer 2 tunnel, and other packets through a normal tunnel.

- **Multiple MTs Share an Interface with Non-Overlapping Address Spaces**: When multiple MTs on an interface do not have overlap in address space, then the destination address space classifies the MT to which a packet belongs. In this case, a router may install routes from different MTs into a shared RIB.

2.13.4.8.3   Some MTs Are Not Used for Packet Forwarding Purposes

An MT in an IS-IS routing domain may still be used even if the resulting RIB associated with it is not used for packet forwarding purposes. For example, routers may perform multicast Reverse Path Forwarding (mRPF) checks on a different RIB (i.e., solely dedicated to mRPF) than the standard unicast RIB, even though it is using an entirely different RIB for multicast packet forwarding. The dedicated mRPF RIB is not used for packet forwarding. However, an incoming packet to a router interface must still be clearly identified as belonging to a unique MT.

2.13.4.9   Multi-Topology Network Management Considerations

When multiple MTs exist within an IS-IS routing domain, some of the routers may be configured to participate only in a subset of the MTs in the domain **[RFC5120]**. This section discusses some of the options available to a network administrator to enable operations on those routers or allow network management stations to access them.

2.13.4.9.1   Create Dedicated Management MT to Include All the Nodes

A network operator may set up a dedicated management topology or 'in-band' management MT for the routing domain. The operator will include in this "mgmt" topology all the routers that need to be managed. The routes that are computed in the "mgmt" topology will be installed in an "mgmt" RIB. If the "mgmt" topology is configured to use a set of address spaces that do not overlap with the default topology, the "mgmt" routes may also be (optionally) installed in the default (topology) RIB.

2.13.4.9.2   Extend the Default Topology to All the Nodes

Even in the case where a network operator does not use the default topology on some of the nodes for IP packet forwarding, the operator may extend the default topology to those nodes for the purpose of network management. However, to prevent IP traffic from being forwarded through those ("management-extended") nodes during network topology changes, it is recommended for network operators to set high costs (metric values) on the links that belong to the extended portion of the default topology.

2.13.5   IS-IS Router Capability TLV

There are some situations where IS-IS routers may need to learn the capabilities of the other routers in their IS-IS area, routing level, or the routing domain, such as, for performing MPLS TE related activities. Reference **[RFC7981]** defines the optional IS-IS Router CAPABILITY TLV (Code 242) which is composed of multiple Sub-TLVs, and allows a router to advertise its capabilities within an IS-IS routing level or the entire routing domain.

TLV 242 consists of 1-byte Type field (with value set to 242), 1-byte Length field that specifies the number of bytes in the Value field, and a variable length Value field that starts with a 4-byte Router ID, indicating the originator of the TLV, followed by 1-byte Flags field (composed of a number of flags), and a set of (0–250 bytes) optional Sub-TLVs which are formatted as described in **[RFC5305]**. More than one TLV 242 from the same source may be present in an LSP.

Reference **[RFC7981]** defined two flag bits:

- **S bit (0x01)**: This is bit 7 (the rightmost bit), and if set to 1, TLV 242 must be flooded across the entire IS-IS routing domain. If set to 0, TLV 242 must not be leaked (redistributed) between routing levels. This bit must not be altered during the leaking of TLV 242.
- **D bit (0x02)**: When TLV 242 is leaked from Level 2 to Level 1, the D bit must be set (1), otherwise, it must be clear (0). A TLV 242 with the D bit set must not be leaked from Level 1 to Level 2 to prevent looping of the TLV.

The originator should set the Router ID in TLV 242 to be identical to the value advertised in the Traffic Engineering Router ID TLV (TLV 134) **[RFC5305]**. If the originating router is not assigned a Traffic Engineering Router ID, the router should set the Router ID to be identical to an IP Interface Address **[RFC1195]** it advertises. If the originating router does not support IPv4, then must use the reserved value 0.0.0.0 in the Router ID field, and the IPv6 TE Router ID Sub-TLV (Type 12) **[RFC5316]** must be present in the TLV. Any TLV 242 that contains a Router ID of 0.0.0.0 and does not carry the IPv6 TE Router ID Sub-TLV must not be used.

Routers that do not support the TLV 242 must silently ignore it and continue processing other TLVs in the same LSP. Routers that do not support specific Sub-TLVs carried within a TLV 242 must silently ignore the unsupported Sub-TLVs and continue processing those Sub-TLVs that are supported in the TLV. In order for LSPs with TLV 242 with IS-IS domain-wide scope originated by Level 1 routers to be flooded across the entire IS-IS domain, at least one Level 1-2 router in every IS-IS area of the domain must support TLV 242.

2.14   IS-IS LINK-STATE DATABASE SYNCHRONIZATION

As soon as an IS-IS router powers up on a network segment, it adds information about its local attachments to its LSDB. The router then sends Hello PDUs on all operational interfaces to discover neighbors. If a neighbor is discovered on an interface, the router and the neighbor attempt to form an adjacency with each other. The routers use the adjacency to advertise their LSDBs to each other. In addition, the routers advertise summary lists of their LSDB information to each other to allow the other router to verify that it has the most up-to-date information in its LSDB. In the event, one of the routers requires an LSP update, it sends a request to that neighbor for that update. All routers in an area follow this exchange process until they all have identical LSDBs.

Having identical LSDBs allows all routers to have a common view of the network topology. Each router then uses the SPF algorithm (based on the Dijkstra Algorithm) to process the information in its identical LSDB to determine a path from itself to each remote destination. Because all routers use the same algorithm to calculate routes, each router must have consistent (synchronized) routing information in its LSDB to get loop-free paths. Each router must ensure that link-state updates are advertised and propagated quickly to provide consistency in LSDBs updates. The concept of having consistent LSDBs is a central tenet of link-state routing protocols like OSPF and IS-IS, and allows these protocols to ensure loop-free routing. Each router then uses its resulting IP Routing Table to make consistent packet forwarding decisions.

2.14.1 IS-IS RELIABLE FLOODING

Proper operation of link-state routing protocols such as OSPF and IS-IS requires a reliable and efficient process for synchronizing the LSDB in each router. In this section we describe the concepts of routing information distribution in IS-IS and LSDB synchronization. IS-IS routers accomplish LSDB synchronization using special LSPs: CSNPs and PSNPs. Routers use CSNPs and PSNPs to ensure that LSPs sent have been received; they function in some way as a reliability mechanism for LSP transmissions. CSNPs and PSNPs contain LSP descriptors or headers describing the LSPs in a router's LSDB and not the actual detailed LSP information.

CSNPs contain a summary list of the LSPs that are maintained in the LSDB of a router similar in function to OSPF Database Description packets described in Chapter 1. PSNPs on the other hand, contain only subset of the router's LSPs, usually a single LSP descriptor block. As described above, a router uses a PSNP to request an LSP that is missing in its LSDB, or to acknowledge the receipt of an LSP.

Routers in an IS-IS Level 1 area maintain identical Level 1 LSDBs, which are synchronized using CSNPs and PSNPs. The LSDB synchronization is to allow all routers within an area to have identical views or maps of the area's topology, which is necessary for routing consistency within the area. A Level 2 LSDB contains area address prefix information that binds all the Level 1 areas together for inter-area (or Level 2) routing.

2.14.2 LSDB UPDATE PROCESS

When network changes occur, IS-IS routers, depending on their routing levels and their configurations (e.g., configured to use routes passed via IS-IS route leaking), may be required to update their LSDBs. A router generates an LSP when an event occurs that would cause the information in its LSDB to change. IS-IS routers generate new LSPs when any of the following events occurs:

- When routers are required to perform periodic updates
- When an IS-IS adjacency or circuit goes up or down
- When an entire IS-IS router goes up or down
- When an IS-IS circuit is assigned a new routing metric value
- When there is a change in System ID
- When there is a change in *manualAreaAddresses*
- When there is change in DIS status
- When there is a changed of an ATT (Attached) Flag
- For Level 1 routing, when intra-area routes change
- For Level 2 routing, when inter-area routes change

When any of these events occur, the router will regenerate LSP(s) reflecting the change with a new Sequence Number. If the event caused the router to generate an LSP which had not previously been generated (e.g., an adjacency goes into the Up state which could not be accommodated in an existing LSP), the Sequence Number is set to 1. The router will then propagate the LSP(s) on every circuit attached to it.

2.14.2.1   LSDB Update Process on Multiaccess Broadcast Network Segments

The DIS creates and sends CSNPs at regular intervals on the broadcast network segment (10 seconds by default in **[ISO10589:2002]**). The CSNPs lists all LSPs present in the DIS's LSDB. Each non-DIS router on the network segment receives the CSNPs and compares them to its local LSDB. The following summarizes possible outcomes of this comparison:

- If the router has an LSP with the same LSP ID in its LSDB, with the same Sequence Number, it performs no further action; both the router and the DIS have the same LSP.
- If the router does not have an LSP with the indicated LSP ID in its LSDB, or the Sequence Number of the received LSP (from the DIS) is higher, the router will request that LSP from the DIS by sending a PSNP onto the network segment. The DIS will receive the PSNP and flood the requested LSP.
- If the router has an LSP with the same LSP ID in its LSDB but the received LSP's Sequence Number is lower, or the newer LSP with this LSP ID in its LSDB is one that the DIS did not advertise in its CSNPs, the router will update the DIS by simply flooding the newer LSP on the network segment right away.

The DIS does not relay LSPs; instead, it serves as a reference point for LSP comparison. If a router detects that it is missing an LSP known by the DIS, or if the local LSP in its LSDB is older than the one known by the DIS, the router will send a PSNP requesting the newer LSP, and the DIS will flood it. If the PSNP sent by the router or the LSP sent by the DIS gets lost during transmission, the process is repeat.

If a router has a newer LSP than the one held by the DIS, or if the DIS does not have it entirely, the router will simply flood the LSP onto the network segment. The

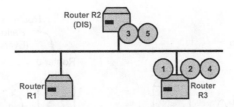

1.	Router R3 (a new router) sends a LAN IS-IS Hello PDU to establish neighbor relationships with the other routers in the broadcast domain.
2.	Router R3 establishes neighbor relationships with Router R1 and Router R2, waits for the timeout of the LSP refresh timer, and then sends its LSP to a multicast address (01-80-C2-00-00-14 in a Level-1 area and 01-80-C2-00-00-15 in a Level-2 area). All neighbors on the network are able to receive the LSP.
3.	The DIS (Router R2) on the network segment adds the received LSP to its LSDB. After the CSNP timer expires, the DIS sends CSNPs to synchronize the LSDBs on the network.
4.	Router R3 receives the CSNPs from the DIS, checks its LSDB, and sends a PSNP to request the LSPs it does not have.
5.	The DIS receives the PSNP and sends Router R3 the required LSPs for LSDB synchronization.

FIGURE 2.24 Process of Updating LSDBs on a Broadcast Network

DIS does not have to send any explicit acknowledgment. If the DIS receives the flooded LSP, it will advertise it in its next CSNP transmission, and this CSNP serves as an implicit acknowledgment of the LSP. If the CSNP sent by the DIS does not advertise the LSP even after it was flooded by the originating router, then the LSP must have been lost in transit, and the router will simply reflood it.

Figure 2.24 is an example that describes the process of updating LSDBs on a broadcast network. The DIS sends CSNPs periodically on the broadcast network segment. The CSNP interval defined by the *CompleteSNPInterval* attribute and the default is 10 seconds **[ISO10589:2002]**. The CSNPs are multicast to all IS-IS routers on the broadcast network segment. An IS-IS router will receive CSNPs and compare the list of LSPs contained in them with its local LSDB, and if there are any missing LSPs, the route will request these by sending a PSNP (Figure 2.24).

Figure 2.25 shows another example of LSDB synchronization on a broadcast network. Router 3 which is elected the DIS transmits a CSNP to all routers on the network. Router 1 compares the list of LSPs in the CSNP with its LSDB and detects that it is missing one LSP. Router 1, therefore, sends a PSNP to the DIS (Router 3) requesting the missing LSP. The DIS receives the PSNP and retransmits the missing LSP. Router 1 receives the PSNP and acknowledges it with a PSNP.

2.14.2.2 LSDB Update Process on Point-to-Points Links

In the IS-IS specification **[ISO10589:2002]**, each IS-IS router on a point-to-point link sends a single IS-IS Hello PDU and these determine whether the adjacency is at Level 1, Level 2 or Level 1-2 (see Reserved/Circuit Type field of the Hello PDU in Figure 2.20). After the two routers have declared the adjacency as "Up", they will attempt to synchronize their LSDBs. The routers may use the three-way handshake mechanism design for point-to-point adjacencies **[RFC5303]**. When the correct routing level adjacency is established, both routers will mark all their LSPs for flooding over the point-to-point link, and each router will send a CSNP describing the contents of its LSDB.

Cisco IOS allows periodic CSNPs to be configured on point-to-point links to further increase the reliability of the LSP flooding process **[CISCTEAPAQ00]**.

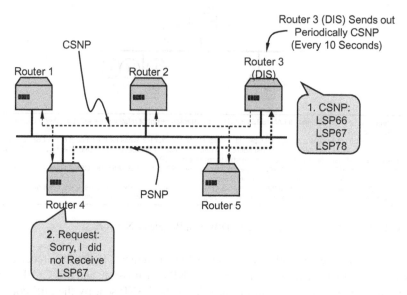

FIGURE 2.25   Link-State Database Synchronization on a Broadcast LAN

According to **[ISO10589:2002]**, each router on point-to-point links sends CSNPs only once and not periodically when the link becomes active. If a router determines from the received CSNP that the neighbor has LSPs that are missing or newer than those in the local LSDB, it will request them using a PSNP. Because CSNPs are not periodically exchanged on point-to-point links in the IS-IS specification **[ISO10589:2002]**, each LSP transmitted must be acknowledged by a corresponding PSNP from the neighbor.

Each router sends a request to the neighbor using PSNPs that specify any missing LSPs, and acknowledges the receipt of the LSPs using again PSNPs. A PSNP serves as an explicit acknowledgment of the receipt of an LSP on a point-to-point link. This process reduces the amount of routing traffic crossing the point-to-point link; each IS-IS router exchanges only the link-state information missing from its LSDB rather than the entire LSDB of its neighbor router.

It is important to note that, every LSP sent over a point-to-point link, whether during the initial LSDB synchronization, or anytime later when the LSDB is updated or an LSP is purged, must be acknowledged using PSNPs or CSNPs. If a neighbor router sends periodic CSNPs, the other router would also accept them as valid acknowledgments of transmitted LSPs.

Figure 2.26 shows an example of LSDB synchronization on a point-to-point link:

- We assume Router 1 establishes an adjacency with Router 2.
- Router 1 and Router 2 each transmit a CSNP to the other. If the LSDB of Router 2, for instance, and the contents of the received CSNP do not match (i.e., are not synchronized), Router 2 will request the required LSPs by sending a PSNP to Router 1.

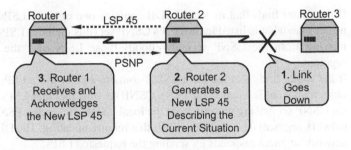

FIGURE 2.26   Link-State Database Synchronization on a Point-to-Point Link

- Router 1 sends the required LSPs to Router 2, and starts the LSP retransmission timer, and waits for Router 2 to send a PSNP acknowledging receipt of the LSPs.
- If Router 1 does not receive a PSNP from Router 2 after the LSP retransmission timer expires, Router 1 retransmits the LSPs until it receives a PSNP from Router 2.

In Figure 2.26, we also assume the link between Routers 2 and 3 fails. The Router 2 router detects this failure and sends a new LSP (LSP 45) announcing this change. Router 1 router receives this LSP, stores it in its LSDB, and then sends a PSNP back to Router 2 to acknowledge receipt of the LSP.

2.14.3   HANDLING LSPs

In an IS-IS domain, LSPs may be locally generated by a router (new LSPs). A router may also receive LSPs from a neighbor on a circuit which may be LSPs generated by another router or may be a copy (duplicate) of an LSP generated by the local router. LSPs received by a router may be older, the same age, or newer than the current contents of the router's local LSDB. This section summaries how a router handles these different LSPs.

2.14.3.1   Handling Newer LSPs

If an IS-IS router receives a newer LSP, it adds it to its local LSDB replacing an older copy of the same LSP currently in the LSDB. The router also marks the newer LSP to be sent (at the routing level associated with the LSP) on all IS-IS circuits on which the router currently has an adjacency in the up state, except the IS-IS circuit on which the LSP was received.

- **Point-to-Point Links**:
  o On point-to-point circuits, the router will flood the LSP periodically until the IS-IS neighbor sends a PSNP to acknowledge receipt of the LSP, or sends an LSP that is the same or newer than the LSP being flooded.
- **Broadcast Multiaccess Network Segments**: On broadcast multiaccess network segments, the router will flood the newer LSP once. The router also examines the set of CNSPs that the DIS sends periodically for the network segment.

o If the router finds that its local LSDB contains one or more LSPs that are newer than what is listed in the DIS's CSNP set (this includes LSPs that are missing from the CSNP set), it refloods those LSPs over the network segment.

o If the router finds that the local LSDB contains one or more LSPs that are older than what is listed in the DIS's CSNP set (this includes LSPs listed in the CSNP set that are missing from the local LSDB), it sends a PSNP on the network segment describing the LSPs that require updating. The DIS for the network segment responds by sending the requested LSPs.

2.14.3.2 Handling Older LSPs

An IS-IS router may receive an LSP that is older than a corresponding copy in its local LSDB. A router may also receive a CSNP or PSNP that describes an LSP that is older than a copy in its local LSDB. If any of these cases happen, the router marks the LSP in its local LSDB to be flooded on the IS-IS circuit on which the older LSP or CSNP/PSNP that contained the older LSP was received. The actions taken by the router at this point are identical to the actions described above after the router finds that its local LSDB contains one or more LSPs that are newer than what is listed in the DIS's CSNP set.

2.14.3.3 Handling Duplicate LSPs

Due to the distributed nature of the LSP flooding process, an IS-IS router may receive copies of an LSP that are identical to the current contents of its local LSDB. Router on a point-to-point link will ignore such LSPs when received. A DIS makes periodic transmission of CSNPs on a broadcast network segment that also serve as an implicit acknowledgement to the sender of an LSP that the LSP has been received. So, in a broadcast network segment, receipt of a duplicate LSP is also always ignored.

2.15 SHORTEST-PATH FIRST AND THE LINK-STATE DATABASE

A router running IS-IS or Integrated IS-IS uses the following process to build its IS-IS Routing Database:

• Applying a SPF algorithm to the LSDB, the router calculates a SPF tree to all know OSI destinations (the OSI addresses or NETs of IS-IS routers and ESs). The link metrics along each path to each destination are totaled to determine which path is the shortest path to the destination.

• The router maintains separate LSDB for Level 1 and Level 2 routes. A Level 1-2 router runs the SPF algorithm twice (once for each routing level) and creates separate SPF trees for each level.

• The router calculates reachability to an ES as a Partial Route Calculation (PRC) based on the Level 1 and Level 2 SPF trees mentioned above. If the network is a pure Integrated IS-IS IP routing environment, there are no OSI ESs. Note that, in Integrated IS-IS, IP subnets attached to an IS-IS router are treated or described as if they are ES.

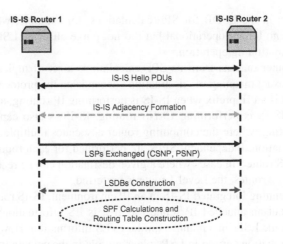

IS-IS Router 1   IS-IS Router 2

IS-IS Hello PDUs

IS-IS Adjacency Formation

LSPs Exchanged (CSNP, PSNP)

LSDBs Construction

SPF Calculations and
Routing Table Construction

1. IS-IS routers send and receive IS-IS Hello PDUs to discover neighbors and establish adjacencies
2. IS-IS adjacency is established (Authentication, IS Type, MTU must match)
3. IS-IS routers build LSPs about their local networks and networks learned from adjacent routers
4. IS-IS routers send the LSPs to the adjacent routers
5. All IS-IS routers construct their LSDBs using the information in the received LSPs
6. Each IS-IS router uses the SPF algorithm to calculate best paths to all known network destinations and then
   installs these in the IS-IS Routing Database

FIGURE 2.27   Summary of IS-IS Protocol Operation

- The router installs the best paths in the IS-IS Routing Database (also referred
  to as the IS-IS Forwarding Database). When Integrated IS-IS is used, the next-
  hop IP address is the IP address of the corresponding next-hop IS-IS neighbor
  router.

Figure 2.27 summarizes the main processing steps in IS-IS protocol operation. Every
IS-IS router requires an NSAP address even if it is performing only IP routing. IS-IS
uses the NSAP address in an LSP to identify the originating router, identify entries in
the LSDB, and construct the underlying IS-IS SPF tree. Integrated IS-IS, even when
used for IP routing, inherits the NSAP node-based addressing scheme for identifying
network nodes. Therefore, even when Integrated IS-IS is used in an IP-only environment, the IS-IS nodes must have NETs.

Each IS-IS router independently reruns a SPF calculation (based on the well-
known Dijkstra algorithm) when the contents of the LSDB change. Each router
determines the shortest paths along a directed graph where the IS-IS routers are represented as the vertices of the graph and the links between the routers are edges having a nonnegative weight. The weight assigned to each branch of the graph can be a
configurable wide metric that has a maximum value of 2^{24} (for each individual link)
and up to 2^{32} for the complete path (root to leaf). A router (the root or single source
vertex) uses Dijkstra algorithm to compute the shortest paths from itself to all other
vertices in this weighted, directed graph.

Before considering a link between any two routers as part of the graph, a two-way
connectivity check is first performed. This prevents routers from using stale routing

information in the LSDB for SPF calculations, for example, when a router in the network is no longer operational but did not purge all of its LSPs that it generated before discontinuing operation.

Each router outputs from its SPF calculations a set of tuples (destination, next hop). The exact meaning of "destination" depends on the protocol mode of IS-IS. A destination is an IP prefix when IS-IS is performing IP routing, and the NSAP of an ES when IS-IS is performing CLNS routing. IS-IS is also capable of supporting ECMP routing, where the computing router associates multiple next hops with the same destination. Independent SPFs are performed for each routing level supported by the IS-IS router. In cases where a given destination can be reached by both Level 1 and Level 2 routes, the Level 1 route is preferred.

When running Integrated IS-IS in an IP environment, IS-IS routers treat IP reachability (about an attached IP subnet) as if it is ES information; IP information is included in the LSPs in a manner similar to ES information. However, this IP reachability information carried in LSPs plays no role in the calculation of the IS-IS SPF tree. This IP reachability information is simply detail information about the leaf connections in the IS-IS SPF tree. Therefore, similar to ES reachability, an IS-IS router updates the IP reachability as only a PRC.

The IS-IS router (running Integrated IS-IS) generates IP routes in the PRC and presents these to the IP Routing Table, where they are accepted based on conventional IP Routing Table route preference rules. For example, when multiple routing protocols present routes to the same destination to the IP Routing Table, their Administrative Distances (also referred to as Route Preferences) are compared to determine which route is to be installed. The route with the lowest Administrative Distance is installed. When IP IS-IS routes are entered in the IP Routing Table, they are marked as belonging to IS-IS Level 1 or Level 2, as appropriate **[CISCTEAPAQ00]**.

It may be argued that separating IP reachability from the core IS-IS network architecture allows Integrated IS-IS to have better scalability than OSPF. In OSPF, LSAs are sent for individual IP subnets, and if an IP subnet fails, an LSA reflecting this change is flooded through the OSPF network and, all routers must run a full SPF calculation.

In an Integrated IS-IS network, the IS-IS routers build the IS-IS SPF tree from CLNS information. If an IP subnet fails in in this environment, LSPs reflecting this change are flooded similar to OSPF. However, if the failed IP subnet is a leaf connection on the IS-IS router (i.e., its loss has no impact on the underlying CLNS network architecture), the IS-IS SPF tree will not be affected and, therefore, the router performs only a PRC **[CISCTEAPAQ00]**.

2.15.1   SPF ALGORITHM HIGHLIGHTS

IS-IS routers exchange IS-IS Hello PDUs which allow them to detect network changes quickly. This also results in faster network convergence. IS-IS uses the SPF algorithm to determine the best routes to network destinations. When network topology changes occur, IS-IS evaluates the changes to determines if a full or PRC is required. Details of the SPF algorithm for IS-IS (a variant of the Dijkstra algorithm) are provided in **[ISO10589:2002]**. Specifically, reference **[RFC1195]** describes the SPF algorithm for IP and Dual environments.

If only the LSP information of an ES has changed, it is not necessary for the attached IS-IS router to recompute the entire SPF tree. If the proper data structures exist in the router, ES may be attached and detached as leaves of the SPF tree, and the IS-IS Routing Database entries modified as appropriate.

All IS-IS routers in an IS-IS routing domain run separate instances of the SPF algorithm independently and concurrently. Intradomain routing of a packet is performed on a hop-by-hop basis. So, the SPF algorithm running in a router determines only the next hop router, and not the complete path, that the packet will take to reach its final destination. To guarantee correct and consistent route computation at every IS-IS router in an IS-IS routing domain, the following properties must be ensured:

- All IS-IS routers in the OSI routing domain use identical topology information (LSDBs) when the routing information distribution converges
- Each IS-IS router in the OSI routing domain generates routes from the same topology information and *set* of routing metrics.

These properties are necessary in order to have consistent, loop-free paths to network destinations.

Each router runs the SPF algorithm to determine a set of legal paths to each destination in the routing domain. The set of legal paths consist of the following:

- A single minimum total metric path: referred to as *minimum or least cost path*.
- A set of minimum total metric paths: referred to as *equal minimum cost paths* (or *equal cost multipaths*).
- A set of paths which will get a packet closer to its final destination than the local router: referred to as *downstream paths*.

An IS-IS router considers paths that do not meet the above conditions as illegal and will not use them. Paths with total metric values exceeding the value of the IS-IS architectural constant *MaxPathMetric* are also illegal and must not be used. For 6-bit metrics (narrow metrics), *MaxPathMetric* is 1,023. The IS-IS parameter *MaxPathMetric* is the maximum total metric value (total path cost) for a complete path to a destination.

Each router, when determining its paths, also ascertains the identity of the adjacent router that lies on the first hop (next-hop router) to the destination on each path. These next-hop routers for each destination are used to create the IS-IS Routing Database (or IS-IS Forwarding Database), which holds the forwarding information for relaying packets at the router. Separate route calculations are performed for each routing level in the IS-IS routing hierarchy (i.e., Level 1 and Level 2) with the supported routing metric.

2.16 OSI ROUTING AND SUBOPTIMAL INTER-AREA ROUTING

A Level 1 router has Level 1 or Level 1-2 neighbors in its own area and is only aware of the topology of its local IS-IS area. It maintains a Level 1 LSDB containing all the information required for intra-area routing. We stated earlier on that, because a Level

1 router uses the closest Level 2-capable router in its local area for sending packets out of the area, suboptimal routing may result. This section clarifies that using a few networking examples.

Routing within an IS-IS Level 1 routing area is based on the System ID carried in the destination ES's OSI NSAP address (see discussion in "NSAP Addresses" section above). Each Level 1-2 router in an IS-IS area sends a default route to its Level 1 routers. The following steps are used when a Level 1 router has a packet that is destined for another IS-IS area in the domain:

- The Level 1 router forwards the packet to the closest Level 1-2 router in the area. The Level 1 router determines the closest Level 1-2 router (which also serves as an exit point from the area) based on the best default route (i.e., the least-cost default route) received from the Level 1-2 routers in its area.
- The selected Level 1-2 router receives the packets, and based on the destination Area ID or Address, routes the packet into the Level 2 backbone. The packet is forwarded hop-by-hop across the Level 2 backbone to the destination area.
- When the packet reaches the destination area, it is forwarded via Level 1 routing again until arrives at its final destination within that area.

A Level 1 router is only aware of the network topology in its area, including all other routers and ESs in its area, but does not know the identity of IS-IS routers or destinations outside of its area (only the destination NSAP address). So, a Level 1 router forwards traffic with destinations outside of their area to a Level 1-2 router in its area. If a destination address is in the same area (Area IDs are equal), a Level 1 router forwards the packet, based on the System ID, to the Level 1 router that advertises the destination address. If the destination address is in a different area (Area IDs are not equal), the Level 1 router forwards the packet to the nearest Level 1-2 router in the local area.

Recall that a Level 1-2 router serves as the interface between a Level 1 and the Level 2 routing subdomain (or backbone). The Level 1-2 router functions as both a Level 1 router (routing packets to Level 1 destinations) and a Level 2 router (routing packets between IS-IS areas). Level 2 routing is based on the Area Address carried in an IS-IS packet. Thus, if a Level 1-2 router receives a packet (from a Level 2 router) that is destined for its local area, it uses the System ID to route the packet at Level 1.

2.16.1 IS-IS SUBOPTIMAL INTER-AREA ROUTING

It should be noted however, that because Level 1 routers in an IS-IS area use a default route to send packets destined for other areas only to the nearest Level 1-2 routers, suboptimal routing can result as illustrated in Figures 2.28 and 2.29 **[CISCISISWP02] [CISCTEAPAQ00]**. Several Level 1-2 routers may inform a Level 1 router about exit points that may lead to a given destination area, but the closest Level 1-2 router may not necessarily provide the optimal path to that destination area. Usually, if there is more than one exit point (multiple Level 1-2 routers), the closest router is selected based on a least-cost criterion. If there are two equal-cost paths, the source Level 1 router may load balance the traffic over these paths.

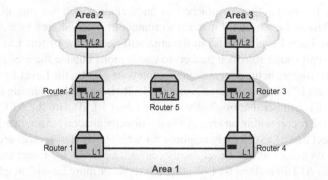

- Let us assume the cost on each link is 10.
- Router 1 (Level 1 router) in Area 1 will send all traffic destined for outside Area 1 to Router 2 (Level 1-2 router) because Router 2 is the closest Level 1-2 neighbor router.
- Router 1 will send packets destined for Area 3 to Router 2.
- Because Router 2, Router 5, and Router 3 are backbone routers, Router 2 will send this packet to Router 3 through Router 5 for delivery into Area 3.
- The more optimal path would be for Router 1 to send the packet directly to Router 3 through Router 4.

FIGURE 2.28   Example 1 of IS-IS Suboptimal Inter-Area Routing

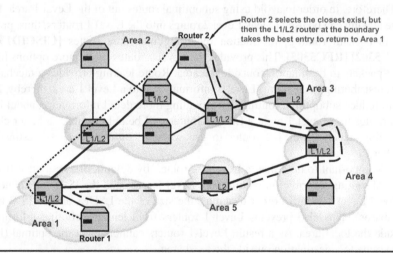

- Router 1 in Area 1 sends a packet destined for Router 2 through Area 1's Level 1–2 router.
- The Level 1-2 router in Area 1 inspects the destination Area ID in the packet and routes it directly into Area 2.
- In Area 2, the packet is routed as Level 1 to Router 2.
- A packet returning from Router 2 to Router 1 is routed by Router 2 to its nearest Level 1–2 router.
- The nearest Level 1-2 router happens to determine the best route to Area as being through Area 4.
- This Level 1-2 router then routes the return packet by a different route than the one taken by the incoming packet. It can be seen that the return path taken is not actually the least-cost one from Router 2 to Router 1.
- Although, asymmetric routing takes place (packets in different directions taking different paths), this is not necessarily harmful to the network and most end-user applications.

FIGURE 2.29   Example 2 of IS-IS Suboptimal Inter-Area Routing

A Level 1-2 router that is attached to another area will set one of the ATT (Attached) bits in Level 1 LSP it sends to routers in its local area (see ATT bit in Figure 2.16). Each Level 1 router in the area will get a copy of this LSP and then discover the exit router to which packets to destinations outside the area can be forwarded. If the routers in the area are running Integrated IS-IS, the Level 1 routers will install a default IP route automatically in their IP Routing Tables pointing toward the nearest Level 1-2 router that originally sent the Level 1 LSP with the ATT bit set.

A Level 1-2 router within an area that is not directly attached to another IS-IS area can also detect that a Level 2-only neighbor router is attached to another area, and set the ATT bit on behalf of this Level 2-only router. The Level 1-2 router can indicate (by setting an ATT in its Level-1 LSP) that it has one or more Level 2 neighbor routers in other areas that can be used by Level 1 routers in the same area as the path of last resort (the default route).

2.16.2   AVOIDING SUBOPTIMAL ROUTING USING IS-IS ROUTE LEAKING

The discussion above shows that the shortest path from an IS-IS Level 1 area to an outside destination may not be through the closest Level 1-2 router. Another problem is that, the closest Level 1-2 router capable of forwarding packets out of an area may not even be operational thereby requiring an approach that provides more exit options. The discussion above shows that sending packets destined for an external destination to the closest Level 1-2 router can lead to suboptimal routing when the shortest path to the destination is through a different Level 1-2 router.

Therefore, in order to avoid using suboptimal routes out of the Level 1 area, IS-IS route leaking can be used to leak Level 2 routes into the Level 1 routers, thus, providing them with routing information regarding inter-area routes [CISCID13796] [RFC5302] [RFC5305]. This provides the Level 1 routers with more options to forward packets to destinations outside the area. Route leaking provides a mechanism for redistributing (or leaking) Level 2 information into a Level 1 area, thereby, helping to reduce suboptimal routing. By providing more detail information about inter-area routes to a Level 1 area, a Level 1 router will be able to make a better choice regarding which Level 1-2 router to use when forwarding packets destined for another IS-IS area.

It is important to note that, a Level 1-2 router, by default, does not advertise the internal routing information within other Level 1 areas and the IS-IS backbone into its local Level 1 area (in order to reduce the size of the LSDBs and IS-IS Routing Databases). This also prevents Level 1 routers from learning routing information outside the local area. As a result, Level-1 routers cannot select the optimal (least-cost) route to a destination outside the local area.

Reference [RFC5302] defined IS-IS route leaking for use with the IS-IS narrow metric TLVs, the IP Internal Reachability Information TLV (TLV 128), and the IP External Reachability Information TLV (TLV 130). Reference [RFC5305] extended route leaking for use with the IS-IS wide metric TLV, the Extended IP Reachability TLV (TLV 135). Both [RFC5302] and [RFC5305] defined an up/down bit (in the TLVs 128, 130, and 135) to indicate to IS-IS routers whether or not the route defined in these TLV has been redistributed or leaked.

If the up/down bit in the TLV is set to 0, this means the route in the TLV was originated within that local Level 1 area. If the up/down bit is set to 1, the route has been redistributed into the Level 1 area from Level 2. According to **[RFC5305]**, if a route is advertised from Level 2 to a Level 1 area, the up/down bit must be set to 1, indicating that the route has traveled down the hierarchy. This provides a way to distinguish Level 2-to-Level 1 inter-area routes from Level 1 intra-area routes. Routes that have their up/down bits set to 1 may only be advertised down the routing hierarchy, i.e., to Level 1. To prevent routing loops from occurring, reference **[RFC5302]** defines the following rules for route leaking:

- Level 1-2 routers must set the up/down bit to 1 for routes that are derived from Level 2 routing and are advertised into Level 1 LSPs. The router must set the bit to 0 for all other routes in Level 1 or Level 2 LSPs.
- Level 1-2 routers must not advertise routes with the up/down bit set (i.e., Level 2-to-Level 1 inter-area routes) that are learned via Level 1 routing back into Level 2.

IS-IS routers use the setting of the up/down bit in the TLVs to prevent routing information and forwarding loops. This is to prevent the looping of routes between routing levels. A network may also use route leaking to implement a simple form of traffic engineering **[CISCID13796]**. Routing information about individual ESs or services can be leaked from specific Level 1-2 routers to control the exit point from a Level 1 area used to reach these addresses.

When route leaking is used in Figure 2.28, Routers 2 and 3 can redistribute routing information about Area 2 and 3 into Area 1. This allows Router 1 and Router 4 to select optimal paths to get to Area 2 and Area 3. Router 1 can now sends traffic destined for Area 3 via Router 3; which reduces the cost to 20, while still forwarding traffic destined for Area 2 through Router 2.

Similarly, Router 4 forwards traffic to Area 2 through Router 2, while still routing to Area 3 via Router 3. By using route leaking on Router 2 and Router 3, Routers 1 and 4 will be able to determine the true routing metrics for reaching Area 2 and Area 3. Essentially, route leaking allows IS-IS to perform "shortest-path exiting" for packets going to other IS-IS areas.

Route filters (e.g., using route maps) and route tagging can be configured on the Level 1-2 routers to filter and tag routes so that routes that have been leaked into a Level 1 area will not be redistributed back into the Level 2 routing subdomain and other areas. This allows a Level 1-2 router to advertise the routing information of other Level 1 areas and the Level 2 backbone area to its Level 1 area, but not allow any Level 1-2 router in the local area to readvertise these routes back to the areas that originated them. Leaked routes will be identified and specially tagged so that they will not be readvertised by any Level 1-2 router into Level 2.

2.16.3 Configuring Default Routes Using the "default information originate" Command

We have seen above that the ATT bit can be set by a Level 1-2 router in its own Level 1 LSPs to indicate to all Level 1 routers within the local IS-IS area that it is a potential exit point of the area. The Level 1 routers will then use the default route to the nearest Level 1-2 router when sending traffic outside the area. Default routing can also be achieved in another way when using Integrated IS-IS.

The **default information originate** command **[CISCISISCOMD20]** can be configured on an IS-IS router running Integrated IS-IS to cause it to generate a default route into an IS-IS area. The router inserts the default route (0.0.0.0/0) in its LSPs (Level 1 or Level 2) and the LSPs are flooded according to the routing level (Level 1 or Level 2). The Level 1 routers in the IS-IS area will always prefer the explicit default route (0.0.0.0) advertised in the LSPs before considering the route advertised via the ATT bit. A Level 2 router does not need to have a default route to originate a default route.

2.16.4   OTHER DRIVERS FOR IS-IS ROUTE LEAKING

Other than enabling Level 1 routers to determine optimal routes out of the local Level 1 area, there are two other significant drivers for leaking routes from the Level 2 backbone and other Level 1 areas into the local area. When a BGP router learns multiple possible routes to external destinations, it selects only the best route to installed in its IP Routing Table. The IGP cost to the BGP next hop address is one of the factors in the BGP best path selection process (see Chapter 3). Many practical BGP networks use this technique, which is known as "shortest exit routing" or "shortest-path exiting". If an IS-IS Level 1 router is not able to determine the exact IGP metric to all BGP routers in *other* Level 1 areas, it cannot perform effective shortest-path exiting **[RFC5302]**.

BGP deployments typically use the IGP metric (in this case IS-IS metric) as part of the BGP Multi-Exit Discriminator (MED) attribute (see Chapter 3). BGP routers advertise the MED value to other domains (autonomous systems) to inform those domains of the optimal entry point into the current domain. When IS-IS is used as the IGP, the IS-IS metric can be taken and inserted as the MED value in the BGP MED attribute. This is to cause external traffic from other domains to enter the local domain through the point closest to the exit router.

Note that a domain that receives the BGP MED attribute may choose to ignore the MED value that is advertised based on local policy. However, current BGP practice is to advertise the IGP metric to other domains in order to provide optimum routing wherever possible. This practice is possible in networks that consist of only a single area, but becomes problematic in IS-IS two-level networks that are based on the traditional "nearest Level 1-2 router" exit policy (using a default route to exit the Level 1 area).

The absence of end-to-end metric information in such networks means that the MED value will not reflect the true cost across the advertising domain. Using route leaking (i.e., providing redistribution of Level 2 and other Level 1 area routing information within the domain) as described above, would alleviate this problem. Route leaking would allow accurate IS-IS metrics across the domain to be computed, resulting in an accurate IGP metric presented in the BGP MED attribute.

2.16.5 SOME PITFALLS OF IS-IS ROUTE LEAKING

Performing route leaking in an unstable environment may carry some risks. Generally, routes that are leaked from Level 2 into an IS-IS Level 1 are not aggregated (or summarized). Route summarization usually prevents instabilities in one area (link or route flapping) from propagating and affecting other areas. Since route leaking is generally used without summarization, each time a topology change occurs in an IS-IS area, all routers in this area may change their routing metrics (because of the change), and the local Level 1-2 router will have to recreate its Level 2 LSP and propagate new routing information into the Level 2 backbone **[CISCISISWP02]**.

This means that route leaking will have to recur in receiving Level 1 areas performing shortest-path exiting. This leads to a situation where, for any topology change in one IS-IS area, Level 1 routers receiving the new routing information will have to recompute (via PRC) shortest-path exit routes to other IS-IS areas as well. This situation therefore requires that route leaking be carefully planned to make shortest-path exiting more effective.

The ultimate goal is to enable Level 1 routers within an IS-IS area to learn routing information to external area without passing, as much as possible, instabilities from one area to another. As explained above, allowing Level 1 routers within an IS-IS area to use least-cost exit points rather than the default nearest Level 1-2 router exit policy, enables many other applications to be deployed in the IS-IS domain.

2.17 IS-IS AUTHENTICATION

To guarantee that only trusted routers can participate in routing in the IS-IS routing domain (or autonomous system), IS-IS protocol exchanges can be authenticated using Simple Password Authentication or Cryptographic Authentication. IS-IS supports Simple Password (or Plaintext) Authentication **[ISO10589:2002] [RFC1195]**, and Cryptographic Authentication **[RFC5304] [RFC5410]** (e.g., Message Digest 5 [HMAC-MD5] authentication). The authentication mechanisms proposed in **[RFC5304]** and **[RFC5410]** provide more secure authentication than plaintext authentication. It should be noted, however, that the authentication mechanisms described in **[RFC5304]** and **[RFC5410]** do not provide confidentiality (similar to other authentication mechanisms in RIPv2, EIGRP, OSPFv2, BGPv4, etc.), since IS-IS PDUs are sent in cleartext. Routing protocols advertise routing topology information, and generally do not provide confidentiality for routing information.

An Authentication TLV option carried in the IS-IS PDU header allows the originator and contents of a PDU to be authenticated by the receiver. The Authentication TLV consists of the tuple <Authentication Type, Authentication Length, Authentication Value>, where the 1-byte *Authentication Type* field specifies the Code for the authentication method, the 1-byte *Authentication Length* field specifies the length of the *Authentication Value* field (Figure 2.30). The variable-length *Authentication Value* field carries the authentication data with length equal to the value specified in the *Authentication Length* field. The Value and its Length in the TLV depends on the authentication algorithm used.

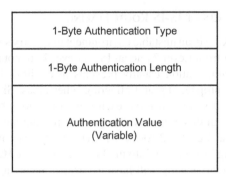

FIGURE 2.30   Authentication TLV Option Fields

IS-IS authentication is configured independently for adjacency formation (via a TLV in IS-IS Hello PDUs) and for LSP authentication (via a TLV in Level 1 and Level 2 LSPs). If only LSP authentication is configured, an IS-IS router can still form an adjacency with an unauthorized IS-IS neighbor, but LSPs cannot be exchanged. The router's LSDB will not contain any entries for this neighbor.

2.17.1   IS-IS SIMPLE PASSWORD AUTHENTICATION

IS-IS as defined in **[ISO10589:2002]** supports only basic security through packet authentication (Plaintext or Simple Password Authentication) using the special TLV 10. Reference **[RFC1195]** also specifies Simple Password Authentication using TLV 133, which removes the password length restrictions imposed by **[ISO10589:2002]**. Both specifications define only simple passwords where routers transmit passwords in cleartext without encryption. A variable-length password is carried in cleartext in the TLV's Authentication Value field.

Authentication passwords can be configured on an IS-IS routing domain on a per-link, per-area, and per-domain basis as follows **[ISO10589:2002]**:

*   IS-IS Hello and ISO 9542 ISH PDUs containing a per-link password
*   Level 1 LSPs containing a per-area password
*   Level 2 LSPs containing a per-domain password
*   Level 1 SNPs containing a per-area password
*   Level 2 SNPs containing a per-domain password

When the per-link, per-area or per-domain authentication is used, each mode is configured with a single transmit password, and a set of receive passwords. The sender always adds the transmit password to a departing PDU. If the transmit password in the PDU matches any of the passwords in the receive password set, the receiver will accept the PDU for further processing. The receive password set allows the receiver to perform graceful password change without temporary loss of connectivity with the sender.

2.17.1.1 Interface Password

An interface authentication can be used to make sure that only IS-IS neighbors that are authorized can establish an IS-IS adjacency over a given interface. Interface authentication is configured on an interface so that neighbor routers can authenticate Level 1 and Level 2 IS-IS Hello PDUs. The interface password is configured on a per-interface basis and must be specified for each routing level (Level 1 and Level 2) independently.

Over point-to-point interfaces, only one IS-IS Hello PDU is sent for Level 1-2 adjacencies, and on these interfaces, the same interface password must be used as a Level 1 and Level 2 password.

When Plaintext Authentication is enabled on a link, a password is included in each IS-IS PDU sent on the link. The IS-IS router supplies the password in all IS-IS PDUs it sends on the interface. The receiving neighbor IS-IS router must supply the correct password that matches the one carried in the received IS-IS PDU on the interface [ISO10589:2002]:

- If an IS-IS router receives a Point-to-Point IS-IS Hello PDU, and the value of *circuitTransmitPassword* or the set of *circuitReceivePasswords* for the link is not null, the router will compare the password in the received Hello PDU with the passwords in the set of *circuitReceivePasswords* for the link on which the Hello PDU was received. If the value in the Hello PDU matches any of these passwords, the router will accept the Hello PDU for further processing. If the value in the Hello PDU does not match any of the circuit-*ReceivePasswords*, then the router will ignore the Hello PDU and generate an *authenticationFailure* event.

2.17.1.2 LSP Authentication (Area- or Domain-Wide)

Area authentication is configured so that IS-IS routers can authenticate Level 1 LSPs, CSNPs, and PSNPs. Domain authentication is configured so that IS-IS routers can authenticate Level 2 LSPs, CSNPS, and PSNPs. Domain-wide and area-wide authentication can be configured separately. One or both types of authentication modes can be run. Area passwords are inserted in Level 1 LSPs to be checked, while domain passwords are inserted and checked for in Level 2 LSPs [ISO10589:2002]:

- If an IS-IS router receives a Level 1 LSP, and the value of *areaTransmitPassword* or the set of *areaReceivePasswords* is not null, the router will compare the password in the received LSP with the passwords in its local set of *areaReceivePasswords*, augmented by the *areaTransmitPassword* value. If the value in the received LSP matches any of these passwords, the router will accept the LSP for further processing. If the value in the LSP does not match any of these password values, then the router will ignore the LSP and generate an *authenticationFailure* event.
- If a Level 2 LSP is received, and the value of *domainTransmitPassword* or the set of *domainReceivePasswords* is not null, the router will compare the

password in the received LSP with the passwords in the set of *domainReceive-Passwords*, augmented by the *domainTransmitPassword* value. If the value in the LSP matches any of these passwords, the router will accept the LSP for further processing. If the value in the LSP does not match any of these values, the router will ignore the LSP and generate an *authenticationFailure* event.

2.17.1.3   Limitations of Simple Password Authentication

Simple password authentication can be used as a way to isolate operator configuration errors (misconfigurations) related to adjacency setups, but does not provide strong protection against malicious attacks on the network. The specifications in **[ISO10589:2002]** and **[RFC1195]** both provide accommodation for more complex and secured authentication using schemes as discussed below.

From packet authentication perspective, IS-IS has a unique security advantage over other IP routing protocols because IS-IS PDUs are directly encapsulated in the Data Link Layer frames and not in IP packets or even CLNP packets. Therefore, in order for a malicious attacker to be able to disrupt the IS-IS routing environment, the attacker has to be physically attached to the Data Link Layer in the IS-IS network, which is a challenging and an inconvenient task to be carried out by most attackers.

Other IP routing protocols, such as RIPv2, EIGRP, OSPFv2, and BGPv4, are susceptible to malicious attacks from a remote IP network location through the Internet, because packets from these routing protocols are carried over IP. This makes these routing protocols more susceptible to remote attacks by intrusive and malicious applications.

2.17.2   IS-IS HMAC-MD5 Cryptographic Authentication

Reference **[RFC5304]** describes Cryptographic Authentication of IS-IS PDUs using the Hashed Message Authentication Codes – Message Digest 5 (HMAC-MD5) algorithm. To do this, this specification proposes an extension to the IS-IS specification in **[ISO10589:2002]** that allows the HMAC-MD5 authentication algorithm to be used in conjunction with the existing authentication mechanisms in **[ISO10589:2002]**.

Reference **[RFC5304]** extends and modifies the mechanisms in **[ISO10589:2002]** by defining a new Authentication Type (TLV Code) for HMAC-MD5 and specifying the algorithms for computing the Authentication Value. The Authentication Type (Code) defined for HMAC-MD5 is 54 (0x36), the Authentication Length field in the TLV is 17 bytes, and the length of the Authentication Value for HMAC-MD5 is 16 bytes.

The HMAC-MD5 algorithm requires a secret key K and "text" T as input as described in **[RFC2104]**. The key secret K is configured on all the IS-IS routers attached to a common network segment requiring IS-IS authentication. For each IS-IS PDU, the secret key K is used to generate/verify a message digest that is carried in the Authentication Value field of the TLV 54. The message digest is a one-way function of the secret key K and the IS-IS PDU. The secret key K is only maintained securely by the IS-IS routers and is never sent across the network segment to protect against eavesdropping on the network and passive attacks. Only the message digest is carried in the Authentication Value field of the TLV 54. The secret key K is the password as specified in **[ISO10589:2002]** for the PDU type and is NEVER sent across the network as in Plaintext Authentication.

The text *T* is the IS-IS PDU to be authenticated but the contents of the Authentication Value field (carried in the Authentication Information TLV 54) is set to zero. The sender sets the Authentication Type to 54 and the Authentication Length field in the TLV to 17 before computing the Authentication Value. Before LSPs are to be sent, the sender sets the Checksum and Remaining Lifetime fields of the PDU to zero (0) before computing the Authentication Value. The sender then places the result of the HMAC-MD5 algorithm in the Authentication Value field.

When the sender is calculating the HMAC-MD5 result for LSPs and SNPs, it uses the Area Authentication (string) password as the secret key for Level 1 LSPs and SNPs, and the Domain Authentication password as secret key for Level 2 LSPs and SNPs. For IS-IS Hello PDUs, the sender uses the Link Level Authentication password as key, which may be different from that of LSPs. The sender calculates the HMAC-MD5 result for the IS-IS Hello PDUs after the packet is padded to the MTU size, if padding is enabled.

When the receiving IS-IS router receives an incoming PDU to be authenticated, it will save the values of the Authentication Value field, the Checksum field, and the Remaining Lifetime field, and then set these fields to zero. After computing the local message digest (using its shared secret key) to be used for authenticating the PDU, it will restore the values of these fields to their original values. The receiver will discard the PDU if the message digest carried in the Authentication Value field does not match the corresponding value computed locally.

2.17.3 IS-IS GENERIC AUTHENTICATION

Reference **[RFC5310]** proposes an extension to the IS-IS authentication mechanism to allow a wide range of cryptographic authentication algorithms to be used in addition to the authentication scheme described in **[RFC5304]**. The proposal uses the Hashed Message Authentication Code (HMAC) construct along with the Secure Hash Algorithm (SHA) as an example, but is developed to be general enough to allow IS-IS to use a wide family of cryptographic hash functions. HMAC (**[RFC2104]**) requires a cryptographic hash function and can use any one of SHA-1, SHA-224, SHA-256, SHA-384, or SHA-512 algorithms to authenticate IS-IS PDUs.

The authentication methods described in **[RFC5304]** and **[RFC5410]** are not used to authenticate the specific originator of an IS-IS PDU, but rather to confirm that the PDU originates from an IS-IS router that has access to either the area or domain password, depending upon the kind of IS-IS PDU being processed. These mechanisms also do not protect IS-IS PDUs against replay attacks and consider this issue out of scope.

2.17.3.1 Authentication TLV and IS-IS Security Association

Reference **[RFC5310]** proposes a new Authentication Type, called the Generic Cryptographic Authentication (CRYPTO_AUTH), to be carried in the IS-IS Authentication TLV 10 as described in Figure 2.31. The new Authentication Type, 3, indicates that the IS-IS adjacency is using the CRYPTO_AUTH mechanism, and it is inserted as the first

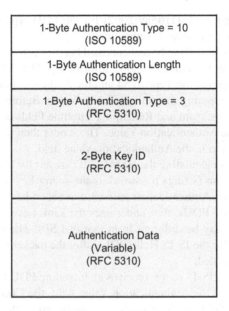

| 1-Byte Authentication Type = 10 (ISO 10589) |
| 1-Byte Authentication Length (ISO 10589) |
| 1-Byte Authentication Type = 3 (RFC 5310) |
| 2-Byte Key ID (RFC 5310) |
| Authentication Data (Variable) (RFC 5310) |

FIGURE 2.31   Generic Authentication TLV Format

byte of the Authentication Value field of the existing IS-IS Authentication TLV 10 described in **[ISO10589:2002]**. The goal of the proposal is to allow any authentication algorithm to be used for authenticating and verifying IS-IS PDUs.

Reference **[RFC5310]** does away with per-interface keys and instead, uses Key IDs that map to unique IS-IS Security Associations (SAs). An IS-IS Security Association holds a set of parameters that are shared by any two authorized or legitimate IS-IS routers in the routing domain:

- **Key Identifier (Key ID)**: This is a two-byte unsigned integer (that is manually configured by the network operator and is) used to uniquely identify an IS-IS Security Association. The receiving IS-IS router determines the active Security Association by examining the Key ID field in the incoming PDU.

  The sending IS-IS router (based on the active Security Associations configured) selects the Security Association to be used, and writes the appropriate Key ID value associated with the selected Security Association in the Key ID field of outgoing IS-IS PDU (Figure 2.31). If the sender, at the time an IS-IS PDU is being sent, sees that multiple valid and active IS-IS Security Associations exist for the outbound interface, it may use any of these existing Security Associations to protect the packet.

Using Key IDs makes it more convenient to change keys while maintaining continuous operation of the IS-IS protocol. Each Key ID in a Security Association specifies two independent components: the Authentication Algorithm (HMAC-SHA-1, HMAC-SHA-224, HMAC-SHA-256, HMAC-SHA-384, or HMAC-SHA-512), and the Authentication Key. In a typical implementation, the network operator would configure a set of Authentication Keys in a Key Chain, with a fixed lifetime specified for each Key in the Key Chain. Each Key ID may indicate an Authentication Key to be used with a different authentication protocol. This allows the use of multiple authentication mechanisms (including the introduction of new authentication mechanisms) and at various times, as needed, without disrupting the IS-IS adjacency.

- **Authentication Algorithm**: This represents the authentication algorithm to be used with the selected IS-IS Security Association. This information is never transmitted over the network segment in cleartext, thereby, allowing a variety of algorithms to be implemented (e.g., HMAC-SHA-1, HMAC-SHA-224, HMAC-SHA-256, HMAC-SHA-384, and HMAC-SHA-512).
- **Authentication Key**: This value represents the cryptographic authentication key associated with the selected IS-IS Security Association. This is a variable-length key and depends upon the Authentication Algorithm specified by the IS-IS Security Association. The Authentication Key is never sent over the network in any form. It is beneficial to change the Authentication Keys used periodically, and in which case, an implementation must be able to store and use more than one key at the same time.

2.17.3.2 Authentication Process

A sender uses the Area Authentication string (password) when calculating the CRYPTO_AUTH results for Level 1 LSPs and SNPs, and the Domain Authentication string when calculating results for Level 2 LSPs and SNPs. For IS-IS Hello PDUs, the sender uses the Link Level Authentication string, which may be different from that of LSPs. The sender calculates the CRYPTO_AUTH result for the IS-IS Hello PDUs after the PDU is padded to the MTU size, if padding is enabled.

To describe the algorithm, the following nomenclature, is used:

- H is the specific hashing algorithm being used (e.g., SHA-256).
- K is the password (Authentication Key) for the IS-IS PDU type (link, area, or domain password) as per **[ISO10589:2002]**.
- Ko is the cryptographic key used with the hash algorithm H.
- B is the internal block size of H (not the hash size), measured in bytes rather than bits. For SHA-1 and SHA-256, $B = 64$; For SHA-384 and SHA-512, $B = 128$.
- L is the length of the hash, measured in bytes rather than bits.
- XOR is the Exclusive-OR operation.
- $Opad$ is a value equal to the hexadecimal value 0x5c replicated B times.
- $Ipad$ is a value equal to the hexadecimal value 0x36 replicated B times.
- $Apad$ is a value equal to the hexadecimal value 0x878FE1F3 replicated $(L/4)$ times.

The processing steps are as follows:

1. **Key Preparation**:
   a. The cryptographic key, Ko, is always L bytes long.

    b. If the Authentication Key (i.e., the IS-IS secret password), K, is L bytes long, then Ko is set equal to Authentication Key, K.

    c. If the Authentication Key, K, is more than L bytes long, then Ko is set equal to the result $H(K)$.

    d. If the Authentication Key, K, is less than L bytes long, then Ko is set to the Authentication Key, K, with zeros appended to the end of K such that Ko is L bytes long.

2. **Compute First Hash**:

First, the IS-IS router fills the Authentication Data field of the IS-IS PDU with the value *Apad*, and the Authentication Type field is set to 0x3. Then, the router computes a first hash (*First-Hash*), also known as the *inner hash*, as follows:

$$First\text{-}Hash = H(Ko \text{ XOR } Ipad \parallel (IS\text{-}IS\ PDU))$$

3. **Compute Second Hash:**

Then the router computes a second hash (*Second-Hash*), also known as the *outer hash*, as follows:

$$Second\text{-}Hash = H(Ko \text{ XOR } Opad \parallel First\text{-}Hash)$$

4. **Result (Authentication Data):**

The router takes the resulting second hash (*Second-Hash*) as the Authentication Data that is written in the Authentication Data field of the outgoing IS-IS PDU (see Figure 2.31). The length of the Authentication Data field (as shown in Figure 2.31) is always equal to the size of the message digest of the specific hash function H chosen to be used. This also implies that if a hash function with larger output size is used, this will also result in the size of the IS-IS PDU to be transmitted being increased.

The sender must fill the Authentication Type and the Authentication Length fields before computing the Authentication Data (see Figure 2.31). The sender computes the Authentication Data as explained above. The Authentication Length of the TLV is set based on the Authentication Algorithm being used: Length is set equal to 23 for HMAC-SHA-1, 31 for HMAC-SHA-224, 35 for HMAC-SHA-256, 51 for HMAC-SHA-384, and 67 for HMAC-SHA-512. Note that 2 bytes have been added to the Authentication Length field value to account for the 2-byte Key ID field, and 1 byte to account for the Authentication Type field size. The sender also fills in the Key ID field.

The sender sets the Checksum and Remaining Lifetime fields of the IS-IS PDU to zero (for the LSPs) before calculating the Authentication Data. The router then writes the result of the Authentication Algorithm in the Authentication Data field (following the 2-byte Key ID field). The router computes the Authentication Data for IS-IS Hello PDU after the PDU has been padded to the MTU size, if padding is enabled. The router sets the Remaining Lifetime field of the LSP to zero before computing the Authentication Data, meaning this field is not authenticated. The Remaining Lifetime field is excluded so that the LSPs may be aged by other intermediate IS-IS routers, without having to recompute the Authentication Data.

The receiving IS-IS router identifies the appropriate IS-IS Security Association by checking the Key ID in the Authentication TLV 10 carried in the incoming IS-IS PDU. The receiver then performs the Authentication Algorithm-dependent receive processing using the algorithm specified by the appropriate IS-IS Security Association for the received IS-IS PDU. The receiver saves the values of the Authentication Value, the Checksum, and the Remaining Lifetime fields before performing any processing.

The receiver then sets the contents of the Authentication Value field of the incoming PDU with *Apad*, and the Checksum and Remaining Lifetime fields with zero, before computing the Authentication Data. The router compares the locally calculated Authentication Data with the Authentication Data in the received PDU, and discards the PDU if the two do not match. In case of a mismatch, the receiver will log an error event.

2.18 SUBNETWORK INDEPENDENT FUNCTIONS OF AN IS-IS ROUTER AND DATA FLOW

Reference **[ISO10589:2002]** groups the IS-IS protocol's routing layer functions into two main categories: subnetwork-independent functions and subnetwork-dependent functions. In IS-IS, the subnetwork-dependent functions relate to the functions of the Data Link Layer. IS-IS recognizes only two types of subnetworks: point-to-point links and multiaccess broadcast network segments.

The subnetwork-dependent functions relate to the capabilities a network entity needs to interface with the Data Link Layer, and include the functions for detecting, establishing, and maintaining routing adjacencies with IS-IS neighbor routers and ESs over various network media types. The operation of the subnetwork-dependent functions relies on certain elements of the CLNP and the ES-IS protocol. The subnetwork-dependent functions work with ES-IS allowing an IS-IS router to determine the NSAP addresses of adjacent neighbors. On multiaccess broadcast network segments like Ethernet networks, a router will be able to obtain SNPAs (i.e., MAC addresses) for all adjacent neighbors and store them in the Adjacency Database.

The subnetwork-independent functions relate to the capabilities a network entity needs to discover, establish, and maintain adjacencies with the routers in an IS-IS domain. It allows an entity to exchange and process routing information and other control information with adjacent IS-IS routers and ESs after validation by the subnetwork-dependent functions. Reference **[ISO10589:2002]** decomposes the subnetwork independent functions of an IS-IS router into the following more specific functional components: Decision Process, Update Process, Forwarding Process, and Receive Process (Figure 2.32). Figure 2.32 is a data-flow diagram that shows the flow of information between these routing functional components.

2.18.1 RECEIVE PROCESS

The Receive Process serves as the entry point for all PDUs, including end-user PDUs, PDUs carrying routing information, control PDUs, and error report PDUs. The Receive Process is responsible for obtaining input information from the following sources:

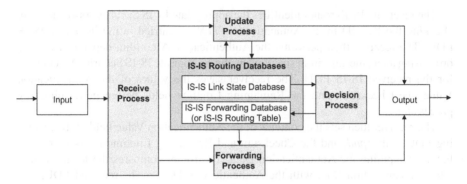

FIGURE 2.32   Subnetwork Independent Functions of an IS-IS Router and Data Flow

- Receive PDUs with their NLPIDs for intradomain routing
- Receive ISO 9542 PDUs exchanged by the ES–IS protocol and derive routing information
- Receive Data PDUs from ESs.
- Perform other actions such as passing the PDU to some other function, for example, to the Forwarding Process for forwarding.

The Receive Process passes routing information and control PDUs (Hello PDUs, LSPs, and Sequence Number PDUs [CSNPs and PSNPs]) to the Update Process, along with an indication of which adjacency they were received on. It also passes user PDUs and error report PDUs to the Forwarding Process.

2.18.2   UPDATE PROCESS

The Update Process is responsible for maintaining the LSDB. The Update Process is also responsible for generating local link-state information (LSPs), based on the Adjacency Database created by the router's subnetwork-dependent functions. The router then advertises these LSPs to all of its neighbors. A router also receives link-state information via LSPs from all of its adjacent neighbor, enters copies of the received LSPs in its local LSDB, and re-advertises them to other neighbors. In a Level 1-2 router, the Update Process manages the Level 1 and Level 2 LSDBs, and floods Level 1 and Level 2 LSPs appropriately.

The LSDB contains a set of information derived from the latest LSPs received from all known routers (within the area, for Level 1 routing, or within the Level 2 subdomain, for Level 2 routing). The Update Process constructs local LSPs, and receives, processes, and propagates LSPs from adjacent routers. Each LSP contains information about the identity of the router that originated the LSP and the routing metric values associated with the adjacency. The Update Process receives LSPs and SNPs (CSNPs and PSNPs) from the Receive Process.

The Update Process is responsible for determining, whether a given LSP represents new, old, or duplicate information with respect to the LSPs stored in the LSDB. It enters new routing information in the LSDB, and also propagates routing information to other routers. The Update Process is responsible for constructing a set of LSPs to inform all the other routers (Level 1 or Level 2) of the state of the links between itself and its neighbors. It is also responsible for assigning a Sequence Number to the LSPs it has created, and for correctly adjusting the Remaining Lifetime field of LSPs originated by other systems in the OSI domain.

The Update Process has the following general characteristics:

- IS-IS Routers generate LSPs as a result of network topological changes, and also periodically. Routers may also generate LSPs indirectly as a result of system management actions (such as when one of the routing metrics for a circuit changed).
- Level 1 LSPs are propagated to all IS-IS routers within an area, but are not propagated out of an area.
- Level 2 LSPs are propagated to all Level 2 routers in the domain.
- LSPs are not propagated outside of an OSI domain.
- Through a set of system management parameters, the Update Process, enforces an upper bound on the amount of routing traffic overhead it generates.

The Update Process also generates one or more new LSPs upon timer expiration, when notified of an Adjacency Database change, and when a change to a system management parameter causes the information in the LSP to change (e.g., a change in *manualAreaAddresses*).

Each LSP in the LSDB has a Remaining Lifetime, Sequence Number, and a Checksum. The LSP's Remaining Lifetime counts down from its time to expiration (with maximum value of *MaxAge*) to 0. The originator of the LSP must periodically refresh its LSPs to prevent their Remaining Lifetime from reaching 0. If the LSP's Remaining Lifetime reaches 0, the expired LSP will be retained in the LSDB for an additional *ZeroAgeLifetime* before it is purged.

If an LSP with an incorrect Checksum is received, the router will initiate a purge of the LSP by setting its Remaining Lifetime value to 0, and reflood it. This triggers the originator of the LSP to send a new LSP. OSPF behaves differently, allowing only the LSP's originator to purge the LSP.

2.18.3 DECISION PROCESS

The Decision Process is responsible for calculating routes to each destination in the ISO domain. If the router supports different routing levels (Level 1 and Level 2), it is run separately for these routing. This process uses the LSDB, which consists of routing information taken from the latest PDUs from every other router in the area, to compute shortest paths from itself to all other systems in the area. Other information in the LSDB include routing metrics for each circuit, and timers that can be adjusted to enhance routing performance.

The Decision Process upon execution, results in the determination of the pair [*circuit, neighbor*], also known as adjacencies, which are stored in the IS-IS Routing Database (also called the Forwarding Database or Forwarding Information Base

[FIB] in IS-IS terminology) of the appropriate routing level. The IS-IS Forwarding Database provides the paths used by the Forwarding Process to forward packets (PDUs) toward their destinations. The Decision Process runs the SPF algorithm on the LSDB, and creates the IS-IS Forwarding Database.

The Decision Process uses the LSDBs containing the link-state information to calculate the best paths for the IS-IS Forwarding Database(s), from which the Forwarding Process can determine the proper next hop for each packet. The router uses the Level 1 LSDB for calculating the Level 1 Forwarding Database(s), and the Level 2 LSDB for calculating the Level 2 Forwarding Database(s).

2.18.4 FORWARDING PROCESS

The Forwarding Process obtains packets to be forwarded from the Receive Process, and uses the IS-IS Forwarding Database to forward these packets toward their destinations. It also performs load sharing when multiple paths exist to a destination, and generates error report PDUs when necessary. The Forwarding Process is responsible both for transmitting PDUs (or packets) originated by the local router system, and for forwarding or relaying packets originated by other routers and ESs. It is also responsible for supplying and managing the buffers necessary to support the relaying of packets to all destinations. It receives packets, via the Receive Process and Update Process to be forwarded on to other nodes.

The Forwarding Process selects an IS-IS Forwarding Database for each packet to be forwarded based on the routing level at which the packet should be forwarded: Level 1 or Level 2. Upon receiving a packet, the Forwarding Process performs a lookup in the appropriate Forwarding Database to determine the best output adjacency to which the packet should be forwarded on its way to the destination.

The Network Layer destination address of a PDU (packet) being forwarded consists of the subfields, Area Address, System ID, and NSEL. Note that the NSEL subfield in the destination address is not examined by IS-IS routers but rather used by ESs (including a destination router) to select the proper Transport Layer entity to which the PDU should be delivered.

2.19 HIGH-LEVEL IS-IS ROUTER ARCHITECTURE, PROCESSES, AND DATABASES

In this section, we describe the processes, interfaces, and databases in a typical IS-IS router implementation. These elements are linked by well-defined asynchronous interfaces as shown in the high-level block diagram of Figure 2.33. This figure provides a simplified diagram of the main IS-IS processes and databases in a typical IS-IS router. We describe these elements in greater detail below.

The discussion here focuses on the relationship between the various subsystems (processes and databases) that provide the subnetwork-independent functions in a typical IS-IS router. This simplified representation highlights only the relevant dependencies between the various processes and databases of IS-IS within the framework of a conventional IS-IS router architecture supporting IP routing.

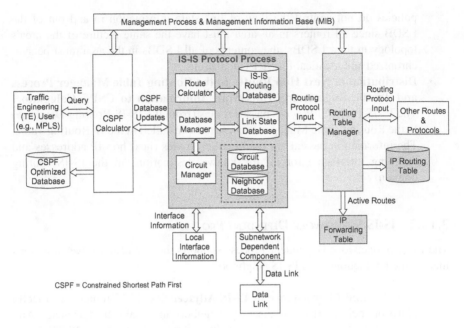

FIGURE 2.33 IS-IS Processes and Databases

2.19.1 IS-IS PROTOCOL PROCESS

The IS-IS Protocol Process supports the core IS-IS protocol functions which includes the following:

- **Maintenance of IS-IS Link-State Database (LSDB):** IS-IS Routers build LSPs describing their local interfaces and address prefixes learned from other adjacent routers. The routers then flood the LSPs to all adjacent neighbors except the neighbor from which they received the LSP. All routers receive the flooded LSPs and construct their LSDB.

- **Calculation of Shortest Paths (Routes) Using the Dijkstra Algorithm:** A router receives all valid LSPs and stores them in its LSDB. For Level 1 routing, for instance, these LSPs describe the topology of a Level 1 area. Each router uses it LSDB to calculate its Shortest-Path Tree (SPT) to all known destinations. The router then uses the SPT to build its IS-IS Routing Database (also called the IS-IS Forwarding Database). Each IS-IS interface has a routing metric that is used in the IS-IS SPT calculation. When multiple routes exist to a destination, the lowest-cost is the one entered in the IS-IS Routing Database.

- **Implementing Routing Policies to Control the Passing of Routes between Routers:** Routing policies can be configured on an IS-IS router running Integrated IS-IS to control the redistribution of IP routing information between IS-IS and other routing protocols (e.g., OSPF and BGP). Note that routing

254 IP Routing Protocols

policies do not control the flow of link-state information in and out of the LSDB since all routers in an area must have the same picture of the area's topology in their LSDBs; the contents of all LSDBs in the area must be synchronized and identical for SPT calculations.

- **Distribution of Next Hop Routes to the Routing Table Manager Process and Traffic Engineering Link-State Information to CSPF Calculator:** The IS-IS Protocol Process passes routes (from the IS-IS Routing Database) to the Routing Table Manager Process to be installed in the IP Routing Table. The information passed are the next hop routes (next hop IP addresses and outgoing interfaces) for all the known destinations in the IS-IS routing domain.

2.19.2 IS-IS SUBNETWORK DEPENDENT PROCESS

The IS-IS Subnetwork Dependent Process supports the Data Link Layer functions that handle interactions with IS-IS neighbors:

- **Creation and Maintenance of IS-IS Adjacencies:** IS-IS routers send Hello PDUs on their interfaces to discover neighbors and establish adjacencies. Any two routers sharing a common network segment will form an IS-IS adjacency if their Hello PDUs contain information that meets the criteria for forming the adjacency. In a multiaccess broadcast network environment, routers report their adjacencies to a DIS, which is responsible for generating the Pseudonode LSPs. The DIS is responsible for creating the Pseudonode, and flooding Pseudonode LSPs, its own LSPs, CSNPs, PSNPs over the network segment, and also for maintaining LSDB synchronization.
- **Forwarding Hello PDUs, LSPs, and SNPs to and from the IS-IS Protocol Process:** The Subnetwork Dependent Process is responsible for passing Level 1 and Level 2 Hello PDUs, LSPs, and SNPs (CSNPs and PSNPs) to and from the IS-IS Protocol Process.
  - o **Intermediate System-to-Intermediate System Hello (IIH):** Sent by IS-IS routers to discover neighbor systems and form adjacencies.
  - o **Link-State Packet (LSP):** Contain information describing the state of adjacencies to neighbor IS-IS systems.
  - o **Complete Sequence Number PDU (CSNP):** Contain a complete list (summaries) of all LSPs in a router's current LSDB, and is used to inform other routers about LSPs that may be missing or outdated from their own LSDBs.
  - o **Partial Sequence Number PDU (PSNP):** Sent by a router to request one or more LSPs, or acknowledge receipt of one or more LSPs.
- **Carries Out IS-IS PDU Verification and Authentication:** IS-IS routers will form an adjacency when they see their MAC addresses (SNPAs) in their neighbor's Hello PDUs and link authentication information matches. This process handles the authentication of all IS-IS routing information (sending and receiving authentication processes).

2.19.3 IS-IS ROUTER DATABASES

The following databases are used by IS-IS routers in a routing domain [ISO10589:2002]:

- **Level 1 Link-State Database**: This LSDB is maintained by Level 1 and Level 1-2 routers, and consists of the latest Level 1 LSPs from each router (or Pseudonode) in the Level 1 area. The Level 1 LSPs lists Level 1 links connected to the routers that originally generated the LSP.
- **Level 2 Link-State Database**: This LSDB is maintained by Level 2 routers, and consists of the latest Level 2 LSPs from each Level 2 router (or Pseudonode) in the routing domain. The Level 2 LSPs lists Level 2 links connected to the routers that originally generated the LSP.
- **Adjacency Database**: This database is maintained by all Level 1 and Level 2 systems in the routing domain. Each system uses this database to keep track of its neighbors. An IS-IS adjacency is a subset of the local routing information which pertains to the reachability of a single neighbor router or ES over a single IS-IS circuit. Adjacencies are used as input to the Decision Process of the IS-IS router when paths are being created through the routing domain. A router creates a separate adjacency for each neighbor on a circuit, and for each routing level (i.e., Level 1 and Level 2). Note that, both a Level 1 and a Level 2 adjacency can exist between two Level 1-2 routers. For IS-IS routers, each entry of the Adjacency Database contains any of the following information:
  o Adjacency on a point-to-point link.
  o Adjacency on a multiaccess broadcast network segment.
  o End System adjacency.
  o Virtual Link adjacency.
- **Circuit Database**: This database is maintained by all Level 1 and Level 2 systems in the routing domain. An IS-IS circuit is a subset of the local routing information maintained by a router pertinent to a single local MAC address (or SNPA). This database contains information about all IS-IS circuit from a router and comprises the array [1, ..., *maximumCircuits*].
- **Level 1 Shortest Paths Database**: This database is maintained by Level 1 and Level 2 routers in the routing domain (unless each circuit is "Level 2 Only"). The contents of this database are computed by the Level 1 Decision Process of the local IS-IS router, using the Level 1 LSDB. The Level 1 IS-IS Forwarding Database is a subset of this Level 1 LSDB.
- **Level 2 Shortest Paths Database**: This database is maintained by Level 2 routers and the contents are computed by the Level 2 Decision Process of the local IS-IS router, using the Level 2 LSDB. The Level 2 Forwarding Database is a subset of this Level 2 LSDB.
- **Level 1 Forwarding Database**: This database is maintained by Level 1 and Level 2 routers in the routing domain (unless each circuit is "Level 2 Only"). IS-IS routers use this database to determine where to forward a data PDUs with destination within the router's own area. It is also used to determine how to

reach a Level 1-2 router within the area, for data PDUs with destinations out-side this router's own area.

- **Level 2 Forwarding Database**: This database is maintained by Level 2 routers in the routing domain. Level 2 routers use this database to determine how to forward a data PDUs over the Level 2 routing subdomain to the destination area.

For practical implementation purposes, the IS-IS Protocol Process in Figure 2.33 builds and maintains three main databases:

- **IS-IS Neighbor Database** which contains a list of all neighbor routers and their attributes. The IS-IS Circuit Database and the Neighbor Database are sim-ply logical abstractions and an IS-IS router may choose to implement these as one integrated databases and not as separate databases.
- **IS-IS Topology Database (or LSDB)** which contains a list of all possible directly attached routers and ES and routes to all known networks within an area (for Level 1 routing).
- **IS-IS Routing (or Forwarding) Database** which contains the best route(s) for each known destination.

2.19.4   ROUTING TABLE MANAGER

When the IS-IS Protocol Process determines the best route to a particular destination, it passes that routing information to the Routing Table Manager. The Routing Table Manager may receive more than one best route to a destination from multiple routing information sources (e.g., EIGRP and OSPF). Because routing metrics from different routing information sources are not comparable, the Routing Table Manager uses the concept of Administrative Distance (or Route Preference) to select the best route for the IP Routing Table. The route with the lowest Administrative Distance value is selected. The best routes from the Routing Table Manager are then installed in the IP Routing Table (also known as the IP RIB).

2.19.5   THE IP ROUTING TABLE

The IS-IS Forwarding Database (referred to as IS-IS Routing Database in Figure 2.33) contains only the best IS-IS routes. These routes are fed into the IP Routing Table of a router to be used in packet forwarding decisions. As discussed above, when multiple routing information sources in a router provide routes to the same destination for the IP Routing Table, including static routes and BGP routes, the router uses the Administrative Distances of the information sources to determine which routes to prefer for the IP Routing Table.

The routing information source with the lowest Administrative Distance is pre-ferred and entered in the IP Routing Table. The accepted best route is installed in the IP Routing Table. In most routers, the IP Routing Table information is processed further into the faster IP Forwarding Table (also called the IP FIB) in order to speed up packet forwarding through the router. The Routing Table Manager Process derives a single active route for each destination from the IP Routing Table, and multiple active routes when ECMP routing is being performed, and enters this information in the IP Forwarding Table. All packet forwarding decisions are then based on the information in the IP Forwarding Table.

Typically, IS-IS being an intradomain routing protocol, is used in conjunction with BGP. BGP learns external routes, while IS-IS, as an IGP, is responsible for learning internal routes (mostly the next-hop information for the BGP routes). Consequently, the routing information maintained by the router has little overlap from either IS-IS or BGP, allowing IS-IS and BGP routes to coexist in the IP Routing Table.

2.19.6   IS-IS Traffic Engineering

Routing protocol traffic engineering extensions enable a router to include traffic engineering information in its routing protocol algorithms in order to calculate the best routes to network destinations. To perform traffic engineering in a network, a detailed knowledge about the network topology and information about the network load and dynamics is required. So, simple extensions to IGPs (IGP-TE) such as OSPF **[RFC3630] [RFC7471]** and IS-IS **[RFC5305] [RFC8570]** have been defined for traffic engineering purposes.

Each router includes link attributes are as part of its LSAs. OSPF extensions for traffic engineering are implemented using OSPF Opaque LSAs, while IS-IS extensions include the definition of new TLVs. The standard flooding algorithm used by OSPF and IS-IS ensures that link attributes are advertised to all routers in the routing domain. A number of traffic engineering extensions can be added to the IGP LSA such as current bandwidth reservation, maximum reserved link bandwidth, maximum link bandwidth, and so on.

Each router maintains network topology information and link attributes in a specialized database called the Traffic Engineering Database (TED). When MPLS is used, the router uses the TED to compute explicit paths over which MPLS Label Switched Paths are created in the network. A router may choose to maintain a separate TED so that the subsequent traffic engineering path computations are can be carried out independent of the IGP and its LSDB. This allows the IGP to continue its operation without interruption, performing the traditional shortest-path computations based on the information contained in the LSDB.

2.19.6.1   Path Computation and Selection

After the IGP-TE has flooded network topology information and link attributes and all routers have entered the routing information in their TEDs, each ingress router (in an MPLS network) uses the TED to compute the paths over which its own set of Label Switched Paths can be created across the routing domain. The egress router can represent the path for each Label Switched Path by either a strict or loose explicit route. An explicit route is configured by predefining the sequence of routers that should form part of the physical path of the Label Switched Path.

A strict explicit route for the Label Switched Path is created if the ingress router specifies *all* the routers that form the Label Switched Path. On the other hand, the Label Switched Path is identified as a loose explicit route if the ingress router specifies only *some* of the routers that form the Label Switched Path. Allowing a network to support strict and loose explicit routes makes it easy for the path selection to be carried out, but still be constrained when necessary.

The ingress router applies a CSPF algorithm to the information in the TED to determine the physical path for each Label Switched Path. CSPF is a SPF algorithm that has been modified to take into consideration specific network constraints like bandwidth and delay when calculating the shortest path across the network. The CSPF algorithm takes as input the following:

- Link-state information about the network topology learned by the IGP (IS-IS) and maintained in the TED.
- Attributes carried by the IGP extensions and associated with the state of network resources. These attributes are stored in the TED (e.g., reserved link bandwidth, available link bandwidth, and total link bandwidth).
- Administrative attributes that serve as constraints for the traffic traversing the proposed Label Switched Path and obtained from user configuration (e.g., maximum hop count, bandwidth requirements, and administrative policy requirements).

The local router running the CSPF algorithm examines each candidate node and link for a new Label Switched Path to determine whether to accept or reject it based on resource availability, or whether choosing that component violates predefined policy constraints. Upon completion, the CSPF outputs an explicit route that consists of a sequence of router addresses that form the shortest path through the network while meeting the network and administrative constraints. This explicit route is then passed to a label distribution protocol (Resource Reservation Protocol – Traffic Engineering [RSVP-TE]), which establishes the forwarding state in the sequence of routers through which the Label Switched Path is created.

2.19.6.2 CSPF Calculator Process

The CSPF Calculator Process in Figure 2.33 handles QoS and traffic engineering queries, and determines the best route to a destination that meets the specified traffic engineering constraints, such as, a specified minimum bandwidth. As discussed above, extensions have been added to IS-IS to provide information about network topology and loading for traffic engineering and MPLS networking purposes.

Specifically, IS-IS supports new TLVs that specify link attributes that are included in the IS-IS LSPs. IS-IS routers use the link-attribute information to populate the TED, which in turn is used by the CSPF algorithm to compute the paths that MPLS Label Switched Paths will take. This path information is used by RSVP-TE to actually set up Label-Switched Paths and reserve bandwidth for them.

2.19.6.3 Signaling and Distributing MPLS Labels

Resource Reservation Protocol (RSVP) is a network control protocol used by end-user devices to request the transport of application data streams or flows from a network with specific qualities of service. Routers also use RSVP to signal the requested QoS to all routers along the path of a flow, allowing them to establish and maintain operational state to provide the requested service. RSVP requests allow each node along the data path to reserve resources.

RSVP-TE is an extension of RSVP for MPLS to support automatic signaling of Label Switched Paths. RSVP-TE is used for signaling the routing of traffic-engineered Label Switched Paths, while LDP (Label Distribution Protocol) is used for signaling the routing of non-traffic-engineered Label Switched Path.

RSVP-TE which relies on a number of extensions to RSVP, is responsible for establishing Label Switched Path state and distributing MPLS labels. It is used only for signaling and establishing Label Switched Paths and not for end-user packet forwarding. OSPF and IS-IS with traffic engineering extensions (IGP-TE) must be working in order for RSVP-TE to function. To work properly, RSVP-TE (which has traffic engineering capabilities) requires an underlying IGP with traffic engineering capabilities (IGP-TE).

The IGP-TE makes it possible for routers to advertise the traffic engineering constraints imposed over the various links in the network. For example, in order for the routers in a network to determine the best link for signaling a Label Switched Path, the IGP must advertise the capacity of a particular link and the amount of reservable capacity. RSVP-TE uses these traffic engineering constraints to request the setup of a Label Switched Path that traverses only those links. Thus, both the IGP-TE (i.e., OSPF-TE or IS-IS-TE) and RSVP-TE must be used and must run simultaneously.

An IGP-TE first identifies the next hop (determine where packets are forwarded), which is information needed by RSVP-TE to assign MPLS labels. RSVP-TE is a signaling protocol (not a routing protocol), and consults the local IGP(-TE) Routing Table to determine how to relay RSVP-TE messages.

RSVP-TE performs the following main functions:

- Specifies an explicit route and sends an RSVP PATH message that travels through an explicit sequence of routers independent of the conventional IP shortest-path routing. RSVP-TE uses PATH messages to signal and request the MPLS label bindings required to establish a Label Switched Path from ingress to egress. The explicit route can be configured either as a strict explicit route or a loose explicit route. A PATH message follows the same path user data take, creating path states in the routers along the path, enabling the routers to learn the previous-hop and next-hop node for setting up the Label Switched Path.
- Permit the RSVP PATH message to request the routers (that form the explicit sequence of routers) to provide MPLS label binding for the Label Switched Path that is being established.
- Send RSVP RESV messages to create and maintain a reservation state in each router through which the Label Switched Path will traverse.

The following steps summarize how RSVP-TE uses the PATH and RESV message to establish a Label Switched Path:

- The ingress Label Edge Router (LER) or sender sends PATH messages toward the egress LER (or receiver) to indicate the Forwarding Equivalence Class (FEC) for which MPLS label bindings are required. Each router along the path (predefined by the IGP-TE) observes the indicated traffic type.

- PATH messages indicate to the routers along the path the necessary bandwidth reservations they need to make, and the MPLS label binding to be distributed to the router upstream.
- The egress LER in response to PATH messages received, sends MPLS label binding information in the RESV messages.
- The Label Switched Path is considered operational when the ingress LER receives the MPLS label binding information.

2.20 SUMMARY OF IS-IS FEATURES

- **Basic Protocol**: The IS-IS protocol is defined in ISO 10589 standard and was originally designed for use as the dynamic routing protocol for ISO CLNS environments **[ISO10589:2002]**.
- **Extensions for IP Routing**: IS-IS was later extended for IP routing in addition to CLNP (the extended protocol also known as Integrated IS-IS or Dual IS-IS).
  o Original IS-IS protocol for IP routing is defined in **[RFC1195]**
  o Differences between the protocol in **[RFC1195]** and the protocol as deployed in practice for IP routing are described in **[RFC3719] [RFC3787]**
  o IS-IS for IPv6 Routing is defined in **[RFC5308]**
- **Other Protocol Features**:
  o ECMP routing
  o IS-IS Cryptographic Authentication **[RFC5304] [RFC5310]**
  o IS-IS Extensions for Traffic Engineering **[RFC5305] [8570]**
  o IPv6 Traffic Engineering in IS-IS **[RFC6119]**
  o Multi-Topology for IS-IS **[RFC5120]**
  o IS-IS Extensions in support of GMPLS **[RFC5307]**
  o Dynamic Hostname Exchange Mechanism for IS-IS **[RFC5301]**

REVIEW QUESTIONS

1. Explain the difference between IS-IS Level 1 and Level 2 routing.
2. Describe the main features of an IS-IS Level 2 backbone.
3. Describe the difference between backbone requirements in OSPF and IS-IS.
4. Explain the differences between the ISO NSAP address, Network Entity Title (NET), and the System ID.
5. What is the function of the NSAP Selector (NSEL) in an OSI NSAP address?
6. What is an SNPA (Sub-Network Point of Attachment)?
7. What are the purposes of NSAP addresses that start with AFI (Address Family Identifier) equal to 49?
8. If the NETs of two IS-IS routers are 49.001a.1122.3344.5566.00 and 49.0001.1921.6800.1001.00, to what IS-IS areas do these routers belong, and what are their 6-byte System IDs?
9. Explain why the following are not valid NETs: 49.0001.1111.1111.1111.02; 4.0002.3333.3333.3333.00.
10. Explain why the following *CAN* be NETs in the same IS-IS area: 49.0001.2222.2222.2222.00; 49.0001.4444.4444.4444.00.

11. Explain why the following *CANNOT* be NETs in the same IS-IS area: 49.0003.5555.5555.5555.00; 49.0007.6666.6666.6666.00.
12. Which part of an NSAP address is used for routing within an IS-IS area and which part for routing between different areas?
13. Explain why the boundary between IS-IS areas cannot lie inside an IS-IS router.
14. What are the two types of network topology supported by IS-IS?
15. Explain the main differences between an IS-IS Level 1 router, Level 2 router, and Level 1-2 router. Describe the Link-State Databases (LSDBs) maintained by a Level 1 router, Level 2 router, and a Level 1-2 router.
16. What are the main functions of IS-IS Hello PDUs?
17. What are the main functions of IS-IS Link-State PDUs (LSPs)?
18. What are the main functions of IS-IS Complete Sequence Numbers PDUs (CSNPs)?
19. What are the main functions of IS-IS Partial Sequence Numbers PDUs (PSNPs)?
20. Why do IS-IS routers send IS Hello PDUs on point-to-point links?
21. Explain why a small random amount of time (jitter) is introduced into the value of a protocol timer.
22. Explain the purpose of the Remaining Lifetime field in an IS-IS LSP.
23. Explain the purpose of the *MaxAge* parameter in IS-IS.
24. Explain the purpose of the *ZeroAgeLifetime* parameter in IS-IS.
25. Explain the purpose of the LSPDBOL (LSP Database Overload) bit in an LSP.
26. Explain the purpose of the ATT (Attached) Flag bit in an LSP.
27. Explain briefly how Level 1 Area Partition Repair is done.
28. Explain the difference between IS-IS narrow metrics and wide (or extended) metrics.
29. Explain why an IS-IS router using narrow metrics can support multiple LSDBs.
30. If two Level 1-2 routers are directly connected and belong to the same IS-IS area, what types of adjacencies and LSDBs can they create?
31. If two Level 1-2 routers are directly connected and belong to different IS-IS areas, what types of adjacencies and LSDBs can they create?
32. What adjacencies do Level 2 routers form with Level 1-2 routers in the same area or another area?
33. What is a Designated IS (DIS) and what are its functions?
34. What is a Pseudonode?
35. Explain the purpose of the Pseudonode ID.
36. What criteria are used in electing the DIS?
37. Explain why IS-IS has no requirement for or concept of a Backup DIS.
38. Explain why there is determinism in DIS election in IS-IS.
39. What type of routes, by default, are advertised by a Level 1-2 router into a Level 1 area?
40. What mechanism can be used to advertise specific routes learned in a Level 2 subdomain into a Level 1 area?
41. Explain why a three-way handshake scheme is necessary for forming IS-IS Point-to-Point adjacencies.

REFERENCES

[CISCID13796].   Cisco Systems, IS-IS Route Leaking Overview, Document ID: 13796, May 13, 2008.

[CISCISISWP02].   Cisco Systems, "Introduction to Intermediate System-to-Intermediate System Protocol", White Paper, Jan. 2002.

[CISCISISCOMD20].   Cisco Systems, Cisco IOS IP Routing: ISIS Command Reference, June 9, 2020.

[CISCTEAPAQ00].   Diane Teare and Catherine Paquet, Building Scalable Cisco Networks, Chapter "Configuring IS-IS Protocol", Cisco Press, October 27, 2000.

[ISO9542:1988].   ISO 9542:1988 – Information processing systems – Telecommunications and information exchange between systems — End system to Intermediate system routing exchange protocol for use in conjunction with the Protocol for providing the connectionless-mode network service (ISO 8473), August 1988.

[ISO10589:2002].   ISO/IEC 10589:2002 – Information technology – Telecommunications and Information Exchange between Systems – Intermediate System to Intermediate System Intra-Domain Routing Information Exchange Protocol for use in Conjunction with the Protocol for Providing the Connectionless-Mode Network Service (ISO 8473)", International Organization for Standardization (ISO). November 2002.

[ISO10747:1994].   ISO/IEC 10747:1994 – Information technology – Telecommunications and information exchange between systems – Protocol for exchange of inter-domain routing information among intermediate systems to support forwarding of ISO 8473 PDUs, October 1994.

[ISO/IEC8348:2002].   ISO/IEC – 8348:2002 Information technology – Open Systems Interconnection – Network service definition, November 2002.

[JUNISRLGTLV20].   Juniper Networks, MPLS Applications User Guide, Section "Shared Risk Link Groups for MPLS ", February 10, 2020.

[RFC792].   J. Postel, "Internet Control Message Protocol", IETF RFC 792, September 1981.

[RFC826].   David C. Plummer, "An Ethernet Address Resolution Protocol or Converting Network Protocol Addresses to 48.bit Ethernet Address for Transmission on Ethernet Hardware", IETF RFC 826, November 1982.

[RFC1195].   R. Callon, "Use of OSI IS-IS for Routing in TCP/IP and Dual Environments", IETF RFC 1195, December 1990.

[RFC1812].   F. Baker, Editor, "Requirements for IP Version 4 Routers", IETF RFC 1812, June 1995.

[RFC1918].   Y. Rekhter, B. Moskowitz, D. Karrenberg, G. J. de Groot, and E. Lear, "Address Allocation for Private Internets", IETF RFC 1918, February 1996.

[RFC2104].   H. Krawczyk, M. Bellare, and R. Canetti, "HMAC: Keyed-Hashing for Message Authentication", RFC 2104, February 1997.

[RFC3630].   D. Katz, K. Kompella, and D. Yeung, "Traffic Engineering (TE) Extensions to OSPF Version 2", IETF RFC 3630, September 2003.

[RFC3719].   J. Parker, Ed., "Recommendations for Interoperable Networks using Intermediate System to Intermediate System (IS-IS)", IETF RFC 3719, February 2004.

[RFC3787].	J. Parker, Ed., "Recommendations for Interoperable IP Networks using Intermediate System to Intermediate System (IS-IS)", IETF RFC 3787, May 2004.
[RFC4203].	K. Kompella and Y. Rekhter, Eds., "OSPF Extensions in Support of Generalized Multi-Protocol Label Switching (GMPLS)", IETF RFC 4203, October 2005.
[RFC4271].	Y. Rekhter, T. Li, and S. Hares, Eds., "A Border Gateway Protocol 4 (BGP-4)", IETF RFC 4271, January 2006.
[RFC4861].	T. Narten, E. Nordmark, W. Simpson, H. Soliman, "Neighbor Discovery for IP version 6 (IPv6)", IETF RFC 4861, September 2007.
[RFC5120].	T. Przygienda, N. Shen, and N. Sheth, "M-ISIS: Multi Topology (MT) Routing in Intermediate System to Intermediate Systems (IS-ISs)", IETF RFC 5120, February 2008.
[RFC5301].	D. McPherson and N. Shen, "Dynamic Hostname Exchange Mechanism for IS-IS", IETF RFC 5301, October 2008.
[RFC5302].	T. Li, H. Smit, and T. Przygienda, "Domain-Wide Prefix Distribution with Two-Level IS-IS", IETF RFC 5302, October 2008.
[RFC5303].	D. Katz, R. Saluja, and D. Eastlake 3rd, "Three-Way Handshake for IS-IS Point-to-Point Adjacencies", IETF RFC 5303, October 2008.
[RFC5304].	T. Li and R. Atkinson, "IS-IS Cryptographic Authentication", IETF RFC 5304, October 2008.
[RFC5305].	T. Li and H. Smit, "IS-IS Extensions for Traffic Engineering", IETF RFC 5305, October 2008.
[RFC5307].	K. Kompella, and Y. Rekhter, Eds., "IS-IS Extensions in Support of Generalized Multi-Protocol Label Switching (GMPLS)", IETF RFC 5307, October 2008.
[RFC5308].	C. Hopps, "Routing IPv6 with IS-IS", IETF RFC 5308, October 2008.
[RFC5310].	M. Bhatia, V. Manral, T. Li, R. Atkinson, R. White, and M. Fanto, "IS-IS Generic Cryptographic Authentication", IETF RFC 5310, February 2009.
[RFC5316].	M. Chen, R. Zhang, and X. Duan, "ISIS Extensions in Support of Inter-Autonomous System (AS) MPLS and GMPLS Traffic Engineering", IETF RFC 5316, December 2008.
[RFC6119].	J. Harrison, J. Berger, and M. Bartlett, "IPv6 Traffic Engineering in IS-IS", IETF RFC 6119, February 2011.
[RFC7471].	S. Giacalone, D. Ward, J. Drake, A. Atlas, and S. Previdi, "OSPF Traffic Engineering (TE) Metric Extensions", IETF RFC 7471, March 2015.
[RFC7775].	L. Ginsberg, S. Litkowski, and S. Previdi, "IS-IS Route Preference for Extended IP and IPv6 Reachability", IETF RFC 7775, February 2016.
[RFC7981].	L. Ginsberg, S. Previdi, and M. Chen, "IS-IS Extensions for Advertising Router Information", IETF RFC 7981, October 2016.
[RFC8570].	L. Ginsberg, Ed., S. Previdi, Ed., S. Giacalone, D. Ward, J. Drake, and Q. Wu, "IS-IS Traffic Engineering (TE) Metric Extensions", IETF RFC 8570, March 2019.

3 Border Gateway Protocol (BGP)

3.1 INTRODUCTION

The Border Gateway Protocol (BGP) is the primary Exterior Gateway Protocol (EGP) used in today's internetworks for distributing routing information between autonomous systems while enforcing policy decisions, and guaranteeing loop-free exchange of routing information. Each route in a BGP is described by a network IP address prefix, a list of autonomous systems the routing information has passed through (also called the autonomous system path), and a variable list of path attributes. An autonomous system represents a group of routers and network address prefixes that share a common routing policy and are managed by a single administrative entity. The autonomous system may run a single or multiple Interior Gateway Protocols (IGPs) but the entire autonomous system is viewed by the outside world as a single entity.

As described in Chapter 2 of Volume 1 of this two-part book, BGP is path-vector routing protocol where each BGP router exchanges network reachability information with its BGP peers (nearest neighbors) by sending to each one of them, the sets of IP network address prefixes that are reachable through it, and the next-hop IP address to which user data should be sent in order to reach those IP network addresses. BGP sends reachability information throughout the internetwork, so that every BGP router has an IP Routing Table that contains IP address prefixes and next-hop addresses that cover the entire internetwork. Any autonomous system that wants to exchange routing information with other autonomous systems will typically have one or more BGP routers. Each BGP router is configured with the IP addresses of the BGP peers with which it is to exchange routing information.

Today's internetworks use BGP version 4 (BGPv4) which supports Classless Inter-Domain routing (CIDR). CIDR helps in the reduction of the size of the IP Routing Tables of routers in the internetwork by allowing the creation and advertisement of aggregate (summary) routes. The use of CIDR allows newer routing protocols like RIPv2, EIGRP, OSPFv2, and Integrated IS-IS to advertise summary or aggregate IP address prefixes. This results in the creation of supernets which also eliminates the concept of classful network addresses in the routing protocols including BGP. Newer routing protocols such as RIPv2, EIGRP, OSPFv2, and Integrated IS-IS all support Variable Length Subnet Masks (VLSMs) and CIDR. The focus of this chapter is BGPv4 **[RFC4271]**. BGP extensions such as Multiprotocol Extensions for BGP, or simply Multiprotocol BGP (MBGP) **[RFC4760]** is considered out of scope and is not discussed here.

3.2 INTERIOR VERSUS EXTERIOR ROUTING

In BGP, a route is defined as a unit of information that consists of a set of network destinations (i.e., one or more network prefixes) that have/share the same set of path attributes **[RFC4098] [RFC4271]**. The set of network destinations are networks whose IP addresses are represented by a single IP address prefix. This single IP address prefix is written in the Network Layer Reachability Information (NLRI) field of a BGP UPDATE message. The set of path attributes to these destinations, on the other hand, is the information written in the Path Attributes field of the BGP UPDATE message.

A BGP route is expressed as an n-tuple <IP Address Prefix, Next Hop, AS-Path, [...other BGP attributes...]> **[RFC4098]**. A BGP router advertises only the routes that it uses itself to its peers. In other words, a BGP router determines the most preferred BGP route among all routes it has received, and uses this route for data forwarding as well as advertise to its peers.

The global public Internet consists of a collection of autonomous systems that are interconnected to allow end-users to exchange information. BGP is the routing protocol that provides the routing between these autonomous systems. Autonomous systems can have routers within their boundaries that run IGPs (such as RIP, EIGRP, OSPF, and IS-IS). The different autonomous systems in turn can be interconnected via an EGP such BGP (Figure 3.1).

Although BGP is an EGP used for loop-free inter-autonomous system routing, it can also be used inside an autonomous system as a conduit for exchanging routing updates between BGP routers in that system. BGP used between routers within the same autonomous system is referred to as internal BGP (iBGP), whereas BGP used between router in different autonomous systems is referred to as external BGP (eBGP) (see Figure 3.2). In Figure 3.2, iBGP is used to provide the iBGP routers with external destination reachability information.

When iBGP is not used, the routes that are learned through eBGP have to be redistributed into an IGP running within an autonomous system, and then redistributed again through eBGP into another autonomous system. However, using iBGP provides a more flexible and efficient way of controlling the exchange of routing information within an autonomous system, and presents a consistent view of the networks within the autonomous system to eBGP neighbors. Also, iBGP provides a number of mechanisms for controlling the exit point from an autonomous system. Thus, iBGP

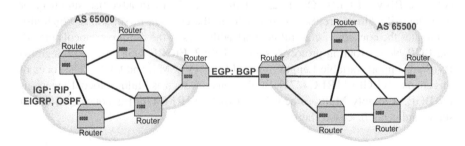

FIGURE 3.1   IGP Versus EGP

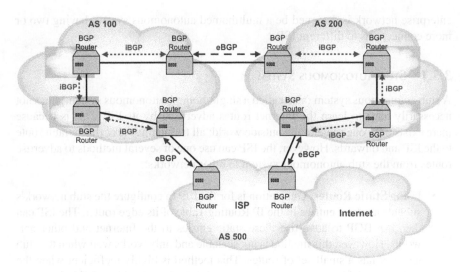

FIGURE 3.2   Comparing iBGP and eBGP

has become the preferred method of exchanging BGP updates within an autonomous system.

3.3   USING BGP

An autonomous system can be classified as belonging to one of the following three types; stub autonomous system, transit autonomous system, and multihomed (non-transit) autonomous system. A stub autonomous system has a single connection to one other autonomous system or the Internet. All data traffic received from or sent to, a destination outside the stub autonomous system must pass over that single connection. A small company branch office network connected to the bigger company network via a single connection is an example of a stub autonomous system.

A transit autonomous system has multiple connections to one or more autonomous systems, and permits data traffic that is not destined for a receiver within that (transit) autonomous system to travel through it. An Internet Service Provider (ISP) network interconnecting two company site networks is an example of a transit autonomous system.

A multihomed (non-transit) autonomous system has multiple connections (i.e., multiple entry points) to one or more other autonomous systems (e.g., ISPs), but it does not permit data traffic received over any one of these connections to be forwarded again out of the (multihomed) autonomous system through another connection. That is, a multihomed autonomous system does not provide transit service to other autonomous systems.

A multihomed autonomous system is similar to a stub autonomous system, except that the ingress and egress points for passing data traffic to or from the autonomous system can be selected from one of multiple connections, depending on which connection provides the best/shortest route to the data destination. Typically, a large enterprise network is designed be a multihomed autonomous system having two or more connections to different ISPs.

3.3.1 STUB AUTONOMOUS SYSTEM

A stub autonomous system (also called a single-homed autonomous system) does not necessarily have to learn the Internet routes advertised by its ISP. This is because there exists only one route to the outside world; all traffic must take this default route to the ISP and onwards. However, the ISP can use one of several methods to advertise routes from the stub autonomous system to other networks:

- **Using Static Routes**: One option is for the ISP to configure the stub network's subnets as static entries in the IP Routing Table of its edge router. The ISP can then use BGP to advertise these static entries to the Internet and other networks. However, this method is not scalable and only works well when the stub network has a small set of routes. This method is highly inefficient when the stub network has many noncontiguous subnets. A preferable solution is to configure a default static route from the stub network to the ISP, and a static route from the ISP toward the stub network (Figure 3.3). The default static route can then be advertised by the ISP to the Internet and other networks via BGP. The default static route provides a path for traffic going outside the stub network.
- **Running an IGP**: An IGP can be used between the stub network and the ISP to advertise the stub network's routes. This provides the benefits of dynamic routing where the stub network can advertise updates when changes in it occur. However, this is an inefficient solution and rarely used because routing instabilities in the stub network will easily be propagated to the ISP and can cause further routing instabilities (see Chapter 2 of Volume 1 of this two-part book for the disadvantages of using an IGP for inter-autonomous system routing).
- **Running BGP**: Although not preferred or recommended, BGP can be used between the stub network and the ISP to learn and advertise the stub network's routes. However, it will be very difficult to get an Autonomous System Number (ASN) from the Regional Internet registries (RIRs) for use in such a simple networking problem because the customer's routing policies are considered an extension of the ISP's routing policies. An alternative solution is for the ISP to assign a private ASN from its pool of ASNs (65,412–65,535), but this assumes that the ISP's routing policies have provisions to support the use of private

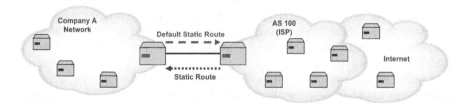

FIGURE 3.3   Illustrating when not to use BGP: Example, in Single-Homed Networks

ASNs as described in **[RFC2270]**. When BGP is used, the stub network can dynamically advertise its public network prefixes to the ISP, and the ISP can propagate a default route to the stub network. To avoid such complications, the best solution is for the ISP to use static routing to the stub network as described above (in Figure 3.3). In this solution, BGP is not needed and the customer (or company) network will not have to learn any Internet routes.

3.3.2 MULTIHOMED AUTONOMOUS SYSTEM

Figure 3.4 illustrates a (non-transit) autonomous system that is multihomed to two ISPs. Here on, multihomed non-transit autonomous system is referred to simply as multihomed autonomous system. A multihomed autonomous system would not re-advertise/propagate routes that it has learned from other autonomous systems but would advertise only its own internal routes. This ensures that traffic generated by network destinations that are outside the multihomed autonomous system would not be directed to it to be forwarded as transit traffic. The connections from the two ISPs to the company network can terminate on different interfaces on the same (single) router, or on different routers to enhance resiliency.

Recall that a multihomed autonomous system does not allow transit traffic to pass through it. So, as a precaution, the edge or boundary routers of the multihomed autonomous system could filter all traffic sent to it and have destinations not belonging to the multihomed autonomous system. Figure 3.4 shows the different methods for passing routing information between the multihomed autonomous system and the two ISPs **[CISCDONSTEW10] [CISCNETACAD17]**:

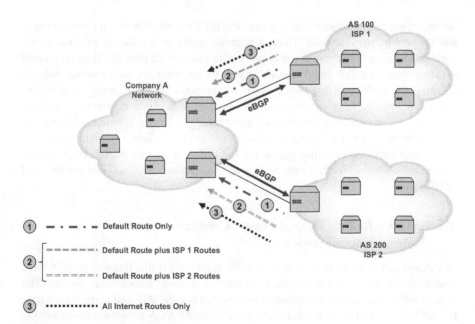

FIGURE 3.4   Illustrating when to use BGP

1. Each ISP configures and propagates only a default route to the customer (or company) network.
   • The default route is propagated to all the internal routers of the customer network. However, because the customer network receives only a default route from each ISP, suboptimal routing may occur. The customer network may choose to use the default route to ISP 1 to send packets to a destination in ISP 2. This problem can be addressed by using appropriate route filtering at the edge or boundary routers of the customer network.
2. Each ISP propagates only a default route and ISP-specific routes to the customer network.
   • These routes may be propagated to all the internal routers of the customer network, or all the boundary routers of the customer network can run BGP to exchange these routes. This option allows the customer network to send traffic to the appropriate ISP for network addresses advertised by that ISP. The customer would choose ISP 1 for network addresses advertised by ISP 1. For all other networks and Internet routes, the customer network may use one of the default routes which can also cause suboptimal routing to occur.
3. Each ISP passes all routes to the customer network.
   • All boundary routers of the customer network run BGP to exchange these routes. Because the customer network receives all Internet routing from both ISPs, it can determine which ISP to use as the best path to send traffic to any destination. This option solves the problem of suboptimal routing; however, the BGP process in the boundary router would require sufficient resources to maintain a large number of Internet routes. This option is typically used by large enterprises and ISPs.

Although not really required, in practice, the ISPs use BGP between their networks and the multihomed network. Here, a network prefix or a group of prefixes in the multihomed network is connected to more than one service provider (i.e., connected to more than one ISP (autonomous system) each with its own routing policy). Creating the multihomed network as an autonomous system running BGP makes sense since the network's prefixes should be part of a single autonomous system, distinct from those of the autonomous systems of its service providers. Doing this allows the customer (the company network) to have a different implementation of route preferences and routing policy for the different service providers. A typical implementation has multiple edge routers on the customer network, one per ISP, and BGP is used.

3.3.3 USING BGP TO PROVIDE TRANSIT CONNECTIVITY WITHIN AN AUTONOMOUS SYSTEM

In a transit autonomous system, transit traffic represents traffic that originates from a source and destined to a destination outside the (transit) autonomous system. The routers within the transit autonomous system that run the IGP and carry transit traffic, are referred to as transit routers (Figures 3.5 and 3.6). A transit autonomous system learns routes from one autonomous system and advertises them to another

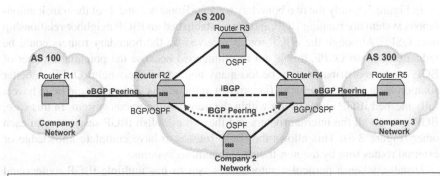

- BGP is typically used within an autonomous system when multiple routing policies exist, or when there is the need to provide transit connectivity between autonomous systems.
- Let us consider the scenario where AS 100 connects to Router R2, and AS 300 connects to R4. AS 200 provides transit connectivity to AS 100 and AS 300.
- Router R2 forms a BGP session directly with R4 to provide transit connectivity between AS 100 and AS 300.
  - However, Router R3 would not know where to route traffic coming from AS 100 or AS 300 when traffic from either autonomous system reaches R3. This is because R3 would not have the appropriate forwarding information to route such traffic.

FIGURE 3.5 Using BGP to Provide Transit Connectivity within an Autonomous System: Single iBGP Peering

autonomous system. This way, the transit autonomous system allows traffic not belonging to it to pass through it.

Transit autonomous systems generally use BGP to connect to other autonomous systems in order to shield their internal non-transit routers from routes advertised by the Internet and other external networks. Not all the routers within the transit autonomous system need to run BGP (Figure 3.5); the internal non-transit routers could use default routing to the BGP routers which allows them to reduce the number of routes they carry. However, in most large ISPs (providing transit services), all the internal routers within usually carry a full set of BGP routes (i.e., the iBGP routers as shown in Figure 3.6).

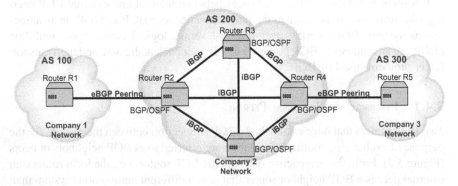

FIGURE 3.6 Using BGP to Provide Transit Connectivity within an Autonomous System: Fully meshed BGP Internetwork

In Figure 3.5, only the two boundary routers (Routers 2 and 4) of the transit autonomous system are running BGP, and have established an iBGP neighbor relationship over OSPF. Although the eBGP routes received by the boundary routers could be redistributed into OSPF, this is not recommended because the potential number of BGP routes redistributed may be too many and could overwhelm OSPF. A better solution for a provider network (i.e., transit autonomous system) would be to have a fully meshed iBGP connectivity within the transit autonomous system. In this case, BGP runs on all the internal routers and they all establish iBGP sessions with each other (Figure 3.6). This allows the iBGP routers to have complete knowledge of external routes sent by the non-transit autonomous systems.

Usually, when a particular autonomous system has multiple iBGP routers and provides transit service for other autonomous systems, then it must ensure that a consistent view of routing within it, is presented to the other (non-transit) autonomous systems. The IGP used within the transit autonomous system (RIPv2, EIGRP, OSPF) provides a consistent view of the interior routes of the autonomous system. However, providing a consistent view of the exterior routes to the transit autonomous system requires having *all* BGP routers within the autonomous system maintain full iBGP connectivity with each other (as shown in Figure 3.6).

It is discussed in the "Understanding the iBGP Full Mesh Requirement" section below that, to prevent routing loops *within* an autonomous system, BGP requires that external routes learned through an iBGP peer must never be propagated to other iBGP peers. It is assumed that the sending eBGP neighbor of the autonomous system (i.e., eBGP Router R2 in Figure 3.6), which also maintains iBGP peering in that system, is fully meshed with all other iBGP peers. This sending eBGP peer (R2) (which appears as an iBGP peer in the autonomous system) is able sends to each iBGP peer the external routes it receives (from other autonomous systems) because of the full mesh connectivity it maintains with all iBGP peers as shown in Figure 3.6. It should be noted that all iBGP routers within a non-transit autonomous system must also have complete knowledge of external routes.

3.4   BGP PEERING

In this section, we discuss the differences between internal and external BGP peering, plus important issues such as the interaction between BGP and IGPs in an autonomous system, BGP peering over physical versus logical connections, multihop eBGP, the advantages of BGP peering using IP loopback addresses, and the Transport Layer protocols used by BGP.

3.4.1   INTERNAL AND EXTERNAL PEERING

Two BGP routers that have established a TCP connection between themselves for the purpose of exchanging routing information are referred to as BGP neighbors or peers (Figure 3.7). From the perspective of a specific BGP router (i.e., the local router), an external peer is a BGP neighbor router that is in a different autonomous system than the local router. Consequently, an eBGP is a BGP connection established between external peers (Figure 3.8). On the other hand, an internal peer is a BGP neighbor

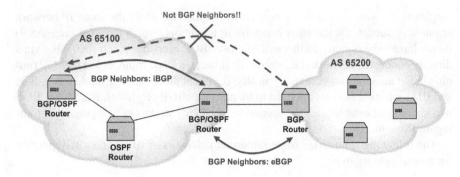

FIGURE 3.7   Identifying BGP Neighbors in a Network

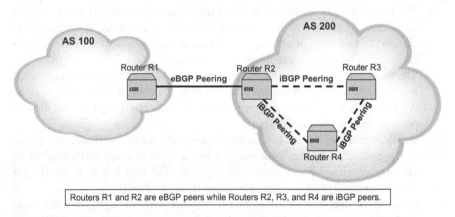

Routers R1 and R2 are eBGP peers while Routers R2, R3, and R4 are iBGP peers.

FIGURE 3.8   Identifying iBGP and eBGP Peers in a Network

router that is in the same autonomous system as the local router. Thus, iBGP is a BGP connection established between internal peers.

For any two BGP routers to be able to establish a BGP session, there must be IP connectivity between them. As in all aspects of IP networking, the TCP connection needed to establish the BGP session depends on the availability of IP connectivity between the two BGP routers; IP reachability must exist before any two BGP routers can peer with one another. BGP depends on other methods such as the use of directly connected interfaces, static routes, and IGPs to provide the connectivity needed between potential peering routers.

Normally eBGP peers are directly connected over a shared network segment (e.g., Ethernet), or over a single link with no intervening routers, and therefore do not require any additional underlying routing mechanism (static route or IGP) to connect them. We discuss below mechanisms for connecting eBGP peers across multiple (router) hops.

When in different autonomous systems (i.e., have different ASNs), eBGP neighbors can be directly connected or reachable via a static route. In many cases, eBGP neighbors are generally directly connected because the share the same IP network segment or subnet. On the other hand, iBGP neighbors (which have the same ASN) do not have to be directly connected. Any two iBGP peers can reach each other via a directly connected network (i.e., shared IP subnet), a static route, or via the IGP running in the autonomous system. Typically, iBGP neighbors are reachable via an IGP, and IP loopback addresses are used to identify the iBGP neighbors. Recall that BGP neighbors must establish a TCP session before they can start exchanging BGP routing information.

The following summarize some important behaviors of iBGP and eBGP peering (in normal operations):

- Except when a mechanism such as Route Reflection is used, routes learned from an iBGP peer are not advertised to other iBGP peers to prevent routing loops within an autonomous system.
- The BGP paths attributes learned from iBGP peers are *not (normally)* changed so as not to impact the path selected to reach networks outside the autonomous system. To prevent routing loops within the autonomous system, the best path chosen throughout the system must be consistent.
- The BGP autonomous system path information (AS-Path) is not manipulated when a route is being advertised to an iBGP peer; the local ASN is added to the AS-Path only when a route is being advertised to an eBGP peer (i.e., at the exit point of the autonomous system).
- In eBGP peering, the BGP next-hop is normally set to the IP address of the sending interface of the local router when a route is being advertised to an eBGP peer. However, in iBGP peering, the BGP next-hop is normally not changed when a route is being advertised to an iBGP peer.
  o In eBGP peering, the BGP next-hop is normally changed, and the local ASN is added to the AS-Path when advertising a route to an eBGP peer. However, in iBGP peering, these attributes are normally left unchanged when advertising a route to an iBGP peer.

Exceptions to these rules are discussed in appropriate sections in this chapter.

3.4.2 INTERACTION BETWEEN IGPs AND BGP

BGP does not discover routes within a network by itself like IGPs do. IGPs are equipped and support mechanisms that allow them to discover IP addresses in a network. Thus, BGP needs to import the routes discovered by IGPs into the BGP Routing Tables so that different autonomous systems can communicate. Furthermore, BGP cannot discover neighbors dynamically, and does not discover neighbors using Hello messages like some IGPs like EIGRP, OSPF and IS-IS do. Any two BGP routers must establish a BGP session over TCP for them to be able to communicate and exchange routing information. This means BGP neighbor adjacencies must be coordinated and also should not change frequently.

IGPs like EIGRP, OSPF, and IS-IS are able to send Hello messages and form adjacencies because any two IGP routers in a routing domain are connected via a common shared network segment (broadcast or point-to-point). The Hello messages do not cross network boundaries (i.e., routers). BGP, on the other hand, uses TCP, which allows BGP to cross network boundaries (as in multihop BGP sessions). This means any two BGP routers that are directly connected, as well as, any two BGP routers that are multiple hops apart, can form an adjacency.

IGPs (like RIP, EIGRP, OSPF, and IS-IS) and BGP use different Routing Table structures to store routing information (see, for example, Chapters 1 and 2). Thus, to enable routers in different autonomous systems to communicate, there is the need to configure the IGPs within the autonomous systems and BGP to interwork so that IGP routes can be imported to BGP Routing Tables, and also, for BGP routes to be imported into IGP Routing Tables. The process of importing or exporting routes from one routing protocol to another is called *route redistribution* (see Chapter 7 of Volume 1 of this two-part book).

When an autonomous system needs to advertise routes to another autonomous system, a boundary router connecting these two systems, also called an eBGP router, imports IGP routes (i.e., routes discovered by IGP within the sending autonomous system) into its local BGP Routing Table. The IGP-acquired routing information is distributed into BGP. Routing policy filters can be used to select which routes and path attributes are manipulated when the IGP routes are being imported into BGP. The BGP Multi-Exit Discriminator (MED) path attribute can also be set to allow eBGP peers to select the best path for traffic entering an autonomous system.

BGP can import routes into the BGP Routing Table using one of the following methods:

- BGP imports IGP routes (RIP, EIGRP, OSPF, and IS-IS routes, including directly connected networks and static routes) from the IGP Routing Table (of the boundary router) into BGP Routing Table.
- BGP imports the routes stored in the IP Routing Table (of the boundary router) one at a time into the BGP Routing Table. In this case, network addresses and masks must exist in the IP Routing Table for them to be imported into BGP. This method of route importation is more accurate and preferable.

The boundary router can also import BGP routes into its IGP Routing Table. Typically, the boundary router uses routing policy filters to select routes and set their path attributes when importing BGP routes. This is to provide a better control on how routes are imported when a large number of BGP routes are available from the sending autonomous system.

A BGP router adds the optimal routes to external destinations and those learned from eBGP peers, plus the routes originated in local autonomous system, to its IP Routing Table. The routes within the local autonomous system are directly connected networks, static routes, or routes originated by an IGP. After establishing a BGP relationship with a neighbor, the BGP router exchanges routes as follows:

- Advertises the BGP routes received from eBGP peers to its iBGP peers and other eBGP peers.
- Advertises the BGP routes received from its iBGP peers only to its eBGP peers and not to other iBGP peers.

- Advertises the best route to its iBGP and eBGP peers when there are multiple valid routes to the same network destination.
- Sends only updated BGP routes to its iBGP and eBGP peers when BGP route changes occur.
- Receives routes from its BGP peers but these routes can be subject to route filtering.

3.4.3 BGP PEERING OVER PHYSICAL VERSUS LOGICAL CONNECTIONS AND MULTIHOP eBGP

For iBGP peering, static routing and an IGP can be used to provide the needed IP connectivity, that is, the routing information needed for establishing the BGP TCP session. However, eBGP peering, generally requires the use of directly connected interfaces between the peering routers. Generally, eBGP peering requires that the eBGP neighbors be physically connected either through directly connected physical interfaces (they share the same physical segment), or share the same IP subnet (i.e., the same broadcast domain as in a Virtual LAN [VLAN]). One restriction of eBGP peering is that, a BGP router receiving a BGP UPDATE message from an eBGP peer that is not physically connected, will drop the message, unless it is configured to do otherwise [CISCHALABS00] [CISCJAINEDGE16].

However, the above eBGP connectivity requirement (where the routers are adjacent to one another) is not always possible in some eBGP peering cases. There are situations where two eBGP neighbors cannot be physically connected but rather, logically connected as illustrated in Figure 3.9. This happens when the two eBGP neighbors (Routers R2 and R4) have interfaces that cannot be physically connected, or are not attach to the same IP subnet [CISCJAINEDGE16]. This scenario also applies to iBGP peers (Routers R1 and R2) as seen in Figure 3.9 but without the physical connectivity restriction in eBGP peering.

Figure 3.9 shows the case where two eBGP peers (e.g., Routers R2 and R4) are connected across a non-BGP router (Router R3). This example shows a peering situation where the eBGP routers are multiple IP hops away from one another, and, therefore, means they can only be logically connected using a combination of static routes and IGP. A BGP session between two eBGP neighbors that are not physically connected is referred to as *multihop eBGP* [CISCHALABS00] [CISCJAINEDGE16].

Any two BGP routers that want to establish a multihop BGP session require an underlying route (via static routing or an IGP) to establish the TCP session between them. BGP routers connected to the same network segment can use ARP (Address Resolution Protocol) to discover the Layer 2 address of the neighbor participating in the peering. Multihop BGP routers require IP Routing Table information to find the IP address of the router to peer with.

Extra configuration is required to indicate to the eBGP peers (R2 and R4) that are not physically connected. Each eBGP peer in the multihop BGP sessions requires appropriate IP Routing Table information to find the IP address of the remote eBGP peer. Router R3 in Figure 3.9 must be provided the appropriate routing information to prevent the formation of forwarding loops and black holes when forwarding packets between the eBGP neighbors.

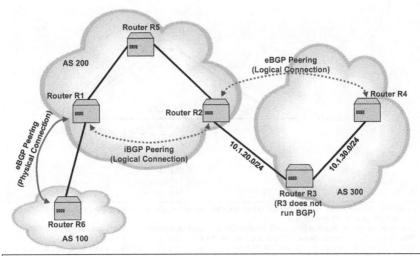

<table>
<tr><td>• Router R1 establishes a direct BGP session with R6 over a physical connection.
    – Routers R1 and R6 setup a directly connected route to reach each other (i.e.,using directly connected
       interfaces).
• Router R2 establishes a logical BGP connection with R4, even though the BGP session passes through R3.
    – R2 uses a static route to reach the network 10.1.30.0/24 attached to R4, while R4 uses a static route or an
       IGP to reach the network 10.1.20.0/24 attached to R2.
    – Meanwhile Router R3 is unaware that R2 and R4 have established a BGP session, even though the packets
       of the eBGP session flow through it.
    – Extra configuration is required to indicate to the eBGP peers (R2 and R4) that they are not physically
       connected to prevent R4 from dropping BGP UPDATE messages sent by R2.
    – Router R3 must learn the appropriate routing information to prevent the formation of forwarding loops and
       blackholes when forwarding packets between the eBGP neighbors.</td></tr>
</table>

FIGURE 3.9   Illustrating Physical and Logical BGP Connections: Multihop BGP Example

Unlike in eBGP peering, in iBGP peering, it does not matter if the iBGP peers in the autonomous system are physically connected or are multiple IP hops apart. Provided there is IP connectivity between the iBGP peers (either through an IGP or static routing), iBGP peering can take place without additional configuration as in multihop eBGP peering. The iBGP peering between Routers R1 and R2 in Figure 3.9 is over a logical connection without additional configuration.

3.4.4   BGP Peering Using IP Loopback Addresses

In BGP peering, a BGP router, by default, sends routing updates over the established BGP session over its outbound interface (carrying that interface's IP address as source address) toward the remote BGP peer's IP address (the destination IP address). Let us consider the three BGP routers (R1, R2, and R3) connected via a full mesh as shown in Figure 3.10. In the event the link between R1 and R3 fails, the BGP session of R3 with R1 will time out and terminate. Router R3 will lose connectivity to R1 even though these two BGP routers could communicate via the multihop path provided by R2. The loss of connectivity between R1 and R3 occurs because iBGP routers generally do not re-advertise routes learned from other iBGP peers.

To address the above connectivity problem, an IGP can be configured on the transit links of the routers to advertise the loopback interface IP addresses of all routers

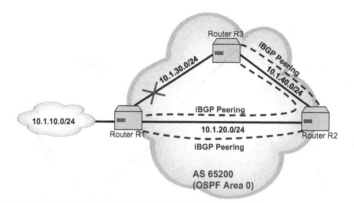

FIGURE 3.10   BGP Peering Using IP Loopback Addresses

(R1, R2, and R3) into the IGP. The BGP neighbor (R1) can then be configured to establish a BGP session to the loopback IP address remote router (R3) **[CISCJAINEDGE16]**. The loopback interface being an internal virtual software interface (with its own IP address), is always operational as long as the router is operational. This means, in the event of link failure between R1 and R3, the BGP session will still be maintained at both endpoints while the IGP discovers another path linking both loopback IP addresses. With this, after a link failure, the single-hop iBGP session between R1 and R3 converts into a multihop iBGP session.

In the above problem, updating only the BGP configuration at Router R1 to have the destination IP address of the BGP session equal to the loopback IP address of the remote router R3 is not enough **[CISCJAINEDGE16]**. The BGP packets leaving R1 will have source IP address still set to the IP address of the outbound interface (of R1). When Router R3 receives a BGP packet (even over an alternative path), it will correlate the source IP address of the packet to the BGP Neighbor Table (of R1). If R3 determines that the source IP address of the BGP packet does not match an entry in the Neighbor Table, it cannot associate the packet with a BGP neighbor and will discard it.

BGP peering using loopback IP addresses allows the BGP session between R1 and R3 to still remain intact even when the link between the two routers fails. Routers R1 and R3 are able to still maintain BGP session connectivity (using their loopback IP addresses) while OSPF learns new routes to allow BGP communication traffic to flow between the routers via R2. R1 performs a recursive Routing Table lookup to identify R2 as the next-hop IP address leading to R3 (see "A Note on BGP Recursive Route Lookup" section below). Using R2 as the next-hop, R1 can still forward packets to R3.

However, BGP peering with the loopback address is sometimes not necessary or effective in some cases. For example, when two eBGP peers are directly connected and a BGP session is set up using their interface IP addresses, the loopback addresses will provide no additional advantage and will not be necessary **[CISCHALABS00]**. This is because, if the directly connecting segment fails, the BGP session will terminate even if loopback interface addresses were used.

3.4.5 BGP TRANSPORT

BGP sends traffic and interacts with neighbors over TCP (Transmission Control Protocol) using TCP port 179. The use of TCP means BGP does not need to implement its own data retransmission, acknowledgement, sequencing, and Application Layer data fragmentation. TCP already supports all the relevant mechanisms and parameters needed for reliable data transmission such as TCP slow start, congestion avoidance, maximum window size, maximum segment size, and various timers.

Upon initialization, a BGP router first chooses a local ephemeral TCP port, or randomly chosen port number greater than 1,024, and tries to contact each configured BGP router on the well-known TCP port 179. The BGP router initiating the TCP session performs a TCP active open, while the BGP peer performs a passive open. Since it is possible for two BGP routers to attempt to initiate a connection to one another at the same time, a connection collision can occur. When a connection collision occurs, each router compares its local Router Identifier (ID) to the Router ID of the colliding neighbor. The BGP router with the higher Router ID value terminates the session on which it performed a passive open, while the BGP router with the lower Router ID value terminates the session on which it performed an active open. This rule is defined to allow only the session initiated by the BGP router with the larger Router ID to be preserved.

A BGP router processes a BGP message only after it is entirely received. RFC 4271 recommends the BGP maximum message size to be 4,096 bytes and all BGP implementations are required to support this size **[RFC4271]**. The smallest BGP message that a router may send is 19 bytes, which consists of the BGP message header without a data portion (see BGP message header below). BGP organizes all multibyte fields in the *network byte order*, where the bytes are ordered from the most significant byte to least significant byte (also called the "big endian" byte order).

BGP's error notification mechanism (using BGP NOTIFICATION messages) relies on the TCP graceful close or termination feature, which allows all outstanding data from either end to be delivered before the TCP connection is closed. TCP graceful close means both ends must indicate that they have "no more data to send" before closing a connection. Both ends must indicate to the other end that it has no more data to send before they start the process of closing the TCP connection.

After two BGP routers have established a TCP connection, they exchange messages to open and confirm the BGP session parameters. Initially, each BGP router sends to the other all candidate BGP routes in its Adj-RIBs-Out database (which represents only is a portion of its IP Routing Table) as allowed by the export policy. Each router then sends incremental updates as network changes occur and the IP Routing Tables change. Unlike routing protocols such as RIP, BGP does not perform periodic advertisement or refresh of the IP Routing Table. When a BGP router makes local policy changes, to allow these changes to have the correct effect without any

BGP connections being reset (i.e., to facilitate non-disruptive routing policy changes), the router performs one of the following actions:

- Retains the current version of the routes that all of its BGP peers have advertised to it for the duration of the connection.
- Implements the Route Refresh Capability feature **[RFC2918]** which allows a BGP router to dynamically request a BGP peer (via a BGP ROUTE-REFRESH message) to re-advertise its Adj-RIBs-Out database (see discussion below).

3.5 BGP MESSAGE TYPES

As illustrated in Figure 3.11, BGP messages are carried within TCP segments and the maximum BGP message size is 4,096 bytes. Each BGP message has a fixed-size header of 19 bytes as described in Figure 3.12. A BGP message may not necessarily carry a data portion after the header, depending on the message type. Figure 3.12 describes the BGP message header structure and the fields within.

BGP messages such as KEEPALIVE messages consist of the BGP message header only and do not have a data portion. A BGP message must not be less than 19 bytes (consisting of 16-byte Marker field, 2-byte Length field, and 1-byte Type field) and not greater than 4,096 bytes.

3.5.1 BGP OPEN MESSAGE

After two BGP routers have established a TCP connection, the first BGP message sent by each side to the other is a BGP OPEN message. If the receiving end finds the

IPv4 Header (20 Bytes)	TCP Fixed Header + Options (if any)	BGP Message

FIGURE 3.11   BGP Message in an TCP/IP Packet

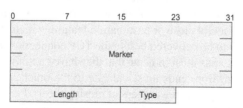

Field	Meaning
Marker (16 Bytes)	This field is included in the BGP message for compatibility reasons and must be set to all ones.
Length (2 Bytes)	This field indicates the total length of the BGP message, including the header in bytes.
Type (1 Byte)	This field indicates the type code of the message. RFC 4271 defines the following message type codes: Type 1 = OPEN; Type 2 = UPDATE; Type 3 = NOTIFICATION; Type 4 = KEEPALIVE. RFC 2918 defines the ROUTE-REFRESH message (Type 5).

FIGURE 3.12   BGP Message Header Format

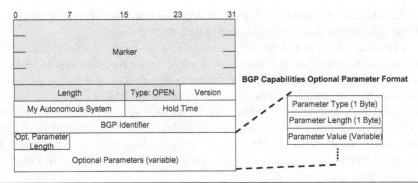

FIGURE 3.13 BGP OPEN Message Format

Field	Meaning
Version (1 Byte)	This indicates the BGP version number of the message. The current BGP version number is 4.
My Autonomous System (2 Bytes)	This indicates the Autonomous System number of the sender.
Hold Time (2 Bytes)	This indicates the number of seconds the message sender proposes for the value of the Hold Timer. This field indicates the proposed maximum time BGP will wait between successive messages (KEEPALIVE and/or UPDATE) from its peer before closing the connection.
BGP Identifier (4 Bytes)	This indicates the unique identifier of the BGP sender which (uniquely) identifies the sender in the network. A given BGP router sets the value of its BGP Identifier to an IP address (typically the Router ID) that is assigned to that BGP router. The value of the BGP Identifier is determined upon system start up and is the same for every local interface and BGP peer.
Optional Parameter Length (1 Byte)	This indicates the total length of the Optional Parameters field in bytes. A value of zero indicates that there are no Optional Parameters present in the message.
Optional Parameters (Variable)	This field contains a list of optional parameters used in the BGP neighborsession negotiation, and where each parameter is encoded as a <Parameter Type, Parameter Length, Parameter Value> triplet: – Parameter Type is a one byte field that unambiguously identifies individual parameters. – Parameter Length is a one byte field that contains the length of the Parameter Value field in bytes. – Parameter Value is a variable length field that is interpreted according to the value of the Parameter Type field.

OPEN message to be acceptable, it sends back a KEEPALIVE message (see below) confirming acceptance of the OPEN message. In addition to the 19-byte fixed-size BGP header, the OPEN message contains a number of fields as described in Figure 3.13. The OPEN message has a minimum length of 29 byes (including the BGP message header).

After the establishment of a BGP session, a participating BGP router upon receiving an OPEN message from the peer, compares the value in the "My Autonomous System" field of the received message to its own local ASN to determine if the two peering routers are in the same autonomous system or not. The outcome of this comparison determines if the BGP peering will be an iBGP or eBGP. This determines the way the BGP peers exchange and process routing updates, and how BGP attributes are carried in the routing updates.

As soon as a BGP router receives an OPEN message, it calculates the value of its local Hold Timer using the smaller of its local configured Hold Time and the Hold Time received from the other end in the received OPEN message. The Hold Time must be either zero or set to a value that is not less than three seconds. A BGP implementation may choose to reject a BGP connection with a neighbor on the basis of the value of the Hold Time specified in the OPEN message.

The value of the actual Hold Time calculated by the BGP router indicates the maximum number of seconds that the router has to wait for incoming KEEPALIVE and/or UPDATE messages from the sender before declaring it as unreachable or unavailable. The parameter *HoldTime* is a mandatory BGP Finite State Machine (FSM) attribute that stores the initial value for the BGP session *HoldTimer*. Reference **[RFC4271]** suggests a default value for the *HoldTime* as 90 seconds. A BGP implementation must allow the *HoldTimer* to be configurable on a per-BGP peer basis, and this flexibility in configuration may also be extended to the other BGP timers discussed below.

The Router ID used in the 4-byte BGP Identifier field is usually determined as the highest IP address in the BGP router, or the loopback IP address used at the time the BGP session is being set up. A loopback address is a virtual software interface address that is always operational at all times the router is operational, independent of the operational state (up/down) of the individual physical router interfaces (see Chapter 1 of **[AWEYA2BK19]**).

The Optional Parameters field in Figure 3.13 carries a list of optional parameters for a BGP router. Also, **[RFC5492]** defines an additional BGP Optional Parameter, called Capabilities, which allows a BGP router to advertise new capabilities graceful without requiring the BGP peering to be terminated. In addition to the Optional Parameter, **[RFC5492]** defines processing rules that allow BGP routers to advertise capabilities in the OPEN message they send. The Capabilities Optional Parameter has the parameter type value equal to 2.

The Capabilities Optional Parameter contains one or more of the following triples <Capability Code, Capability Length, Capability Value> **[RFC5492]**:

- **Capability Code**: This is a one-byte integer that unambiguously identifies an individual capability.
- **Capability Length**: This is a one-byte integer that contains the length (in byte) of the Capability Value field.
- **Capability Value**: This is a variable length field that is interpreted according to the value carried in the Capability Code field.

When a BGP router receives an OPEN message carrying the Capabilities parameter, it examines the information contained within it to determine which capabilities the remote peer supports. If the router determines that it also supports the indicated capabilities, it can start using those capabilities. If the router does not support a capability, it sends a NOTIFICATION message (see below) containing the Error Subcode "Unsupported Optional Parameter" to the remote BGP peer. When this occurs, the remote router should attempt to re-establish the BGP connection without including the previously rejected capability.

Any two BGP routers that supports the **[RFC5492]** specification can establish a BGP peering even when they are presented with unrecognized capabilities, so long as they both support all the capabilities required to establish the BGP peering. Reference **[RFC4271]** which is the base BGP-4 specification, requires that a BGP router that receives an OPEN message with one or more BGP Optional Parameters that are unrecognizable, must terminate the BGP peering. This requirement tends to make it difficult for the introduction of new capabilities in BGP, and so, **[RFC5492]** was developed to remedy this.

3.5.2 BGP UPDATE Message

BGP routers advertise routes between themselves using UPDATE messages. BGP routers in the internetwork use the routing information in the UPDATE messages to construct a network topology map that describes the relationships between the various autonomous systems that make up the internetwork. **[RFC4271]** discusses rules that BGP routers use to detect and remove routing information loops, and other routing anomalies that may arise from inter-autonomous system routing.

An UPDATE message always carries the fixed-size 19-byte BGP header, plus other fields (Figure 3.14), some of which may not be present in every UPDATE message. The minimum length of the UPDATE message is 23 bytes. This is composed of 19 bytes for the fixed BGP header, plus 2 bytes for the Withdrawn Routes Length field, plus 2 bytes for the Total Path Attribute Length field (assuming the Withdrawn Routes and the Path Attribute fields carry no data).

3.5.2.1 Using the BGP UPDATE Message

A BGP UPDATE message is a routing advertisement that may contain (simultaneously) one or multiple withdrawal of unfeasible routes, and a single NLRI field that carries, possibly, multiple feasible routes (i.e., network destination addresses prefixes) that share common path attributes to a BGP peer (Figure 3.14). A feasible route is a route that has been advertised and is available for use by the receiving node. An unfeasible route is a feasible route that was previously advertised and is no longer available for use.

A BGP router can advertise multiple routes that have the same path attributes in a single UPDATE message by including the network prefixes of these multiple routes in the NLRI field of the UPDATE message. As shown in Figure 3.14, the NLRI in an UPDATE message consists of one or more IP address network prefixes all having the same set of path attributes. Each IP address prefix in the NLRI field of the UPDATE message is combined/paired with the (common) set of path attributes to form a BGP route. In other words, the NLRI represents a set of network destinations to which packets can be forwarded (from a particular point [router] in the network) along a common path described by the shared set of path attributes.

Each UPDATE message a BGP router sends contains a variable length sequence of path attributes. As illustrated in Figure 3.14, each path attribute of variable length is expressed as a triple <Attribute Type, Attribute Length, Attribute Value>. When a BGP router decides to advertise a route that it had previously received, it may modify, or add to the path attributes of the route before advertising it to a peer.

A BGP router may receive an UPDATE message from a peer only when its BGP FSM in the Established State (see discussion below). Receiving an UPDATE message when the router's BGP FSM is in any other state results in an error. When the

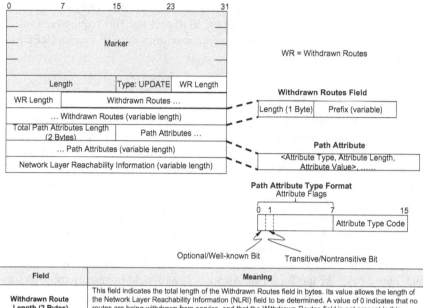

FIGURE 3.14 BGP UPDATE Message Format

Field	Meaning
Withdrawn Route Length (2 Bytes)	This field indicates the total length of the Withdrawn Routes field in bytes. Its value allows the length of the Network Layer Reachability Information (NLRI) field to be determined. A value of 0 indicates that no routes are being withdrawn from service, and that the Withdrawn Routes field is not present in this UPDATE message.
Withdrawn Routes (variable length)	This is a variable-length field that contains a list of IP address prefixes for the routes that are being withdrawn from service. Each IP address prefix is encoded as a 2-tuple of the form <length, prefix>: – The Length field indicates the length in bits of the IP address prefix. A length of zero indicates a prefix that matches all IP addresses (with prefix, itself, of zero bytes). – The Prefix field contains an IP address prefix, followed by the minimum number of trailing bits needed to make the end of the field fall on an byte boundary. Note that the value of any trailing bits is irrelevant.
Total Path Attribute Length (2 Bytes)	This field indicates the total length of the Path Attributes field in bytes. Its value allows the length of the NLRI field to be determined. A value of 0 indicates that neither the NLRI field nor the Path Attribute field is present in this UPDATE message.
Path Attributes (variable length)	Every UPDATE message carries a variable-length sequence of path attributes, except an UPDATE message that carries only the withdrawn routes. Each path attribute is a triple <attribute type, attribute length, attribute value> of variable length: Attribute Type is a two-byte field that consists of the Attribute Flags bytes, followed by the Attribute Type Code byte.
Network Layer Reachability Information (variable length)	This variable length field contains a list of IP address prefixes. These IP prefixes are considered reachable by the advertising router and are advertised to the remote peer. The Reachability information is encoded as one or more 2-tuples of the form <length, prefix>: – The Length field indicates the length in bits of the IP address prefix. A length of zero indicates a prefix that matches all IP addresses (with prefix, itself, of zero bytes). – The Prefix field contains an IP address prefix, followed by enough trailing bits to make the end of the field fall on an byte boundary. Note that the value of any trailing bits is irrelevant.

router receives an UPDATE message, it checks each field in the message for validity, as specified in the "NOTIFICATION message" section below.

BGP routers store routes in their Routing Information Bases (RIBs). A RIB has the following logical subcomponents: the Adj-RIBs-In, the Loc-RIB, and the Adj-RIBs-Out. Each BGP router maintains the Adj-RIBs-In and Adj-RIBs-Out on a per-peer basis. Each of these databases is logically associated with each peer of the BGP router. These databases are described in the "BGP RIB Manager Process" section below.

Initially, when a set of BGP routers establish BGP sessions between themselves, they exchange all candidate BGP routes they have learned. After this initial route exchange, each BGP router only sends incremental updates when network changes occur. The incremental updates (as done in other protocols such as EIGRP and OSPF) reduces processing overheads in the BGP route processors as well as reduce network bandwidth consumption by routing updates.

3.5.2.2   Withdrawn Route Field and Handling BGP Route Withdrawals

In the event that a given BGP route becomes unavailable or unreachable, a BGP router noticing this declares the route as invalid (i.e., unfeasible route) and informs its peers by withdrawing the route. If there is only a change in the routing information associated with a particular network prefix, or a new path to that prefix has be found, then the BGP router does not need to withdraw the route, instead the router can advertise a replacement route.

A BGP router can list multiple routes that are to be withdrawn from service in a single UPDATE message. Each route listed is identified by its destination IP network prefix, which unambiguously identifies the route to the BGP router to receive the advertisement. When an UPDATE message advertises only routes that are to be withdrawn from service, it does not include path attributes or NLRI. Also, when the message advertises only a feasible route, then it need not carry the Withdrawn Routes field.

An UPDATE message that a BGP router sends should not list the same network address prefix written in the Withdrawn Routes field in the NLRI field. However, BGP routers that receive UPDATE message that do otherwise must be able to process such messages. A BGP router should handle such UPDATE messages as if the Withdrawn Routes field does not contain the double listed network address prefix.

There are a number of methods available for a BGP router to inform its peers that a route that it had previously advertised is no longer reachable or available for use. A BGP router can use any one of the following three methods to indicate that a route has been withdrawn from service:

- The router can include in the Withdrawn Routes field in the UPDATE message (Figure 3.14), the IP prefix associated with a previously advertised route. This indicates to the BGP peer that the route listed in Withdrawn Routes field is no longer available for use.
    - o   The router can include explicitly in the Withdrawn Routes field, one or more prefixes that are no longer reachable or unfeasible (the UPDATE message may also contain one or more new routes). The router does not need to include additional information for the routes being withdrawn such as associated path attributes (e.g., AS-Path).
- The router can replace the (withdrawn) route with another route with the same NLRI and advertised it.
    - o   Because a BGP router only advertises a single best route for each destination that is reachable, sending a BGP UPDATE message that contains a prefix that was previously advertised by the router, but with a new set of path attributes, serves an implicit withdrawal of the previously advertised prefix.

- The BGP router can close the BGP connection, which implicitly removes from service all routes that the router and its peer had advertised to each other.

Advertising a replacement route also allows the router to change the path attribute(s) of a route. The replacement route has the same network address prefix as the original route, but has associated with it new (changed) path attributes.

If a BGP router receives an UPDATE message that contains a non-empty Withdrawn Routes field, then the router has to remove from its Adj-RIBs-In database, these previously advertised routes (or destinations IP address prefixes contained in the Withdrawn Routes field). The BGP router then has to run its BGP path selection process (referred to as Decision Process in **[RFC4271]**) again because these previously advertised routes are no longer available for use.

3.5.2.3   UPDATE Message Path Attribute Field

Each UPDATE message carries a variable length sequence of BGP path attributes, except those UPDATE messages that contain only withdrawn routes. The two-byte Attribute Type field (Figure 3.14) consists of the Attribute Flags byte, followed by the Attribute Type Code byte. The remaining bytes of the Path Attribute field (with format <Attribute Type, Attribute Length, Attribute Value>) carry the attribute value which are interpreted according to the settings of the Attribute Flags and the Attribute Type Code.

A BGP UPDATE message can advertise just a single set of path attributes, applicable to multiple network destinations, as long as these destinations share these path attributes. This means all path attributes carried in a given BGP UPDATE message apply to all the network destinations listed in the NLRI field of that message.

3.5.2.3.1   Attribute Flags

The Attribute Flags are described as follows:

- Bit 0 (the high-order bit) of the Attribute Flags byte (called the *Optional bit*) indicates whether the path attribute is well-known (if set to 0) or optional (if set to 1).
- Bit 1 (the second high-order bit) of the Attribute Flags byte (called the *Transitive bit*) indicates whether an optional path attribute is non-transitive (if set to 0) or transitive (if set to 1).
  - o   BGP routers must set the transitive bit to 1 for well-known path attributes.
- Bit 2 (the third high-order bit) of the Attribute Flags byte (called the *Partial bit*) specifies whether the information carried in the optional transitive path attribute is complete (if set to 0) or partial (if set to 1).
  - o   BGP routers must set the Partial bit to 0 for well-known path attributes and for optional non-transitive path attributes.
- Bit 3 (the fourth high-order bit) of the Attribute Flags byte (called the *Extended Length bit*) indicates whether the Attribute Length is one byte (if set to 0) or two bytes (if set to 1).
  - o   When a BGP router sets the Extended Length bit of the Attribute Flags byte to 0, the *third* byte of the Path Attribute carries the length of the path attribute data in bytes.

o When a BGP router sets the Extended Length bit of the Attribute Flags byte to 1, the *third* and *fourth* bytes of the path attribute carry the length of the attribute data in bytes.

• Remaining 4 bits (the lower-order four bits) of the Attribute Flags byte are unassigned/unused. BGP routers must set these to all zero when sent and must ignore them when received.

3.5.2.3.2 Attribute Type Code

The Attribute Type Code bytes contains the defined numerical value (or code) assigned to a particular Attribute Type. Reference **[RFC4271]** defines a number of Attribute Types and Attribute Type Codes. However, there exist many non-standard or proprietary Attributes Types defined by router vendors such as the Cisco-specific WEIGHT attribute (which has only local significance in the BGP router on which it is configured).

Reference **[RFC4271]** defines the following Attribute Types and Attribute Type Code values:

• ORIGIN Attribute = Type Code 1
• AS_PATH Attribute = Type Code 2
• NEXT_HOP Attribute = Type Code 3
• MULTI_EXIT_DISC Attribute = Type Code 4
• LOCAL_PREF Attribute = Type Code 5
• ATOMIC_AGGREGATE Attribute = Type Code 6
• AGGREGATOR Attribute = Type Code 7

If a BGP router receives an UPDATE message with an optional non-transitive attribute that is not recognized, the router quietly ignores the attribute. If an optional transitive attribute in the UPDATE message is not recognized, the router sets the Partial bit (the third high-order bit) in the Attribute Flags byte to 1, and retains the attribute for propagation to other BGP routers. If the BGP router recognizes an optional attribute and it has a valid value, then, depending on the type of the optional attribute, the router processes the attribute locally, retains and updates it, if necessary, for possible propagation to other BGP routers.

3.5.2.4 Network Layer Reachability Information (NLRI) Field

The NLRI field (Figure 3.14) is a variable length field that contains a list of IP network address prefixes. The length of this field, in bytes, is not encoded explicitly in the UPDATE message, but can be calculated as:

$$NLRI\ Field\ Length = UPDATE\ Message\ Length - 23$$
$$-Total\ Path\ Attributes\ Length$$
$$-Withdrawn\ Routes\ Length$$

The UPDATE message Length, Withdrawn Routes Length, and Total Path Attribute Length, are all encoded in the BGP UPDATE message as shown in Figure 3.14. The UPDATE message has a minimum length of 23 bytes.

If a BGP router receives an UPDATE message that contains a feasible route, the router updates the Adj-RIBs-In with this route as follows:

- If the router determines that the NLRI of the new route is identical to one of the routes the router currently has stored in its Adj-RIBs-In, then the router has to replace the older route in the Adj-RIBs-In with the new route, thus implicitly withdrawing the older route from service.
- Otherwise, if the router determines that the Adj-RIB-In has no route with identical NLRI to the new route in the UPDATE message, then it has to enter the new route in the Adj-RIBs-In. Once the BGP route has updated the Adj-RIBs-In, it has run again its BGP path selection process (Decision Process in [RFC4271]).

3.5.2.5 Frequency of UPDATE Messages

In this section, we describe two parameters that can be used to control the rate at which UPDATE messages can be sent.

3.5.2.5.1 Frequency of Route Advertisement

The parameter *MinRouteAdvertisementIntervalTimer* in [RFC4271] specifies the minimum amount of time that must elapse before a BGP router sends the next advertisement to a peer about new, modified, and/or withdrawn routes to a particular network destination. The rate of route advertisement applies on a per-destination basis, although the setting of *MinRouteAdvertisementIntervalTimer* parameter is done on a per-BGP peer basis.

This means any two sequential UPDATE messages that a BGP router sends to a peer advertising feasible routes and/or withdrawal of unfeasible routes to a particular common set of network destinations, must be spaced apart by at least the setting of the *MinRouteAdvertisementIntervalTimer* parameter. It is obvious that the router can only achieve this by maintaining a separate timer for each common set of destinations. However, it must be recognized that doing this would entail unwarranted system implementation overhead.

For this reason, a BGP implementation may use any technique that ensures that the spacing between any two sequential UPDATE messages sent to a BGP peer (advertising feasible routes and/or withdrawal of unfeasible routes to a common set of destinations) is at least *MinRouteAdvertisementIntervalTimer*, plus, also ensuring that an acceptable constant upper bound on the interval is maintained.

Given that within any given autonomous system fast routing convergence is a key requirement, it is desirable for a BGP implementation to meet one of the following conditions:

a. The value of the *MinRouteAdvertisementIntervalTimer* used for iBGP peers should be smaller than the value of the *MinRouteAdvertisementIntervalTimer* used for eBGP peers, or
b. The route advertisement rate limiting procedure describe above should not be applied to routes advertised to iBGP peers.

Reference **[RFC4271]** suggests a default value for the *MinRouteAdvertisement IntervalTimer* on eBGP connections as 30 seconds, and for iBGP connections as 5 seconds. It should be noted that the route advertisement rate limiting procedure does not limit the rate at which a BGP router does route selection, but limits only the rate at which it does route advertisement. If the router selects new routes multiple times while waiting for the *MinRouteAdvertisementIntervalTimer* to expire, then the last route the router has selected is advertised at the end (expiration) of *MinRouteAdvertisementIntervalTimer*.

3.5.2.5.2   *Frequency of Route Origination*

The parameter *MinASOriginationIntervalTimer* in **[RFC4271]** specifies the minimum amount of time that must elapse before a BGP router sends the next advertisements of UPDATE messages that report changes within the router's own autonomous systems. This parameter determines the minimum separation between successive advertisements the router sends about routes originating from its own autonomous system. Reference **[RFC4271]** suggests a default value for the *MinASOriginationIntervalTimer* as 15 seconds.

3.5.3   BGP NOTIFICATION Message

A BGP router sends NOTIFICATION messages (Figure 3.15) to its peer when it detects errors or special conditions. When a router detects an error condition, it sends a NOTIFICATION message and then closes the BGP connection. The NOTIFICATION message has a minimum length of 21 bytes (i.e., the BGP message header + Error Code + Error Subcode). Using this minimum message length, the Data field length can be determined from the following formula:

$$NOTIFICATION\ Message\ Data\ Length = BGP\ Message\ Length - 21$$

The Length field in Figure 3.15 represents the entire BGP messages length (see Figure 3.12).

Reference **[RFC4271]** defines the following Error Codes (see Figure 3.15):

- 1 – Message Header Error
- 2 – OPEN Message Error
- 3 – UPDATE Message Error
- 4 – Hold Timer Expired
- 5 – Finite State Machine Error
- 6 – Cease

The Error Subcodes (Figure 3.15) are shown in Table 3.1.

When a BGP router detects any of the conditions described here, it sends a NOTIFICATION message, with the correct Error Code, Error Subcode, and Data fields, and closes the BGP connection (unless the router is explicitly configured to not send a NOTIFICATION message and close the BGP connection). If no Error

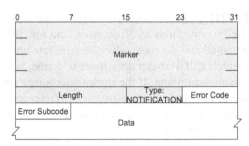

Field	Meaning
Error Code	This field indicates the type of NOTIFICATION. RFC 4271 defines a number of Error Codes. For example, Error Code 1 represents Message Header Error, and Error Code 2 represents OPEN Message Error.
Error Subcode	This field provides more specific information about the nature of the reported error. Each Error Code may have one or more Error Subcodes associated with it. If no appropriate Error Subcode is defined, then a zero (Unspecific) value is used for the Error Subcode field. For example, Error Subcode 1 represents Connection Not Synchronized, and Error Subcode 2 represents Bad Message Length.
Data	This variable-length field is used to diagnose the reason for the NOTIFICATION. The contents of the Data field depend upon the Error Code and Error Subcode.

FIGURE 3.15   BGP NOTIFICATION Message Format

TABLE 3.1
BGP Error Subcodes

Message Header Error Subcodes	OPEN Message Error Subcodes	UPDATE Message Error Subcodes
1 – Connection Not Synchronized	1 – Unsupported Version Number	1 – Malformed Attribute List
2 – Bad Message Length	2 – Bad Peer AS.	2 – Unrecognized Well-known Attribute
3 – Bad Message Type	3 – Bad BGP Identifier	3 – Missing Well-known Attribute
	4 – Unsupported Optional Parameter	4 – Attribute Flags Error
	5 – Deprecated	5 – Attribute Length Error
	6 – Unacceptable Hold Time	6 – Invalid ORIGIN Attribute
		7 – Deprecated
		8 – Invalid NEXT_HOP Attribute
		9 – Optional Attribute Error
		10 – Invalid Network Field
		11 – Malformed AS_PATH

Subcode is specified in the NOTIFICATION message, then the field must contain all zero value. Unless specified explicitly, the BGP router must leave the Data field (of the NOTIFICATION message that is sent to the peer to indicate an error) empty.

Closing the BGP connection means, the peering BGP routers will close the TCP connection, clear the associated Adj-RIBs-In databases, and deallocate all resources for that BGP connection. The BGP router marks the entries in the Loc-RIB database associated with the remote peer as invalid. The router then recalculates the best routes for the destinations of the routes marked as invalid. Before the BGP router deletes invalid routes from its system, it advertises to its BGP peers, either withdrawals for the routes marked as invalid, or the new recalculated best routes for the routes marked as invalid.

3.5.3.1 Handling Message Header Error

When a BGP router detects an error while processing the BGP Message Header, it must indicate that error by sending a NOTIFICATION message with the Error Code "Message Header Error". The router adds an Error Subcode to elaborate on the specific nature of the error. The Marker field of the BGP message header is expected to have a value of all ones. If a peer router receives a message with the Marker field of the message header not all ones, then a synchronization error has occurred and it must set the Error Subcode (in the NOTIFICATION message) to "Connection Not Synchronized".

If a BGP router detects at least one of the following conditions to be true, then it must set the Error Subcode to "Bad Message Length":

- If a BGP message header has the Length field having a value less than 19 bytes or greater than 4,096 bytes.
- If an OPEN message has the Length field value less than the minimum length of the OPEN message.
- If an UPDATE message has the Length field value less than the minimum length of the UPDATE message.
- If a KEEPALIVE message has the Length field value not equal to 19 bytes.
- If a NOTIFICATION message has the Length field value less than the minimum length of the NOTIFICATION message.

The BGP router must also fill the Data field of the NOTIFICATION message with the erroneous Length field.

If the receiving BGP router does not recognize the Type field of the BGP message header, then it must set the Error Subcode to "Bad Message Type". It must also fill in the Data field with the erroneous Type field.

3.5.3.2 Handling OPEN Message Error

When a BGP router detects an error while processing an OPEN message, it must indicate that error by sending a NOTIFICATION message with the Error Code "OPEN Message Error". The router also adds an Error Subcode to elaborate on the specific nature of the error. If the BGP router detects that the version number in the Version field of the received OPEN message (Figure 3.13) is not supported, then it must set the Error Subcode to "Unsupported Version Number". The router also performs the following actions:

- If the BGP router finds the OPEN message's My Autonomous System field to be unacceptable, then it must set the Error Subcode to "Bad Peer Autonomous System".
- If the BGP router finds the OPEN message's Hold Time field to be unacceptable, then it must set the Error Subcode to "Unacceptable Hold Time".

- If the BGP router finds the OPEN message's BGP Identifier field to be syntactically incorrect, then it must set the Error Subcode to "Bad BGP Identifier". Syntactic incorrectness means that the BGP Identifier field represents an invalid unicast IP host address.
- If the BGP router detects one of the Optional Parameters in the OPEN message as not recognizable, then it must set the Error Subcode to "Unsupported Optional Parameters".
- If the BGP router detects one of the Optional Parameters in the OPEN message to be recognizable, but is malformed, then it must set the Error Subcode to 0 (meaning, "Unspecific").

3.5.3.3 Handling UPDATE Message Error

When a BGP router detects an error while processing an UPDATE message, it must indicate that error by sending a NOTIFICATION message with the Error Code "UPDATE Message Error". The router also includes the Error Subcode to elaborate on the specific nature of the error. The router begins error checking of an UPDATE message by examining the path attributes contained in it:

- If the router determines that *Withdrawn Routes Length* or *Total Attribute Length to be too large* (i.e., if Withdrawn Routes Length + Total Attribute Length + 23 bytes exceeds the UPDATE message Length), then the router must set the Error Subcode to "Malformed Attribute List".
- If the router finds any recognized BGP path attribute in the UPDATE message has associated with it *Attribute Flags that conflict with the Attribute Type Code* (see Figure 3.14), then the router must set the Error Subcode to "Attribute Flags Error". The router must also write the erroneous attribute (type, length, and value) into the Data field (in Figure 3.15).
- If the router finds any recognized BGP path attribute in the UPDATE message has an *Attribute Length that conflicts with the expected length* (based on the Attribute Type Code; see Figure 3.14), then the router must set the Error Subcode to "Attribute Length Error". The router must also write the erroneous attribute (type, length, and value) into Data field.
- If the router finds any of the w*ell-known mandatory BGP path attributes are not present* in the UPDATE message, then the router must set the Error Subcode to "Missing Well-known Attribute". The router must also write the Attribute Type Code of the missing, well-known path attribute into Data field.
- If the router finds any of the *well-known mandatory BGP path attributes in the UPDATE message are not recognized*, then the router must set the Error Subcode to "Unrecognized Well-known Attribute". The router must also write the unrecognized path attribute (type, length, and value) into Data field.
- If the router finds the *ORIGIN attribute in the UPDATE message has an undefined value*, then the router must set the Error Subcode to "Invalid Origin Attribute". The router must also write the unrecognized attribute (type, length, and value) into the Data field.
- If the router determines the NEXT_HOP attribute field in the UPDATE message is syntactically incorrect, then the router must set the Error Subcode to

"Invalid NEXT_HOP Attribute". The router must also write the incorrect attribute (type, length, and value) into the Data field. Syntactic incorrectness means
that the NEXT_HOP attribute received represents an invalid IP host address.
To be considered semantically correct, the IP address in the NEXT_HOP must
satisfy the following criteria:

o The IP address received must not be the address of the receiving BGP router.

o In the case of eBGP peering, where the sending and receiving BGP routers
   are one IP hop away from each other, the IP address in the NEXT_HOP
   attribute must be either the sending router's IP address that is used to establish the BGP connection, or the receiving BGP router interface and the
   interface associated with the NEXT_HOP IP address must both share a
   common subnet.

If the router determines the NEXT_HOP attribute to be semantically incorrect,
then the router should log the error, and ignore the route. In this case, the router
does not send a NOTIFICATION message to the sender, and does not close the
BGP connection.

- If the router checks the AS_PATH attribute and finds the attribute to be syntactically incorrect, then the router must set the Error Subcode to "Malformed
  AS_PATH".

- If the router receives an UPDATE message from an eBGP peer, it may check
  whether the leftmost ASN in the AS_PATH attribute (with respect to the position of bytes in the UPDATE message) is equal to the ASN of the BGP peer
  that sent the message. If the receiving router determines this not to be the case,
  it must set the Error Subcode to "Malformed AS_PATH".

- If the router recognizes an optional path attribute, then it must check the value
  of this attribute. If the router detects an error, it must discard the attribute, and
  set the Error Subcode to "Optional Attribute Error". The router must also write
  the attribute (type, length, and value) into the Data field.

  o If the router finds any attribute has appeared more than once in the UPDATE
     message, then the router must set the Error Subcode to "Malformed
     Attribute List".

- The router must check the NLRI field in the UPDATE message for syntactic
  validity. If the router finds the value in this field to be syntactically incorrect,
  then it must set the Error Subcode to "Invalid Network Field".

  o If the router determines a network address prefix in the NLRI field to be
     semantically incorrect (e.g., an unexpected multicast IP address), it should
     log an error locally, and ignore the prefix.

The router treats an UPDATE message that contains correct path attributes, but with
no NLRI, as a valid UPDATE message.

3.5.3.4 Handling NOTIFICATION Message Error

If a BGP router sends a NOTIFICATION message to a peer, and the peer detects an
error in that message, the receiving peer cannot send a NOTIFICATION message to
report this error back to the sender. Instead, the receiving peer should log locally any
such errors (e.g., an unrecognized Error Code or Error Subcode), and bring the error

to the attention of the administration of the sending router. **[RFC4271]** does not specify actions to be taken when this happens and considers the response of the peer to be outside the scope of the specification.

3.5.3.5   Handling Hold Timer Expired Error

If a BGP router does not receive successive KEEPALIVE, UPDATE, and/or NOTIFICATION messages from a peer within the period specified in the Hold Time field of the OPEN message, then it sends a NOTIFICATION message to the peer with the Error Code "Hold Timer Expired", and closes the BGP connection.

3.5.3.6   Handling BGP Finite State Machine Error

If a BGP router detects any error in the BGP FSM (e.g., receipt of an unexpected event), it indicates the error to the peer by sending a NOTIFICATION message with the Error Code "Finite State Machine Error".

3.5.3.7   Sending the Error Code Cease

In the absence of any fatal errors (as listed in this section), a BGP router may choose, at any given time, to send a NOTIFICATION message with the Error Code "Cease" to close its BGP connection. However, the router must not use the NOTIFICATION message with the "Cease" Error Code when a fatal error indicated (as indicated below) does exist.

A BGP router may support the ability to set a locally configured, upper bound on the number of network address prefixes it is willing to accept from a peer. When the router detects that the upper bound is reached, then, under control of local configuration, the router may perform one of the following:

a. The router discards new network address prefixes received from the BGP peer (while continuing to maintain the BGP connection with the peer).
b. The router terminates the BGP connection with the peer.

If the BGP router decides to terminate its BGP connection with the peer because the number of address prefixes received from that peer has exceeded the locally configured, upper bound, then the router must send a NOTIFICATION message with the Error Code "Cease" to the peer. The router also logs this event locally.

3.5.3.8   BGP Connection Collision Detection

If two BGP routers try to establish a BGP connection with each other at the same time, then two parallel BGP connections will be formed. If the source IP address used by the first connection is the same as the destination IP address used by the second connection, and the destination IP address of the first connection is the same as the source IP address of the second connection, then connection collision has occurred. When this happens (a connection collision occurs), then one of the connections must be closed.

Based on the value of the BGP Identifier, **[RFC4271]** establishes a convention for determining which BGP connection is to be preserved when a connection collision occurs. To do this, the BGP Identifiers of the BGP peers involved in the connection collision are compared and only the connection initiated by the BGP router with the higher BGP Identifier value is retained.

A BGP router upon receiving an OPEN message, must examine all of its BGP connections that are in the BGP OpenConfirm State. A BGP router may also examine BGP connections in an OpenSent State if it knows the peer's BGP Identifier using means outside of the BGP specifications [RFC4271]. If any of these connections is a connection to a remote BGP router whose BGP Identifier is the same as the one in the received OPEN message, and this connection collides with the connection over which the router received the OPEN message, then the router performs the following collision resolution procedure:

1. The BGP router compares its BGP Identifier to the BGP Identifier of the remote peer (as specified in the OPEN message). The router does the comparison by converting the BGP Identifiers to host byte order and treating them as 4-byte unsigned integers.
2. If the router determines that the value of its own BGP Identifier is less than the remote router's Identifier value, then it closes the BGP connection that already exists (i.e., the BGP connection that is already in the OpenConfirm State), and accepts the BGP connection initiated by the remote BGP router.
3. Otherwise, the router closes the newly created BGP connection (i.e., the BGP connection associated with the newly received OPEN message), and continues using the existing BGP connection (i.e., the BGP connection that is already in the OpenConfirm State).

Unless BGP is configured to allowed (i.e., behave) otherwise, a BGP connection collision that is occurring with an existing BGP connection that is already in the Established State, will cause the newly created connection to be closed. It is important to note that a BGP router cannot detect connection collision with connections that are in the Idle, Connect, or Active BGP states. A BGP router accomplishes the closing of the BGP connection (that results from the collision resolution procedure described above) by sending a NOTIFICATION message to the remote router with the Error Code "Cease".

3.5.4   BGP KEEPALIVE Message

A BGP router sends KEEPALIVE messages to its peer periodically to ensure that the BGP connection between them is kept alive. As shown in Figure 3.16, the KEEPALIVE message has a length of 19 bytes and consists of only the BGP message header. BGP has its own internal keepalive mechanism to determine if BGP peers are available/reachable. It does not use any TCP-based keepalive mechanism. Instead, a BGP router exchanges KEEPALIVE messages with its peers frequently enough to not cause the Hold Timer to expire.

Reference [RFC4271] recommends a reasonable maximum interval between KEEPALIVE messages (i.e., the Keepalive Time) to be one third of the Hold Time interval. Also, BGP routers must not send KEEPALIVE messages more frequent than one per second. A BGP router may adjust the frequency of sending KEEPALIVE

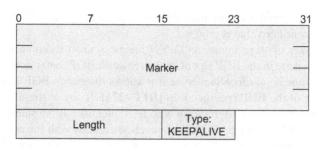

FIGURE 3.16 BGP KEEPALIVE Message Format

messages as a function of the Hold Time interval. If the Hold Time interval negotiated between two BGP routers is zero, then they must not send periodic KEEPALIVE messages **[RFC4271]**. The parameter *KeepaliveTime* is a mandatory BGP FSM attribute that stores the initial value of the BGP session *KeepaliveTimer*.

3.5.5 BGP ROUTE-REFRESH Message

The BGP-4 specification **[RFC4271]** does not define a mechanism for a BGP router to dynamically request the re-advertisement of the Adj-RIBs-Out database from a BGP peer. When the inbound routing policy in a BGP router for routes coming from a particular peer changes, all the network address prefixes from that peer must still be made available so that the BGP router can re-examined them against the new policy. To accomplish this, the BGP router uses a commonly used approach, called "soft-reconfiguration", where the router stores an unmodified copy of all routes from that peer at all times, even when its routing policies do not change. The BGP router requires additional memory and CPU processing to maintain the unmodified copy of the routes from the peer (and all peers).

Reference **[RFC2918]** defines an alternative solution that allows a BGP router to avoid the additional maintenance cost of using more memory and CPU processing for storing peer unmodified routes. The new BGP capability called "Route Refresh Capability" in **[RFC2918]**, allows a BGP router to dynamically send route refresh requests to a BGP peer for re-advertisement of its Adj-RIBs-Out database. A BGP router uses the BGP Capabilities Advertisement defined in **[RFC5492]** to advertise the BGP Route Refresh Capability to a peer. The BGP router advertises this capability to a peer in a BGP OPEN message using the Capability Code 2 and Capability Length 0.

A BGP router advertises the Route Refresh Capability to a peer to indicate to the peer that the sending router is capable of receiving and properly processing ROUTE-REFRESH messages from the peer. The ROUTE-REFRESH message (Figure 3.17), defined in **[RFC2918]**, is a new BGP message type defined in addition to those in **[RFC4271]**. A BGP router may send a ROUTE-REFRESH message to its peer only if the peer sends the Route Refresh Capability to it.

The <AFI, SAFI> carried in the ROUTE-REFRESH message should be one of the <AFI, SAFI> that the BGP peer has advertised to the BGP router at the time the BGP session was being established (via a BGP OPEN message containing the Route

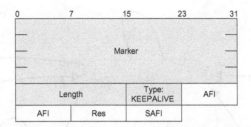

Field	Meaning
Address Family Identifier (AFI) (2 Byes)	This field carries the identity of the Network Layer protocol associated with the NLRI that follows.
Res (1 Byte)	This field is reserved and should be set to 0 by the sender and ignored by the receiver.
Subsequent Address Family Identifier (SAFI) (1 Byte)	This field provides additional information about the type of the Network Layer Reachability Information (NLRI) carried in the attribute.

FIGURE 3.17   BGP ROUTE-REFRESH Message Format

Refresh Capability advertisement). If a BGP router receives a ROUTE-REFRESH message with the <AFI, SAFI> from its peer that the router did not advertise at the time the BGP session was being established (via the capability advertisement), the router is required to ignore that ROUTE-REFRESH message. Otherwise, the BGP router is required to re-advertise to that peer the Adj-RIBs-Out associated with the <AFI, SAFI> carried in the ROUTE-REFRESH message, based on its outbound route filtering policy.

3.6   BGP SESSION STATES AND FINITE STATE MACHINE

A BGP router maintains a separate FSM for each peer it is configured to communicate with (Figure 3.18). A BGP neighbor negotiation process goes through different phases (or FSM stages) before the BGP connection is fully established. When two BGP routers are paired to create a potential BGP peering connection, each BGP router will attempt to establish a connection to the other, unless the router is configured to remain in the idle state, or configured to remain passive.

Here, the connecting side of the TCP connection (i.e., the side of a TCP connection that sends the first TCP SYN packet) is referred to as the active or outgoing connection. The listening side of the TCP connection (i.e., the side of the TCP connection that sends the first TCP SYN-ACK) is referred to as the passive side or incoming connection. One side (active side) performs the active open by sending a TCP SYN to the other side. In response, the passive side replies with a TCP SYN-ACK.

A BGP router must connect to a peer and also listen on TCP port 179 for replies from the passive side (or incoming connection). The router uses port 179 to connect to a peer but also listens on that port for peers trying to connect to it. The terms active and passive have slightly different meanings in BGP peering. In TCP usage, a TCP connection has only one active side and one passive side (as described above). However, when a BGP router is configured to participate in a peering as active, it

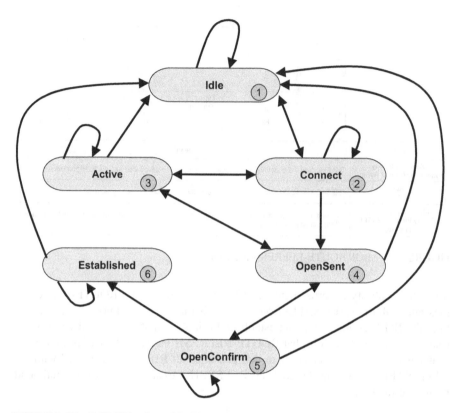

FIGURE 3.18   BGP Finite State Machine

may end up being either on the active side or passive side of the peering TCP connection that eventually gets established. But, once the two BGP peers have completed the TCP connection, it becomes inconsequential which router ends up being on the active or passive side. The only difference here is which side of the TCP connection (or equivalently, which BGP peer) has TCP port number 179.

The BGP router must instantiate a FSM for each incoming connection. There might be instances where a router knows the identity of the peer on the other end of an incoming connection, but does not know the BGP Identifier of the peer. When this happens, there may exist both an incoming and outgoing connection for the same configured BGP peering, a situation referred to as a BGP connection collision as discussed above. When a BGP connection collision occurs before the peering routers can determine which connection a peer is associated with, there may exist two connections for one peer. After the peering routers have successfully resolved the connection collision, they have to discard the BGP FSM for the connection that is closed.

A BGP router maintains, at most, one BGP FSM for each configured BGP peering. The router also maintains one more BGP FSM for each incoming TCP connection used for peering with a potential neighbor who has not yet been identified

[RFC4271]. BGP FSM the router maintains corresponds to exactly one TCP connection – one FSM per BGP connection. A BGP router may have more than one connection with a peer if these connections are configured to use a different pair of IP addresses – a configuration referred to as multiple "configured peerings" to the same BGP peer **[RFC4271]**.

3.6.1 IDLE STATE

The initial or starting state of the FSM of a BGP router is the Idle State (see Figure 3.18). In the Idle State, the BGP FSM in the router rejects all incoming BGP connections. Also, the router has not yet allocated any resources to the BGP FSM. In response to a FSM mandatory administrative event "*ManualStart*" (Event 1 in **[RFC4271]**), or an optional administrative event "*AutomaticStart*" (Event 3), the router performs the following:

- Initializes all BGP resources for the FSM associated with the incoming BGP peer connection.
- Initializes the mandatory session *ConnectRetryCounter* to zero. This counter indicates the number of times the BGP router has attempted to establish a BGP peer session.
- Starts the mandatory session *ConnectRetryTimer* from its initialized value. This timer indicates the length of time the BGP router must wait before retrying to establish a BGP peer session.
- Initiates the establishment of a TCP connection to the remote BGP peer.
- Listens for a TCP connection that may have been initiated by the remote BGP peer.
- Changes the state of the FSM to the *Connect State* (see Figure 3.18).

A FSM administrative event is an event through which the system administrator of the BGP router uses an operator's interface and BGP Policy Engine to signal to the BGP FSM to start or stop. In *ManualStart*, the administrator of the BGP router manually starts the peer BGP connection, while in *AutomaticStart*, the router itself automatically starts the BGP connection. In the Idle State, the BGP FSM ignores the mandatory event "*ManualStop*" (Event 2) and the optional event "*AutomaticStop*" (Event 8).

One reason for including the optional *AutomaticStop* event in a BGP implementation is that, a BGP router may receive UPDATE messages with a number of network address prefixes from a given BGP that result in the total number of address prefixes exceeding the maximum number of address prefixes configured on the router **[RFC4271]**. In such a case, the *AutomaticStop* event feature allows the router to automatically disconnect the remote BGP peer.

Mandatory attribute/event here means that all BGP implementations must understand this type of attribute/event. Optional attributes/events are defined as discretionary and a BGP implementation may choose to include these if it deems necessary.

3.6.2 Connect State

In the Connect State, the BGP FSM in the local router is waiting for the TCP connection with the remote BGP peer to be completed. In this state, the router ignores the FSM administrative start events (Events 1, 3 to 7).

- In response to the mandatory administrative event *"ManualStop"* (Event 2) initiated by the operator, the router performs the following:
  o Drops the TCP connection associated with the remote BGP peer (by sending a TCP FIN).
  o Releases all BGP resources for the FSM associated with the BGP peer.
  o Sets the mandatory session *ConnectRetryCounter* to zero.
  o Stops the mandatory session *ConnectRetryTimer* and resets it to zero.
  o Changes the state of the FSM to *Idle State*.
- In response to the mandatory timer event *"ConnectRetryTimer_Expires"* (Event 9), the router performs the following:
  o Drops the TCP connection associated with the remote BGP peer (by sending a TCP FIN).
  o Restarts the mandatory session *ConnectRetryTimer*.
  o Stops the optional session *DelayOpenTimer* and resets it to zero. This optional session timer is used to delay the sending of an OPEN message on a BGP connection.
  o Initiates the establishment of a TCP connection to the remote BGP peer.
  o Continues to listen for a TCP connection that may have be initiated by the remote BGP peer.
  o Allows the BGP FSM to remain in the *Connect State*.
- If the optional timer event *"DelayOpenTimer_Expires"* (Event 12) occurs while the FSM is in the Connect State, the router performs the following:
  o Sends a BGP OPEN message to the remote peer,
  o Sets the mandatory session *HoldTimer* to a large value (**[RFC4271]** suggests a value of 4 minutes).
  o Changes the state of the FSM to *OpenSent State*.
- If the BGP FSM receives the optional TCP connection-based event *"TcpConnection_Valid"* (Event 14), the FSM processes the TCP connection, and the FSM for the connection remains in the *Connect State*.
  o This event indicates that the router has received a TCP connection request with a valid source IP address, TCP port, destination IP address, and TCP Port. The BGP's destination TCP port number should be 179. A TCP connection request is signaled to the router upon receiving a TCP SYN from the remote peer.
- If the BGP FSM receives the optional TCP connection-based event *"Tcp_CR_Invalid"* (Event 15), the router rejects the TCP connection, and the FSM for the connection remains in the *Connect State*.
  o The *Tcp_CR_Invalid* event indicates that the router has received a TCP connection request with either an invalid source IP address or TCP port number, or an invalid destination IP address or TCP port number.

- If the BGP router succeeds in making a TCP connection (i.e., the mandatory TCP connection-based event *"Tcp_CR_Acked"* (Event 16) or *TcpConnectionConfirmed* [Event 17] occurs), the router checks the optional session attribute *DelayOpen* prior to processing. This attribute allows a BGP implementation to be configured to delay for a specific time period (i.e., the *DelayOpenTime*), the sending of an OPEN message. This delay is meant to allow the remote BGP peer sufficient time to send the first OPEN message. If the *DelayOpen* attribute is set to TRUE, the router performs the following:
  - o Stops the mandatory session *ConnectRetryTimer* (if running) and resets it to zero.
  - o Sets the optional session *DelayOpenTimer* to the initial value.
  - o Allows the BGP FSM to remain in the **Connect State**.

    The mandatory session *Tcp_CR_Acked* event (Event 16) indicates that the router has sent a request to establish a TCP connection with the remote peer. The router has sent a TCP connection request via TCP SYN, received a TCP SYN-ACK message from the remote peer, and sent in response to this, a TCP ACK.

    The mandatory session *TcpConnectionConfirmed* event (Event 17) indicates that the router has received a confirmation from the remote peer that it has established the TCP connection. The remote peer sent a TCP SYN. The local router sent a TCP SYN, TCP ACK message and now has received a final TCP ACK.

- If the optional session attribute *DelayOpen* is set to FALSE, the router performs the following:
  - o Stops the mandatory session *ConnectRetryTimer* (if running) and resets it to zero.
  - o Completes the initialization of BGP
  - o Sends a BGP OPEN message to the remote peer,
  - o Sets the mandatory session *HoldTimer* to a large value (e.g., 4 minutes).
  - o Changes the state of the FSM to **OpenSent State**.

- If the BGP FSM receives the mandatory TCP connection-based event *TcpConnectionFails* (Event 18), the router checks the *DelayOpenTimer*. If this timer is running, the router performs the following:
  - o Restarts the mandatory session *ConnectRetryTimer* from its initialized value.
  - o Stops the optional session *DelayOpenTimer* and resets it to zero.
  - o Continues to listen for a TCP connection that may have been initiated by the remote BGP peer.
  - o Changes the state of the FSM to **Active State**.

The *TcpConnectionFails* event indicates that the router has received a TCP connection failure notice. The remote BGP could have done this by sending a TCP FIN. The local router would respond to this with a TCP FIN-ACK. Another possibility is that, the local BGP router experienced a timeout in the TCP connection and then closed the connection.

- If the optional session *DelayOpenTimer* is not running, the router performs the following:
  - o Stops the mandatory session *ConnectRetryTimer* (if running) and resets it to zero,
  - o Drops the TCP connection associated with the remote BGP peer (by sending a TCP FIN).
  - o Releases all BGP resources for the FSM associated with the remote BGP peer
  - o Changes the state of the FSM to *Idle State*.
- If the BGP router receives a BGP OPEN message while the *DelayOpenTimer* is running (i.e., the optional BGP message-based event "*BGPOpen with DelayOpenTimer running*" [Event 20] occurs), it performs the following:
  - o Stops the *ConnectRetryTimer* (if running) and resets it to zero.
  - o Completes the initialization of BGP.
  - o Stops the *DelayOpenTimer* and resets it to zero.
  - o Sends a BGP OPEN message to the remote BGP peer.
  - o Sends a BGP KEEPALIVE message to the remote BGP peer.
  - o If the initial value of the mandatory session *HoldTimer* is non-zero, the router performs the following:
    - Starts the *KeepaliveTimer* from its initialized value. Typically, a BGP router initializes the *KeepaliveTimer* to the specified Keepalive Time value and counts down to zero, and then sends out another BGP KEEPALIVE message to the peer.
    - Resets the session *HoldTimer* to the value negotiated with the BGP peer.
  - o Else, if the initial value of the *HoldTimer* is zero, the router performs the following:
    - Resets the session *KeepaliveTimer* to zero.
    - Resets the session *HoldTimer* value to zero.
  - o Changes the state of the FSM to *OpenConfirm State*.

  If the BGP router detects the value of the My Autonomous System field in the BGP OPEN message (Figure 3.13) to be the same as the local ASN, it sets the connection status to an "internal" connection; otherwise, it sets the status to "external" connection.
- If the BGP router checks the BGP message header and detects an error (i.e., the mandatory BGP message-based event "*BGPHeaderErr*" (Event 21) occurs) or detects a BGP OPEN message error (i.e., the mandatory BGP message-based event "*BGPOpenMsgErr*" [Event 22] occurs), the router performs the following:
  - o (Optionally) If the optional attribute *SendNOTIFICATIONwithoutOPEN* is set to TRUE, then the router first sends a BGP NOTIFICATION message to the remote BGP peer with the appropriate Error Code.
  - o Stops the *ConnectRetryTimer* (if running) and resets it to zero.
  - o Releases all BGP resources for the FSM associated with the remote BGP peer.
  - o Drops the TCP connection associated with the remote BGP peer (by sending a TCP FIN).

o   Increments the connection's *ConnectRetryCounter* by 1.

o   (Optionally) performs peer oscillation damping if the optional session attribute *DampPeerOscillations* is set to TRUE.

o   Changes the state of the FSM to **Idle State**.

The *BGPHeaderErr* event is generated when the BGP router detects that a received BGP message header is not valid. The *BGPOpenMsgErr* event is generated when the BGP router detects that a received BGP OPEN message has errors. The optional BGP message-based attribute *SendNOTIFICATIONwithoutOPEN* allows a BGP router to send a BGP NOTIFICATION to the remote peer without first sending a BGP OPEN message. Without this optional session attribute, the BGP peer at the end of the BGP connection assumes that the router must send a BGP OPEN message prior to sending a BGP NOTIFICATION message. The BGP router sets the optional session attribute *DampPeerOscillations* to TRUE to indicate that it is using logic that damps BGP peer oscillations in the **Idle State**.

- If the BGP router receives a BGP NOTIFICATION message with a version error (i.e., the mandatory BGP message-based event "*NotifMsgVerErr*" [Event 24] occurs), the router checks the *DelayOpenTimer*. If this timer is running, the router performs the following:

o   Stops the *ConnectRetryTimer* (if running) and resets it to zero.

o   Stops the *DelayOpenTimer* and resets it to zero.

o   Releases all BGP resources for the FSM associated with the remote BGP peer.

o   Drops the TCP connection associated with the remote BGP peer (by sending a TCP FIN).

o   Changes the state of the FSM to **Idle State**.

The *NotifMsgVerErr* event is generated when the BGP router receives a BGP NOTIFICATION message with "Version Error".

- If the optional session *DelayOpenTimer* is not running, the router performs the following:

o   Stops the *ConnectRetryTimer* (if running) and resets it to zero.

o   Releases all BGP resources for the FSM associated with the remote BGP peer.

o   Drops the TCP connection associated with the remote BGP peer (by sending a TCP FIN).

o   Increments the connection's *ConnectRetryCounter* by 1.

o   (Optionally) performs peer oscillation damping if the optional session attribute *DampPeerOscillations* is set to TRUE.

o   Changes the state of the FSM to **Idle State**.

- The BGP also responds to any other events as listed below (see details in **[RFC4271]**):

o   Events 8 (*AutomaticStop*)

o   Events 10 and 11 (*HoldTimer_Expires* and *KeepaliveTimer_Expires*)

o   Event 13 (*IdleHoldTimer_Expires*)

o   Event 19 (*BGPOpen*)

o   Events 23 (*OpenCollisionDump*)

o Events 25 to 28 (*NotifMsg*, *KeepAliveMsg*, *UpdateMsg*, and *UpdateMsgErr*)
In response to these other events, the BGP router performs the following:
o Stops the *ConnectRetryTimer* (if running) and resets it to zero.
o Releases all BGP resources for the FSM associated with the remote BGP peer.
o Drops the TCP connection associated with the remote BGP peer (by sending a TCP FIN).
o Increments the connection's *ConnectRetryCounter* by 1.
o (Optionally) performs peer oscillation damping if the optional session attribute *DampPeerOscillations* is set to TRUE.
o Changes the state of the FSM to *Idle State*.

3.6.3 ACTIVE STATE

In the Active State, BGP FSM in the local router is trying to establish a BGP connection with a remote peer by listening for a TCP connection and accepting it.

- In this state, the router ignores the FSM start events, which are **[RFC4271]**:
  o Event 1: *ManualStart*
  o Events 3: *AutomaticStart*
  o Events 4: *ManualStart_with_PassiveTcpEstablishment*
  o Events 5: *AutomaticStart_with_PassiveTcpEstablishment*
  o Events 6: *AutomaticStart_with_DampPeerOscillations*
  o Events 7: *AutomaticStart_with_DampPeerOscillations_and_PassiveTcpEstablishment*
- In response to the mandatory *ManualStop* event (Event 2) in the Active State (initiated by the operator), the router performs the following:
  o If the optional session *DelayOpenTimer* is running and the optional session attribute *SendNOTIFICATIONwithoutOPEN* is set (TRUE), the router sends a BGP NOTIFICATION message with Error Code "Cease" to the remote BGP peer.
  o Releases all BGP resources for the FSM associated with the remote BGP peer including stopping the *DelayOpenTimer*.
  o Drops the TCP connection associated with the remote BGP peer (by sending a TCP FIN).
  o Resets the *ConnectRetryCounter* to zero.
  o Stops the *ConnectRetryTimer* and resets it to zero.
  o Changes the state of the FSM to *Idle State*.
- In response to a *ConnectRetryTimer_Expires* event (Event 9), the router performs the following:
  o Restarts the session *ConnectRetryTimer* from its initialized value.
  o Initiates a TCP connection to the remote BGP peer,
  o Continues to listen for a TCP connection that may have been initiated by a remote BGP peer
  o Changes the state of the FSM to *Connect State*.

- If the router receives the optional timer event *DelayOpenTimer_Expires* (Event 12), it performs the following:
  - o Resets the *ConnectRetryTimer* to zero.
  - o Stops and resets the *DelayOpenTimer* to zero.
  - o Completes the initialization of BGP.
  - o Sends a BGP OPEN message to the remote BGP peer.
  - o Sets the session *HoldTimer* to a large value (e.g., 4 minutes).
  - o Changes the state of the FSM to *OpenSent State*.
- If the BGP router receives the optional TCP connection-based *TcpConnection_ Valid* event (Event 14), it processes the TCP connection flags and stays in the *Active State*.
- If the BGP router receives the optional TCP connection-based *Tcp_CR_Invalid* event (Event 15), it rejects the TCP connection and stays in the *Active State*.
- If the BGP router succeeds in making a TCP connection (i.e., events *Tcp_CR_ Acked* (Event 16) or *TcpConnectionConfirmed* [Event 17] occur), it checks the optional *DelayOpen* attribute prior to processing the following:
  - o If the *DelayOpen* attribute is set to TRUE, the router performs the following:
    - ■ Stops the *ConnectRetryTimer* and resets it to zero.
    - ■ Sets the initial value of the session *DelayOpenTimer* to *DelayOpenTime*.
    - ■ Allows the BGP FSM to remain in the *Active State*.
  - o If the *DelayOpen* attribute is set to FALSE, the router performs the following:
    - ■ Resets the session *ConnectRetryTimer* to zero.
    - ■ Completes the initialization of BGP.
    - ■ Sends a BGP OPEN message to the remote BGP peer,
    - ■ Sets the session *HoldTimer* to a large value (e.g., 4 minutes)
    - ■ Changes the state of the FSM to *OpenSent State*.
- If the router receives the mandatory TCP connection-based *TcpConnectionFails* event (Event 18), it performs the following:
  - o Restarts the *ConnectRetryTimer* from the initialized value.
  - o Stops and resets the *DelayOpenTimer* value to zero.
  - o Releases all BGP resources for the FSM associated with the remote BGP peer.
  - o Increments the session *ConnectRetryCounter* by 1.
  - o (Optionally) performs peer oscillation damping if the optional session attribute *DampPeerOscillations* is set to TRUE.
  - o Changes the state of the FSM to *Idle State*.
- If the BGP router receives a BGP OPEN message while the *DelayOpenTimer* is running (i.e., the optional BGP message-based event *"BGPOpen with DelayOpenTimer running"* [Event 20] occurs), it performs the following:
  - o Stops the *ConnectRetryTimer* (if running) and resets it to zero.
  - o Stops and resets the *DelayOpenTimer* to zero.
  - o Completes the initialization of BGP.
  - o Sends a BGP OPEN message to the remote BGP peer.
  - o Sends a BGP KEEPALIVE message to the remote BGP peer.

o If the session *HoldTimer* value is non-zero.
  ▪ Starts the session *KeepaliveTimer* from its initialized value.
  ▪ Resets the session *HoldTimer* to the value negotiated with the BGP peer.
o Else if the session *HoldTimer* is zero.
  ▪ Resets the session *KeepaliveTimer* to zero.
  ▪ Resets the session *HoldTimer* to zero.
o Changes the state of the FSM to **OpenConfirm State**.
If the BGP router detects the value of the My Autonomous System field in the BGP OPEN message (Figure 3.13) to be the same as the local ASN, it sets the connection status to an "internal" connection; otherwise, it sets the status to "external" connection.

- If the BGP router checks the BGP message header and detects an error (i.e., the mandatory BGP message-based event "*BGPHeaderErr*" [Event 21] occurs) or detects a BGP OPEN message error (i.e., the mandatory BGP message-based event "*BGPOpenMsgErr*" [Event 22] occurs), the router performs the following:
  o (Optionally) If the optional attribute *SendNOTIFICATIONwithoutOPEN* is set to TRUE, then the router first sends a BGP NOTIFICATION message to the remote BGP peer with the appropriate Error Code.
  o Stops the *ConnectRetryTimer* (if running) and resets it to zero.
  o Releases all BGP resources for the FSM associated with the remote BGP peer.
  o Drops the TCP connection associated with the remote BGP peer (by sending a TCP FIN).
  o Increments the connection's *ConnectRetryCounter* by 1.
  o (Optionally) performs peer oscillation damping if the optional session attribute *DampPeerOscillations* is set to TRUE.
  o Changes the state of the FSM to **Idle State**.
- If the BGP router receives a BGP NOTIFICATION message with a Version Error (i.e., the mandatory BGP message-based event "*NotifMsgVerErr*" [Event 24] occurs), the router checks the *DelayOpenTimer*. If this timer is running, the router performs the following:
  o Stops the *ConnectRetryTimer* (if running) and resets it to zero.
  o Stops the *DelayOpenTimer* and resets it to zero.
  o Releases all BGP resources for the FSM associated with the remote BGP peer.
  o Drops the TCP connection associated with the remote BGP peer (by sending a TCP FIN).
  o Changes the state of the FSM to **Idle State**.
- If the optional session *DelayOpenTimer* is not running, the router performs the following:
  o Stops the *ConnectRetryTimer* (if running) and resets it to zero.
  o Releases all BGP resources for the FSM associated with the remote BGP peer.
  o Drops the TCP connection associated with the remote BGP peer (by sending a TCP FIN).

o Increments the connection's *ConnectRetryCounter* by 1.
o (Optionally) performs peer oscillation damping if the optional session attribute *DampPeerOscillations* is set to TRUE.
o Changes the state of the FSM to ***Idle State***.
• In response to all other events such as Events 8, 10 to 11, 13, 19, 23, 25 to 28, the router performs the following:
o Sets the session *ConnectRetryTimer* to zero.
o Releases all BGP resources for the FSM associated with the remote BGP peer.
o Drops the TCP connection associated with the remote BGP peer (by sending a TCP FIN).
o Increments the connection's *ConnectRetryCounter* by 1.
o (Optionally) performs peer oscillation damping if the optional session attribute *DampPeerOscillations* is set to TRUE.
o Changes the state of the FSM to ***Idle State***.

3.6.4 OPENSENT STATE

In the OpenSent State, the BGP FSM in the local router waits for a BGP OPEN message from its remote BGP peer.

• In this state, the router ignores the FSM start events (Events 1, 3 to 7).
• In response to the mandatory *ManualStop* event (Event 2) in the OpenSent State (initiated by the operator), the router performs the following:
o Sends a BGP NOTIFICATION message with Error Code "Cease" to the remote BGP peer.
o Resets the *ConnectRetryCounter* to zero.
o Releases all BGP resources for the FSM associated with the remote BGP peer.
o Drops the TCP connection associated with the remote BGP peer (by sending a TCP FIN).
o Resets the *ConnectRetryTimer* to zero.
o Changes the state of the FSM to ***Idle State***.
• In response to an *AutomaticStop* event (Event 8) in the OpenSent State (initiated by the system), the router performs the following:
o Sends a BGP NOTIFICATION message with Error Code "Cease" to the remote BGP peer.
o Sets the session *ConnectRetryTimer* to zero.
o Releases all BGP resources for the FSM associated with the remote BGP peer.
o Drops the TCP connection associated with the remote BGP peer (by sending a TCP FIN).
o Increments the connection's *ConnectRetryCounter* by 1.
o (Optionally) performs peer oscillation damping if the optional session attribute *DampPeerOscillations* is set to TRUE.
o Changes the state of the FSM to ***Idle State***.

- If the session timer event *HoldTimer_Expires* (Event 10) occurs, the router performs the following:
  - o Sends a BGP NOTIFICATION message to the remote BGP peer with the Error Code "Hold Timer Expired".
  - o Sets the session *ConnectRetryTimer* to zero.
  - o Releases all BGP resources for the FSM associated with the remote BGP peer.
  - o Drops the TCP connection associated with the remote BGP peer (by sending a TCP FIN).
  - o Increments the connection's *ConnectRetryCounter* by 1.
  - o (Optionally) performs peer oscillation damping if the optional session attribute *DampPeerOscillations* is set to TRUE.
  - o Changes the state of the FSM to *Idle State*.
- If the BGP router receives a *TcpConnection_Valid* (Event 14), *Tcp_CR_Acked* (Event 16), or a *TcpConnectionConfirmed* event (Event 17), it is possible that a second TCP connection may be in progress. The router's BGP FSM tracks the second TCP connection according to the BGP Connection Collision Detection process **[RFC4271]** until it receives a BGP OPEN message.
- The BGP FSM ignores a TCP Connection Request for an Invalid TCP port number (*Tcp_CR_Invalid* [Event 15]).
- If the router receives a *TcpConnectionFails* event (Event 18) is, it performs the following:
  - o Closes the BGP connection with the remote peer.
  - o Restarts the session *ConnectRetryTimer*.
  - o Continues to listen for a TCP connection that may have been initiated by the remote BGP peer
  - o Changes the state of the FSM to *Active State*.
- When the BGP router receives a BGP OPEN message from a peer, it checks all fields in the message for correctness. If the router detects no errors in the BGP OPEN message, it performs the following:
  - o Resets the session *DelayOpenTimer* to zero.
  - o Sets the session *ConnectRetryTimer* to zero.
  - o Sends a BGP KEEPALIVE message to the remote BGP peer.
  - o Sets the session *KeepaliveTimer* (as described below).
  - o Sets the session *HoldTimer* according to the value negotiated with the BGP peer.
  - o Changes the state of the FSM to *OpenConfirm State*.
  If the Hold Time value negotiated with the peer is zero, then the router does not start the *HoldTimer* and *KeepaliveTimer*. If the BGP router detects the value of the My Autonomous System field in the BGP OPEN message (Figure 3.13) to be the same as the local ASN, it sets the connection status to an "internal" connection; otherwise, it sets the status to "external" connection.
- If the BGP router checks the BGP message header and detects an error (i.e., the mandatory BGP message-based event "*BGPHeaderErr*" [Event 21] occurs) or detects a BGP OPEN message error (i.e., the mandatory BGP message-based

event *"BGPOpenMsgErr"* [Event 22] occurs), the router performs the following:

o Sends a BGP NOTIFICATION message to the remote BGP peer with the appropriate error code.

o Sets the session *ConnectRetryTimer* to zero.

o Releases all BGP resources for the FSM associated with the remote BGP peer.

o Drops the TCP connection associated with the remote BGP peer (by sending a TCP FIN).

o Increments the connection's *ConnectRetryCounter* by 1.

o (Optionally) performs peer oscillation damping if the optional session attribute *DampPeerOscillations* is set to TRUE.

o Changes the state of the FSM to *Idle State*.

- The BGP router must apply an appropriate BGP connection collision detection mechanism when it receives a valid BGP OPEN message (see discussion above). If the router determines that a connection in the OpenSent State is the one that must be closed, it signals the optional BGP message-based event *"OpenCollisionDump"* (Event 23) to its FSM. If such an event occurs when the FSM is in the OpenSent State, the router performs the following:

o Sends a BGP NOTIFICATION message with the Error Code "Cease" to the remote BGP peer.

o Sets the session *ConnectRetryTimer* to zero.

o Releases all BGP resources for the FSM associated with the remote BGP peer.

o Drops the TCP connection associated with the remote BGP peer (by sending a TCP FIN).

o Increments the connection's *ConnectRetryCounter* by 1.

o (Optionally) performs peer oscillation damping if the optional session attribute *DampPeerOscillations* is set to TRUE.

o Changes the state of the FSM to *Idle State*.

- If the BGP router receives a BGP NOTIFICATION message with a Version Error (i.e., the mandatory BGP message-based event *"NotifMsgVerErr"* [Event 24] occurs), the router performs the following:

o Sets the session *ConnectRetryTimer* to zero.

o Releases all BGP resources for the FSM associated with the remote BGP peer.

o Drops the TCP connection associated with the remote BGP peer (by sending a TCP FIN).

o Changes the state of the FSM to *Idle State*.

- In response to all other events such as Events 9, 11 to 13, 20, 25 to 28, the router performs the following:

o Sends a BGP NOTIFICATION message to the remote BGP peer with the Error Code FSM Error.

o Sets the session *ConnectRetryTimer* to zero.

o Releases all BGP resources for the FSM associated with the remote BGP peer.

o Drops the TCP connection associated with the remote BGP peer (by send-ing a TCP FIN).
o Increments the connection's *ConnectRetryCounter* by 1.
o (Optionally) performs peer oscillation damping if the optional session attri-bute *DampPeerOscillations* is set to TRUE.
o Changes the state of the FSM to **Idle State**.

3.6.5 OPENCONFIRM STATE

In the OpenConfirm State, the BGP FSM in the local router waits for a BGP KEEPALIVE or NOTIFICATION message from the remote BGP peer.

- In this state, the router ignores the FSM start events (Events 1, 3 to 7).
- In response to the mandatory administrative event "*ManualStop*" (Event 2) initiated by the operator, the router performs the following:
  o Sends a BGP NOTIFICATION message with the Error Code Cease to the remote BGP peer.
  o Releases all BGP resources for the FSM associated with the BGP peer.
  o Drops the TCP connection associated with the remote BGP peer (by send-ing a TCP FIN).
  o Sets session *ConnectRetryCounter* to zero.
  o Stops the session *ConnectRetryTimer* to zero.
  o Changes the state of the FSM to **Idle State**.
- In response to an *AutomaticStop* event (Event 8) initiated by the system, the router performs the following:
  o Sends a BGP NOTIFICATION message with Error Code Cease to the remote BGP peer.
  o Sets the session *ConnectRetryTimer* to zero.
  o Releases all BGP resources for the FSM associated with the remote BGP peer.
  o Drops the TCP connection associated with the remote BGP peer (by send-ing a TCP FIN).
  o Increments the connection's *ConnectRetryCounter* by 1.
  o (Optionally) performs peer oscillation damping if the optional session attri-bute *DampPeerOscillations* is set to TRUE.
  o Changes the state of the FSM to **Idle State**.
- If the session timer event *HoldTimer_Expires* (Event 10) occurs before the router receives a BGP KEEPALIVE message, the router performs the following:
  o Sends a BGP NOTIFICATION message to the remote BGP peer with the Error Code Hold Timer Expired.
  o Sets the session *ConnectRetryTimer* to zero.
  o Releases all BGP resources for the FSM associated with the remote BGP peer.
  o Drops the TCP connection associated with the remote BGP peer (by send-ing a TCP FIN).

- o Increments the connection's *ConnectRetryCounter* by 1.
- o (Optionally) performs peer oscillation damping if the optional session attribute *DampPeerOscillations* is set to TRUE.
- o Changes the state of the FSM to *Idle State*.
- If the BGP router receives a *KeepaliveTimer_Expires* event (Event 11), it performs the following:
  - o Sends a BGP KEEPALIVE message to the remote BGP peer.
  - o Restarts the session *KeepaliveTimer* associated with the BGP peer.
  - o Allows the FSM to remain in the *OpenConfirmed State*.
- In response to a *TcpConnection_Valid* event (Event 14), or the success of a TCP connection (Tcp_CR_Acked [Event 16], or a *TcpConnectionConfirmed* event [Event 17]) while in OpenConfirm State, the router needs to track the second TCP connection.
  - o If a TCP connection is attempted with an invalid TCP port (*Tcp_CR_Invalid* event [Event 15]), the router ignores the second TCP connection attempt.
- If the router receives a *TcpConnectionFails* event (Event 18) from the underlying TCP engine, or a BGP NOTIFICATION message (i.e., the mandatory BGP message-based event "*EventNotifMsg*" [Event 25] occurs), it performs the following:
  - o Sets the session *ConnectRetryTimer* to zero.
  - o Releases all BGP resources for the FSM associated with the remote BGP peer.
  - o Drops the TCP connection associated with the remote BGP peer (by sending a TCP FIN).
  - o Increments the session *ConnectRetryCounter* by 1.
  - o (Optionally) performs peer oscillation damping if the optional session attribute *DampPeerOscillations* is set to TRUE.
  - o Changes the state of the FSM to *Idle State*.
- If the BGP router receives a BGP NOTIFICATION message with a Version Error (i.e., the mandatory BGP message-based event "*NotifMsgVerErr*" [Event 24] occurs), the router performs the following:
  - o Sets the session *ConnectRetryTimer* to zero.
  - o Releases all BGP resources for the FSM associated with the remote BGP peer.
  - o Drops the TCP connection associated with the remote BGP peer (by sending a TCP FIN).
  - o Changes the state of the FSM to *Idle State*.
- If the BGP router receives a valid BGP OPEN message (*BGPOpen* [Event 19]), it processes the BGP collision detect function. If the router detects that the connection is to be dropped due to connection collision, it performs the following:
  - o Sends a BGP NOTIFICATION message with Error Code "Cease" to the remote BGP peer.
  - o Sets the session *ConnectRetryTimer* to zero.
  - o Releases all BGP resources for the FSM associated with the remote BGP peer.

o Drops the TCP connection associated with the remote BGP peer (by sending a TCP FIN).

o Increments the connection's *ConnectRetryCounter* by 1.

o (Optionally) performs peer oscillation damping if the optional session attribute *DampPeerOscillations* is set to TRUE.

o Changes the state of the FSM to **Idle State**.

• If the BGP router receives a BGP OPEN message, it checks all fields for correctness. If router detects an error in the BGP message header (*BGPHeaderErr* event [Event 21]) or BGP OPEN message (*BGPOpenMsgErr* event [Event 22]), it performs the following:

o Sends a BGP NOTIFICATION message to the remote BGP peer with the appropriate error code.

o Sets the session *ConnectRetryTimer* to zero.

o Releases all BGP resources for the FSM associated with the remote BGP peer.

o Drops the TCP connection associated with the remote BGP peer (by sending a TCP FIN).

o Increments the connection's *ConnectRetryCounter* by 1.

o (Optionally) performs peer oscillation damping if the optional session attribute *DampPeerOscillations* is set to TRUE.

o Changes the state of the FSM to **Idle State**.

• If during the processing of another BGP OPEN message, the BGP router detects that a BGP connection collision has occurred and the connection is to be closed, the BGP FSM will issue the optional BGP message-based event "*OpenCollisionDump*" (Event 23). This event will cause the router to perform the following:

o Sends a BGP NOTIFICATION message to the remote BGP peer with the Error Code Cease.

o Sets the session *ConnectRetryTimer* to zero.

o Releases all BGP resources for the FSM associated with the remote BGP peer.

o Drops the TCP connection associated with the remote BGP peer (by sending a TCP FIN).

o Increments the connection's *ConnectRetryCounter* by 1.

o (Optionally) performs peer oscillation damping if the optional session attribute *DampPeerOscillations* is set to TRUE.

o Changes the state of the FSM to **Idle State**.

• If the BGP router receives a BGP KEEPALIVE message (i.e., the mandatory BGP message-based event "*KeepAliveMsg*" [Event 26] occurs), it performs the following:

o Restarts the session *HoldTimer*.

o Changes the state of the FSM to **Established State**.

The *KeepAliveMsg* event is generated when the BGP router receives a BGP KEEPALIVE message.

• In response to all other events such as Events 9, 12 to 13, 20, 27 to 28, the router performs the following:

o Sends a BGP NOTIFICATION message to the remote BGP peer with the Error Code "Finite State Machine Error".
o Sets the session *ConnectRetryTimer* to zero.
o Releases all BGP resources for the FSM associated with the remote BGP peer.
o Drops the TCP connection associated with the remote BGP peer (by sending a TCP FIN).
o Increments the connection's *ConnectRetryCounter* by 1.
o (Optionally) performs peer oscillation damping if the optional session attribute *DampPeerOscillations* is set to TRUE.
o Changes the state of the FSM to *Idle State*.

3.6.6 ESTABLISHED STATE

The Established State is the final state of the BGP neighbor session process. In this state, the BGP FSM in the local router is now ready to exchange BGP UPDATE, KEEPALIVE, and NOTIFICATION messages with its remote BGP peer.

* In this state, the router ignores the FSM start events (Events 1, 3 to 7).
* In response to the mandatory administrative event *"ManualStop"* (Event 2) initiated by the operator, the router performs the following:
  o Sends a BGP NOTIFICATION message with the Error Code "Cease" to the remote BGP peer.
  o Sets session *ConnectRetryCounter* to zero.
  o Deletes (from its IP Routing Table) all routes associated with the BGP connection with the remote BGP peer.
  o Releases all BGP resources for the FSM associated with the BGP peer.
  o Drops the TCP connection associated with the remote BGP peer (by sending a TCP FIN).
  o Stops the session *ConnectRetryTimer* to zero.
  o Changes the state of the FSM to *Idle State*.
* In response to an *AutomaticStop* event (Event 8) initiated by the system, the router performs the following:
  o Sends a BGP NOTIFICATION message with Error Code Cease to the remote BGP peer.
  o Sets the session *ConnectRetryTimer* to zero.
  o Deletes (from its IP Routing Table) all routes associated with the BGP connection with the remote BGP peer.
  o Releases all BGP resources for the FSM associated with the remote BGP peer.
  o Drops the TCP connection associated with the remote BGP peer (by sending a TCP FIN).
  o Increments the connection's *ConnectRetryCounter* by 1.
  o (Optionally) performs peer oscillation damping if the optional session attribute *DampPeerOscillations* is set to TRUE.
  o Changes the state of the FSM to *Idle State*.

- If the session timer event *HoldTimer_Expires* (Event 10) occurs, the router performs the following:
  o Sends a BGP NOTIFICATION message to the remote BGP peer with the Error Code "Hold Timer Expired".
  o Sets the session *ConnectRetryTimer* to zero.
  o Releases all BGP resources for the FSM associated with the remote BGP peer.
  o Drops the TCP connection associated with the remote BGP peer (by sending a TCP FIN).
  o Increments the connection's *ConnectRetryCounter* by 1.
  o (Optionally) performs peer oscillation damping if the optional session attribute *DampPeerOscillations* is set to TRUE.
  o Changes the state of the FSM to **Idle State**.
- If the BGP router receives a *KeepaliveTimer_Expires* event (Event 11), it performs the following:
  o Sends a BGP KEEPALIVE message to the remote BGP peer
  o Restarts the *KeepaliveTimer*, unless the negotiated value of the *HoldTime* is zero.
  Each time the router sends a BGP UPDATE or KEEPALIVE message, it restarts the *KeepaliveTimer*, unless the *HoldTime* value negotiated is zero.
- A *TcpConnection_Valid* event (Event 14) occurring upon the receipt of a valid TCP port, will cause the BGP FSM to track the second connection.
  o The router will ignore an invalid TCP connection (indicated by the occurrence of the *Tcp_CR_Invalid* event [Event 15]).
  o In response to an indication that the TCP connection to the remote BGP peer has been successfully established (*Tcp_CR_Acked* event [Event 16] or *TcpConnectionConfirmed* event [Event 17]), the router will track the second connection until it sends a BGP OPEN message.
- If the BGP router receives a valid BGP OPEN message (*BGPOpen* event [Event 19]), and if the optional attribute *CollisionDetectEstablishedState* is set to TRUE, the router will check the BGP OPEN message to see if it collides with any other BGP connection. If the BGP router (implementation-dependent) determines that this BGP connection needs to be terminated, it will process the optional BGP message-based event *OpenCollisionDump* (Event 23). If the BGP connection needs to be terminated, the router performs the following:
  o Sends a BGP NOTIFICATION message with Error Code "Cease" to the remote BGP peer.
  o Sets the session *ConnectRetryTimer* to zero.
  o Deletes (from its IP Routing Table) all routes associated with the BGP connection with the remote BGP peer.
  o Releases all BGP resources for the FSM associated with the remote BGP peer.
  o Drops the TCP connection associated with the remote BGP peer (by sending a TCP FIN).
  o Increments the connection's *ConnectRetryCounter* by 1.

     o (Optionally) performs peer oscillation damping if the optional session attribute *DampPeerOscillations* is set to TRUE.

     o Changes the state of the FSM to **Idle State**.

- If BGP router receives a BGP NOTIFICATION message (*NotifMsgVerErr* [Event 24] or *NotifMsg* [Event 25] event) or a *TcpConnectionFails* event [Event 18] from the underlying TCP engine, it performs the following:

     o Sets the session *ConnectRetryTimer* to zero.

     o Deletes (from its IP Routing Table) all routes associated with the BGP connection with the remote BGP peer.

     o Releases all BGP resources for the FSM associated with the remote BGP peer.

     o Drops the TCP connection associated with the remote BGP peer (by sending a TCP FIN).

     o Increments the connection's *ConnectRetryCounter* by 1.

     o Changes the state of the FSM to **Idle State**.

- If the BGP router receives a BGP KEEPALIVE message (*KeepAliveMsg* event [Event 26]), it performs the following:

     o Restarts session *HoldTimer*, if the negotiated value of the Hold Time is non-zero.

     o Allows the FSM to remain in the **Established State**.

- If the BGP router receives a BGP UPDATE message (*UpdateMsg* event [Event 27]), it performs the following:

     o processes the BGP UPDATE message.

     o Restarts the session *HoldTimer*, if the negotiated value of the Hold Time is non-zero.

     o Allows the FSM to remain in the **Established State**.

- If the BGP router receives a BGP UPDATE message, and the processing of the message detects an error (i.e., the mandatory BGP message-based event "*UpdateMsgErr*" [Event 28] occurs), the router performs the following:

     o Sends a BGP NOTIFICATION message with Error Code "Update Message Error" to the remote BGP peer.

     o Sets the session *ConnectRetryTimer* to zero.

     o Deletes (from its IP Routing Table) all routes associated with the BGP connection with the remote BGP peer.

     o Releases all BGP resources for the FSM associated with the remote BGP peer.

     o Drops the TCP connection associated with the remote BGP peer (by sending a TCP FIN).

     o Increments the connection's *ConnectRetryCounter* by 1.

     o (Optionally) performs peer oscillation damping if the optional session attribute *DampPeerOscillations* is set to TRUE.

     o Changes the state of the FSM to **Idle State**.

- In response to all other event such as Events 9, 12 to 13, 20 to 22, the router performs the following:

     o Sends a BGP NOTIFICATION message with Error Code "Finite State Machine Error" to the remote BGP peer.

o Sets the session *ConnectRetryTimer* to zero.
o Deletes (from its IP Routing Table) all routes associated with the BGP connection with the remote BGP peer.
o Releases all BGP resources for the FSM associated with the remote BGP peer.
o Drops the TCP connection associated with the remote BGP peer (by sending a TCP FIN).
o Increments the connection's *ConnectRetryCounter* by 1.
o (Optionally) performs peer oscillation damping if the optional session attribute *DampPeerOscillations* is set to TRUE.
o Changes the state of the FSM to *Idle State*.

3.7   BGP VERSION NEGOTIATION

Two BGP routers may negotiate the version of BGP used (see Version field in the OPEN message format in Figure 3.13) by making multiple attempts at opening a BGP connection between them, starting with the highest BGP version each router supports **[RFC4271]**. If opening a BGP connection fails and ends up with the sending of an Error Code "OPEN Message Error" and an Error Subcode "Unsupported Version Number", then the BGP router will have available information about the BGP Version it tried using, the BGP Version its peer tried using, the BGP Version number indicated by its peer in the received NOTIFICATION message, and the BGP Version it supports locally. If the two routers do support one or more common BGP Versions, then they can use this information to quickly determine the highest BGP Version they both support.

3.8   BGP PATH ATTRIBUTES

When a BGP routers receives routing updates from multiple autonomous systems that describe different paths to the same network destination, it selects a single best path for reaching that destination. The BGP router then propagates the best path to its neighbors.

A BGP router uses BGP path attributes as pieces of information to describe the different network prefixes it includes in the BGP UPDATE messages it sends to neighbors. Every UPDATE message (Figure 3.14) contains a variable sequence of BGP attributes except those BGP UPDATE messages that carry only withdrawn routes. Each BGP attribute is described as a TLV representing an attribute *type*, attribute *length*, and attribute *value*.

This section describes the most commonly used BGP path attributes and how BGP uses them in the BGP route decision-making process. The route selection process is based on the values of the BGP attributes that the UPDTAE messages contain and other BGP-configurable factors in the local router. The list of path attributes carried in the BGP UPDATE messages are used by every BGP router along the path to compare different network paths, if multiple paths exist, and to the select the best paths to be installed in a router's IP Routing Table.

3.8.1 CATEGORIES OF BGP PATH ATTRIBUTES

The section describes the main categories of BGP path attributes (see Figures 3.19 and 3.20).

BGP defines four different categories of path attributes (see Figures 3.19 and 3.20): Well-known mandatory; Well-known discretionary; Optional transitive; Optional non-transitive. These path attributes are carried in BGP UPDATE messages (Figure 3.14) a router sends. All BGP implementations must be capable of recognizing all the well-known path attributes. When a BGP router updates any of the well-known path attributes, it must propagate these path attributes to its peers in the UPDATE messages it sends.

- **Well-Known Mandatory**: All BGP implementations must understand this type of BGP path attributes – these attributes must be included in all BGP UPDATE messages a BGP router sends. As shown in Figures 3.19 and 3.20, a BGP router

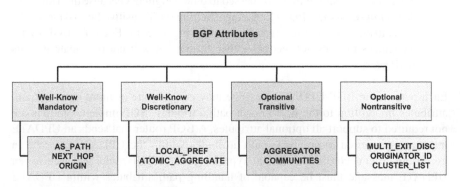

FIGURE 3.19 BGP Attributes

BGP Attribute	eBGP	iBGP	
AS_PATH	Well-known Mandatory	Well-known Mandatory	Automatically included in UPDATE message
NEXT_HOP	Well-known Mandatory	Well-known Mandatory	
ORIGIN	Well-known Mandatory	Well-known Mandatory	
LOCAL_PREF	**Not allowed**	Well-known Discretionary	
ATOMIC_AGGREGATE	Well-known Discretionary	Well-known Discretionary	Can be configured to help provide path control
AGGREGATOR	Optional Transitive	Optional Transitive	
COMMUNITY	Optional Transitive	Optional Transitive	
MULTI_EXIT_DISC	Optional Nontransitive	Optional Nontransitive	

FIGURE 3.20 List of BGP Attributes

must include the well-known mandatory path attributes in every UPDATE message that it sends that contains NLRI in both iBGP and eBGP routing exchanges.
BGP uses well-known mandatory attributes to ensure that all BGP implementations understand a minimum standard set of path attributes.

- **Well-Known Discretionary**: Other BGP path attributes are defined as discretionary which means a BGP router may choose to include these in a particular
  UPDATE message if it deems necessary. All BGP implementations must
  understand this type of path attributes but they do not have to be included in all
  BGP UPDATE messages a BGP router sends to its neighbors.
- **Optional Transitive**: These BGP path attributes do not need to be understood
  by all BGP implementations, but they will be passed to other neighbors since
  the Transitive flag bit (in the UPDATE message) is set to 1 (see Figure 3.14). A
  BGP implementation will accept and advertise this attribute even if it does not
  recognize it.
- **Optional Non-Transitive**: Unlike the optional transitive BGP path attributes,
  these path attributes will not be passed to other neighbors because the Transitive
  flag is *not* set in the UPDATE message. Setting the Transitive bit to 0 means the
  path attribute is non-transitive as shown in Figure 3.14. Even if a BGP router
  recognizes or does not recognize this attribute, it will not propagate it to its
  BGP peers.

Each path in the BGP UPDATE message may contain one or more optional path
attributes in addition to the well-known path attributes. BGP implementations are
not required to support all optional attributes. A BGP router that sends an UPDATE
message is required to order the path attributes within the UPDATE message in
ascending order according to attribute type. However, a BGP router that receives an
UPDATE message must be capable of handling path attributes within UPDATE
messages that are placed out of order **[RFC4271]**. Also, an UPDATE message must
not contain the same attribute type more than once within its Path Attributes field.

The originator and any other BGP router on the BGP path is allowed to attach
new, optional transitive attributes to an UPDATE message. If the BGP router attach-
ing a new attribute is not the originator, it will set the Partial bit in the Attribute Flags
byte to 1. The rules governing the addition of new non-transitive optional attributes
depends on the specific path attribute being modified. Also, the BGP routers on a
path are allow to update all optional path attributes (both transitive and non-transitive), if they see this to be necessary.

3.8.1.1 Handling Unrecognized Optional BGP Path Attributes

The setting of the Transitive bit in the BGP UPDATE message's Attribute Flags byte
determines how a BGP router will handle an unrecognized optional BGP path attribute (see Figure 3.14):

- BGP routers should accept paths with unrecognized optional transitive
  attributes.
- If a BGP router accepts a path with an unrecognized optional transitive attribute and propagates the path to other BGP peers, then it must pass that

unrecognized optional transitive attribute, along with the Partial bit (see Figure 3.14) in the Attribute Flags byte set to 1.

- If a BGP router accepts a path and propagates it to other BGP peers with a recognized optional transitive attribute, and the Partial bit in the Attribute Flags byte is set to 1 by a previous autonomous system, it must not set the bit back to 0.
- A BGP router that receives an unrecognized optional non-transitive attribute must quietly ignore it and not propagate it to other BGP peers.

3.8.2 ORIGIN ATTRIBUTE (ORIGIN)

The Origin attribute (Type Code 1) is a well-known mandatory BGP path attribute that defines the network entity that originated the path information in an UPDATE message. The data byte can take one of the following values:

- Value = 0: Meaning IGP – NLRI is interior to the originating autonomous system.
- Value = 1: Meaning EGP – NLRI learned via EGP.
- Value = 2: Meaning INCOMPLETE – NLRI learned by some other means.

The Origin attribute is generated by the BGP router that originated the associated NLRI in the UPDATE message and is not to be changed by any other router along the path. A BGP router considers the Origin attribute when running the BGP best path process to determine the best path to a network destination when multiple paths exist. BGP prefers the path with the lowest Origin attribute type value (type 0 is the lowest followed by type 1).

3.8.3 AS-PATH ATTRIBUTE (AS_PATH)

The AS-Path attribute (Type Code 2) is a well-known mandatory BGP path attribute that consists of a sequence of autonomous system path segments. The BGP AS-Path attribute lists the autonomous systems (i.e., ASNs) through which the reachability information in the UPDATE message has traversed (Figure 3.21). Whenever a routing update (i.e., UPDATE message) enters an autonomous system, BGP prepends the ASN to the AS-Path attribute before it is advertised to the next eBGP peer. This enables BGP to track routes, and also enables it to detect and prevent routing loops. In turn, the routing information that BGP advertises will list the autonomous systems through which the information has passed through. The information stored in the AS-Path is one of the primary criteria that a BGP router uses to evaluate routes for inclusion in its IP Routing Table.

Each autonomous system path segment is expressed as the triple <Path Segment Type, Path Segment Length, Path Segment Value>.

- **Path Segment Type**: This is a 1-byte field that can take the following values:
  - o Value = 1: Segment Type is AS_SET, which is an unordered set of autonomous systems a route in the UPDATE message has passed through.

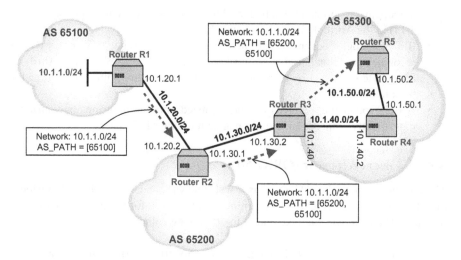

FIGURE 3.21 Illustrating the Use of the BGP AS-Path Attribute (AS_PATH)

    o Value = 2: Segment Type is AS_SEQUENCE which is an ordered set of autonomous systems a route in the UPDATE message has passed through.

- **Path Segment Length**: This is a 1-byte field containing the number of autonomous systems (not the number of bytes) in the Path Segment Value field.
- **Path Segment Value**: This field contains one or more ASNs, each encoded as a 2-byte field.

The components of the list in the AS-Path attribute can be in the form of an AS_SET or AS_SEQUENCE. The route aggregation rules described in the "BGP Path Attributes and Route Aggregation" section below show how AS_SETs are used. AS_SETs reduce the size of the AS-Path information because they list each ASN only once, regardless of how many times an ASN may have been written in multiple AS-Paths that were aggregated.

3.8.3.1 Using the AS-Path Attribute

An AS_SET implies that the network destinations listed in the UPDATE message's NLRI field can be reached through paths that pass through at least some of the AS_SET's constituent autonomous systems. The significance of AS_SETs is to provide sufficient routing information to avoid the creation of routing information loops. However, the use AS_SETs may cause potentially feasible paths to be pruned because such paths are no longer listed individually as seen in AS_SEQUENCEs. In practice, the use of AS_SETs is not likely to cause major routing problems because once an IP packet arrives at the edge of a group of autonomous systems, the receiving BGP router is likely to have maintained more detailed path information to be able to distinguish individual paths for the available network destinations.

Within an autonomous system, routes are passed between iBGP peers with their AS-Path attributes unaltered. However, when a route is being sent to an eBGP peer, the ASN of the originating eBGP peer is added to the route. Each eBGP peer that receives the route, prepends its own ASN to the route before passing it on to other eBGP routers. At the final eBGP router, the AS-Path of the route will list all the ASNs the route has passed through (including the local ASN). The ASN of the originating eBGP router will be at the end of the list (the rightmost position). This ordered list is an AS_SEQUENCE path segment.

BGP uses the ASNs in the AS-Path attribute to prevent routing loops from forming in the internetwork (of autonomous systems). Given that each routing update that is exchanged between eBGP routers contains a list of all ASNs the update has traversed, an eBGP router that receives an update can check the AS-Path attribute to see if its local ASNs is already listed. If the local ASN is already listed, then this means the update was seen in the past but has somehow found its way back to the receiving eBGP router's autonomous system. So, to avoid creating a routing loop, the receiving router simply ignores that update (and will not pass it on to other BGP routers).

The information carried in the AS-Path attribute is part of the information a BGP router uses in its BGP best path selection process to determine the best paths to network destinations. When comparing different paths to the same destination and assuming all other higher priority path attributes are equal, BGP prefers a shorter AS-Path over longer ones. In the event of a tie, BGP will consider other path attributes as discussed in the "BGP Best Path Selection Algorithm" section below.

3.8.3.2   Originating a BGP Route

The following explain the actions a BGP router will take when originating a route:

1. The BGP router that is originating the route will write its own ASN in a path segment of type AS_SEQUENCE, and will place it in the AS-Path attribute of all UPDATE messages it sends to an external (eBGP) peer.
   a. At the originating router, the ASN of the local autonomous system will be the only entry of the AS_SEQUENCE path segment, and this path segment will be the only segment in the AS-Path attribute.
2. The originating router will add an empty AS-Path attribute in all UPDATE messages it sends to internal (iBGP) peers. An empty AS-Path attribute has a length field containing the value zero.

3.8.3.3   Propagating a BGP Route

When a BGP router receives an UPDATE message from another BGP router and propagates a route it has learned from it, it will modify the AS-Path attribute of the route based on the location of the BGP router to which the route will be sent:

1. When the BGP router advertises the route to an internal (iBGP) peer, the AS-Path attribute associated with the route is not to be modified.
2. When the BGP router advertises the route to an external (eBGP) peer, it will update the AS-Path attribute as follows:
   a. If the router determines the first path segment of the AS-Path attribute to be of type AS_SEQUENCE, it will prepend its own AS Number (ASN) as the last element of the sequence. That is, it will place the ASN in the leftmost

position with respect to the position of bytes in the UPDATE message. If prepending this information will cause an overflow in the AS-Path attribute segment (i.e., exceed 255 autonomous systems), the router will generate and prepend a new segment of type AS_SEQUENCE and then write its own ASN to this new segment.

b. If the first path segment of the AS-Path attribute is of type AS_SET, the router will generate and prepend a new path segment of type AS_ SEQUENCE to the AS-Path attribute, and write its own ASN in that new segment.

c. If the AS-Path attribute is empty, the router will generate a path segment of type AS_SEQUENCE, write its own ASN into that segment, and place that segment into the AS-Path attribute.

Whenever a BGP router modifies the AS-Path attribute and finds that there is the need to include or prepend its own ASN, it may include or prepend more than one instance of its own ASN in the AS-Path attribute. The BGP router has to be properly configured to do this.

3.8.4   Next-Hop Attribute (NEXT_HOP)

The Next-Hop attribute (Type Code 3) is a well-known mandatory BGP path attribute, and in the typical case, specifies the (unicast) IP address of the next best router from the local router (i.e., the next-hop) that leads to the set of network destinations listed in the NLRI field of the UPDATE message (Figure 3.14). Although, this describes the general use of the Next-Hop attribute, the application of this attribute varies slightly depending on the networking situation as described in the following subsections (see also Figures 3.22 to 3.24).

3.8.4.1   Advertising a Route to an Internal BGP Peer

For an iBGP peer, the local router calculates the Next-Hop attribute as follows [RFC4271]:

1. When a BGP router sends an UPDATE message to an internal peer and the route is not locally originated, the router should not modify the Next-Hop attribute unless the router has been explicitly configured to advertise its own IP address as the Next-Hop (see Figure 3.23).

2. When a given BGP router advertises a locally originated route to an internal peer, the router should use as the Next-Hop, the interface IP address of the router through which it can reach the advertised network.

3. If the route being advertised to an internal peer is directly connected to the local BGP router, or if the interface IP address of the router through which the local router can reach the advertised network is the IP address of another internal peer, then the BGP router should use its own interface IP address as the Next-Hop. That is, the IP address of the interface that the local BGP router uses to reach the internal peer to receive the advertised route.

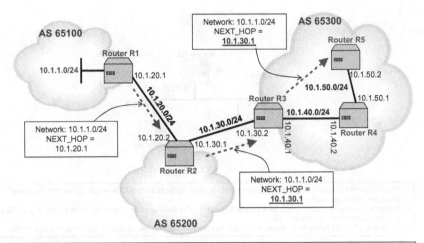

• When Router R3 sends an update to an internal peer R5 about network 10.1.1.0/24, R3 does not modify the Next-Hop attribute unless it is explicitly configured to do so using the `next-hop-self` command (see. also Example 2)
• The Next-Hop is updated between eBGP peers (eBGP peering between R1 and R2, and eBGP peering between R2 and R3) but not between iBGP peers (iBGP peering between R3 and R5).

FIGURE 3.22 Example 1: Illustrating the use of the BGP Next-Hop Attribute (NEXT_HOP)

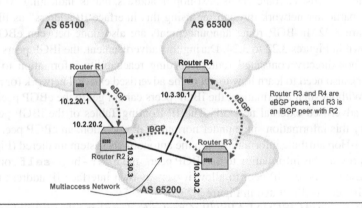

Scenario 1:
• BGP allows the BGP Next-Hop to be set when advertising a route between iBGP peers.
• Router R1 sends an update to Router R2, listing its interface 10.2.20.1 as a Next-Hop. When R2 forwards that update to R3, the Next-Hop IP address is still listed as 10.2.20.1. Router R3 (which has a BGP peering with R2) needs to have a route to the 10.2.2.0/24 network to have a valid Next-Hop.
• So to address this, the BGP configuration mode on R2 can be changed (e.g., using the `next-hop-self` command in Cisco IO) to allow R2 to advertises its IP address (10.3.30.3) to R3 as the Next-Hop for networks from AS 65100, rather than the IP address of R1 (10.2.20.1). This means, R3 does not have to know about the external network between R1 and R2 (i.e., network 10.2.20.0/24).

Scenario 2:
• On multiaccess networks (e.g., Ethernet), the BGP Next-Hop attribute can be changed to avoid traffic travelling an extra hop.
• When R3 sends an update to R4 about network 10.2.20.0/24, it will normally list its interface IP address (10.3.30.2) as the Next-Hop for R4 to use. However, given that R2, R3, and R4 are all on the same multiaccess network, it is inefficient for R4 to forward traffic to R3, and then for R3 to forward it on to R2. This situation unnecessarily adds an extra hop to the traffic path.
• So, to resolve this issue, R3 can be configured (using the `next-hop-self` command) to advertise a Next-Hop of 10.3.30.3 (R2's interface) for the network 10.2.20.0 instead of listing its own interface IP address (10.3.30.2).

FIGURE 3.23 Example 2: Illustrating the use of the BGP Next-Hop Attribute

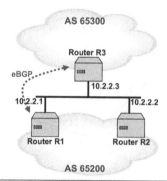

The BGP Next-Hop attribute can also beuse as away to direct traffic to another BGP router, rather than the router advertising the route itself.
Router R3 can be instructed to use R2 as the BGP Next-Hop for destinations in AS 65200, even though Router R3 is learning these routes directly from R2.
To do this, when R1 forwards an update to R3, the Next-Hop IP address will be setto10.2.2.2 (the interface of R2).

FIGURE 3.24 Example 3: Illustrating the use of the BGP Next-Hop Attribute: BGP Third Party Next-Hop

Most BGP implementations support commands (such as the **next-hop-self**) that allow the network administrator to set the BGP Next-Hop when a route is being advertised between iBGP peers **[CISCBGPCOMD19] [CISCMASCOMD14]**. In BGP, an router advertises a route via its next-hop IP address, that is, indicating "you can reach this particular network through me using this interface IP address" as illustrated in Figure 3.22. In BGP, route announcements are also done between eBGP peers as shown in Figures 3.22 to 3.24. During this advertisement, the iBGP peers (which are the non-directly connected routers needing reachability information to the eBGP peer), also need to learn how to get to the advertised external network (or route).

Without such information, the iBGP peers cannot get to the eBGP peers and then the advertised external network. The IP Routing Tables of the iBGP peers do not carry this information since under normal BGP operation, an eBGP peer passes the Next-Hop attribute information into the autonomous system unaltered (Figure 3.22). To provide this information to the iBGP peers, the **next-hop-self** command can be used at the eBGP peer to allow it pass its own interface IP address to the next iBGP peer as illustrated in Figure 3.23.

Reference **[CISCHALABS00]** (Chapter 6) describes other example networking scenarios where the **next-hop-self** command can be used. To configure a local BGP router to be the next-hop for a given BGP neighbor, the minimum command syntax is "**neighbor** *ip-address* **next-hop-self**", where *ip-address* is the IP address of the BGP neighbor to which advertisements will be sent with the local BGP router acting as the next-hop router **[CISCBGPCOMD19]**. The command forces the local BGP router to advertise itself to the BGP neighbor as being the next-hop of the advertised routes.

3.8.4.2 Advertising a Route to an External BGP Peer One Hop Away

When a given BGP router sends an UPDATE message to an external peer advertising a route, and the peer is one IP hop away, it performs the following actions (see Figures 3.22 to 3.24) **[RFC4271]**:

1. If the route being advertised was learned from an iBGP peer, or is one that is locally originated, the router can use in the Next-Hop attribute, the interface IP address of the iBGP peer through which it can reach the advertised network, provided that BGP peer shares a common subnet with this IP address. This represents one form of "third party" Next-Hop attribute (see Figure 3.24).

2. Otherwise, if the route being advertised was learned from an eBGP peer, the router can use in the Next-Hop attribute, the interface IP address of any adjacent router, provided that the eBGP peer shares a common subnet with the selected interface IP address. This represents another form of "third party" Next-Hop attribute (see Figure 3.24).

3. Otherwise, if the BGP router determines that eBGP peer to which the route is being advertised shares a common subnet with one of its own interfaces, it may use the IP address associated with this interface in the Next-Hop attribute. This situation is called a "first party" Next-Hop attribute.

4. By default (i.e., if none of the above conditions apply), the BGP router should use in the Next-Hop attribute, the IP address of the interface that the router itself uses to establish the BGP connection to the eBGP peer.

3.8.4.3   Advertising a Route to an External BGP Peer Multiple Hops Away

When a given BGP router sends an UPDATE message to an external peer advertising a route, and the peer is multiple IP hops away from the router (i.e., "multihop eBGP") **[RFC4271]**, it will perform the following:

1. If the BGP router is configured to propagate the Next-Hop attribute, then when advertising a route that it has learned from one of its peers, it does not modify the Next-Hop attribute. Instead, it will set the Next-Hop attribute of the advertised route to be exactly the same as the Next-Hop attribute of the learned route.

2. By default (i.e., if the above does not hold), the BGP router should use in the Next-Hop attribute, the IP address of the interface that it uses to establish the BGP connection to the eBGP peer.

From the Next-Hop discussion thus far, we derive the following key points:

- For eBGP peering, when a BGP router receives and advertises routes that originated from an eBGP peer, it sets the Next-Hop to be the IP address of that eBGP peer that announced the route.
- For iBGP peering, when a BGP router receives and advertises routes that originated from inside its autonomous system, it sets the Next-Hop to be the IP address of the iBGP peer that announced the route.
- When a BGP router propagates a route into an autonomous system via eBGP, the Next-Hop learned by eBGP is propagated unaltered into iBGP. The router leaves the Next-Hop to be the IP address of the eBGP peer from which the route was learned.
- When a BGP router advertises a route on a multiaccess medium (e.g., Ethernet), usually, the router sets the Next-Hop to be the IP address of the router connected to the medium that originated the route.

3.8.4.4   Other Rules Governing the Advertisement of a Route

Under normal circumstances, a BGP router chooses the Next-Hop attribute in the UPDATE message such that it takes the shortest available path. A BGP router must be able to disable the advertisement of third party Next-Hop attributes in order to handle media/networks that are imperfectly bridged (imperfections in a broadcast media/network).

A BGP router must not originate a route to be advertised to a peer using an IP address of that peer as Next-Hop. Also, BGP router must not install a route in its IP Routing Table with itself as the next-hop. A BGP router uses the Next-Hop attribute to determine the actual outbound interface and immediate next-hop IP address that should be used to forward packets that are in transit to the associated network destinations.

A BGP router determines an immediate next-hop IP address by performing a recursive route search in its IP Routing Table for the IP address carried in the Next-Hop attribute, and selects one entry if multiple entries of equal cost exist.

- The entry in the router's IP Routing Table that resolves the IP address carried in the Next-Hop attribute always specifies the outbound router interface.
- If the IP Routing Table entry specifies a directly attached subnet, but does not specify a next-hop IP address, then the router should use the IP address in the Next-Hop attribute as the immediate next-hop IP address.
- If the entry also specifies the next-hop IP address, then the router should use this address as the immediate next-hop address for packet forwarding.

3.8.5   Multi-Exit Discriminator (MED) Attribute (MULTI_EXIT_DISC)

The MED attribute (Type Code 4) is an optional non-transitive BGP path attribute that contains a four-byte unsigned integer (called a metric) **[CISCWHMCDA04]** **[RFC4271]** **[RFC4451]**. A BGP router may use the MED attribute value (sometimes called the metric) in its BGP path selection process to determine the preferred entry point when multiple entry points exist to a neighboring autonomous system (Figure 3.25). The MED attribute value is used to reflect the preferred route into an autonomous system that has multiple entry points.

As illustrated in Figure 3.25, BGP routers use the MED attribute on external (inter-autonomous system) links to discriminate among multiple exit/entry points to a neighboring autonomous system. All other factors being equal, the BGP routers prefer the exit point with the lower MED attribute value.

A BGP router sends the MED attribute to its eBGP peers to be propagated to BGP routers within their autonomous systems as shown in Figure 3.25. This means, if an eBGP router receives the MED attribute, it can only propagate it to iBGP routers within its autonomous system. The router must not propagate the MED attribute received from a neighboring autonomous system to other neighboring autonomous systems. The MED attribute is only exchanged between autonomous systems, and when an autonomous system receives a MED attribute, that attribute stays within that system and is not passed on to the next autonomous system. When a routing update

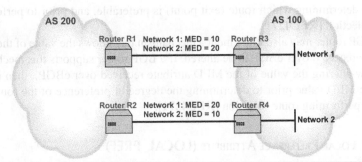

• Let us assume that AS 100 and AS 200 are connected by two separate BGP links: one to Routers R3 and the other to R4.
• We assume that Network 1 in AS 100 is located nearer Router R3, and Network 2, also in AS 100, is located nearer Router R4.
• Because the AS-Paths are equivalent, two routes exist for each network, one through Router R3 and the other through Router 4.
• To force all traffic destined for Network 1 through Router R3, AS 100 assigns a MED metric for each router to Network 1 atits exit point. The route to Network 1 through Router R3 is assigned a MED metric of 10, while the route to Network 1 through Router R4 is assigned a MED metric of 20.
• In this scenario, BGP routers in AS 200 will select the route with the lower MED metric for their IP routing tables.
However, it is recommended that the overall network be carefully evaluated for preferred routes (i.e., the BGP Local Preference attribute should also be considered) before configuring the MED options as explained later.

FIGURE 3.25 Illustrating Use of the BGP Multi-Exit Discriminator (MED) Attribute (MULTI_EXIT_DISC)

is passed on to another autonomous system, the eBGP router doing so will set the MED attribute value back to the default of 0.

Furthermore, when an autonomous system receives an UPDATE message containing a MED attribute, the BGP routers in that system use the MED attribute value received in their BGP best path selection process (see "BGP Best Path Selection Algorithm" section below). It should be noted that the BGP routers compare MED attributes for routes from eBGP peers that are in the same autonomous systems. This is because MED attributes from different autonomous systems are not comparable. The MED attribute value from a particular autonomous system is typically assigned to reflect the internal topology of the autonomous system, the routing protocol employed, and the routing policies used.

When two autonomous systems have multiple connections (routes) between them, it is not uncommon, in practice, for the MED attribute value sent by one of the autonomous systems to the other, to be based on the IGP metric in that system. In this case, the IGP metric reflects how close/far a particular network (prefix) is to one of the entry points to the autonomous system. The closest entry point for the network will have the lowest IGP metric (and consequently the lowest MED attribute value). This allows BGP to prefer the route that has the lowest MED attribute value or, equivalently, the route that leads to the entry point that is closest to the network prefix. Different MED attribute values can also be assigned to the entry points in order to load balance traffic between the two autonomous systems as illustrated in Figure 3.25.

A BGP router must support a locally configured/implemented mechanism that allows the MED attribute to be removed from a route. If a BGP router implements a mechanism to remove the MED attribute from a route, then it must do this removal prior to determining which route (exit point) is preferable, and prior to performing route selection **[RFC4271]**.

A BGP router may also implement a mechanism that allows the value of the MED attribute received over eBGP to be altered. If a BGP router supports this mechanism (to allow altering the value of the MED attribute received over eBGP), then it must alter the MED value prior to determining the degree of preference of the route, and prior to performing route selection **[RFC4271]**.

3.8.6   Local Preference Attribute (LOCAL_PREF)

The Local Preference attribute (Type Code 5) is a well-known discretionary BGP path attribute that contains a four-byte unsigned integer. A BGP router uses this path attribute to inform its other iBGP peers about its degree of preference for an advertised route (Figures 3.26 and 3.27). A given BGP router includes the Local Preference attribute in all UPDATE messages that it sends to other iBGP peers. When presented with multiple routes, a BGP router prefers the route with the highest Local Preference value.

The Local Preference attribute has significance only within an autonomous system and is exchanged only between iBGP peers within that autonomous system. A BGP router in an autonomous system may receive routing updates from several other autonomous systems describing routes to a given network destination. The BGP router communicates the Local Preference attribute to all other internal peers within

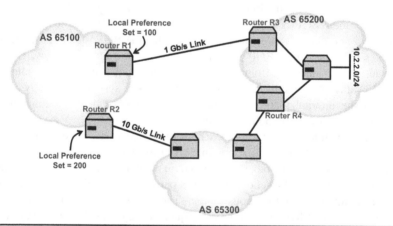

- We assume that AS 65100 is receiving two possible paths to the network 10.2.2.0/24, one of which is received through AS 65200, and the other through AS 65300.
- Although the path through AS 65200 is shorter (has one autonomous system hop rather than two), AS 65100 prefers to send traffic destined to the 10.2.2.0/24 prefix along the outbound 10Gb/s high-speed link, rather than along the slower outbound 1 Gb/s link.
- So, setting the BGP Local Preference on the 102.2.0/24 prefix as it is received on Router R1 to the value 100, and on Router R2 to 200, forces all of the BGP routers within AS 65100 to prefer the path through Router R2 - the higher speed 10 Gb/s link is thus preferred.

FIGURE 3.26   Illustrating the Use of the BGP Local Preference Attribute (LOCAL_PREF)

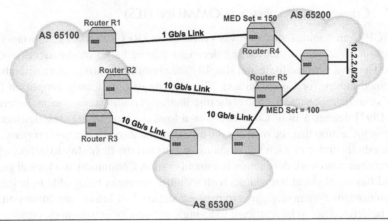

- We assume that AS 65200 sets the MED on its 1Gb/s exit point to the value 150, and the MED on its 10 Gb/s exit point to 100, with the intention of making the 10 Gb/s connection preferred. However, using the MED in this scenario will not yield the intended result as explained here.
    - First, AS 65100 will receive three paths to the 10.2.2.0/24 network, one through AS 65300, and two through AS 65200.
    - The MED of the path through AS 65100 and the paths through AS 65200 will not be compared, since their AS Paths are not the same.
    - If we assume that AS 65100 has set its BGP Local Preferences on Routers R1, R2, and R3, so that the path through AS 65300 will be preferred, then the MED advertised from AS 65200 (by Routers R4 and R5) will have no impact on traffic routing decision from AS 65100, since BGP MED is considered after BGP Local Preference during the execution of the BGP routing decision algorithm.
- If the path through AS 65300 had no texisted, or, for some other reason, was not preferred over the path through AS 65200, then the MEDs advertised by Routers R4 and R5 might have some impact on the best path decision made by AS 65100.
- On the other hand, if AS 65100 sets some BGP metric (such as the BGP Local Preference) to have a higher degree of preference in the BGP decision algorithm with the intended outcome that one path should be prefer over the other, then the MED would never be considered in the BGP decision algorithm.

FIGURE 3.27 Illustrating the Scenario where Using of the BGP Multi-Exit Discriminator (MED) Attribute has no Impact on Routing Decisions

its autonomous system to give them a common view on how to exit the autonomous system. The Local Preference attribute provides an indication to the iBGP routers within the autonomous system about which path is preferred to exit the autonomous system.

A BGP router first calculates the degree of preference for each external route based on the locally configured routing policy. It then includes the calculated degree of preference in the Local Preference attribute when advertising the route to its iBGP peers. The BGP router also uses the degree of preference it has learned via Local Preference attribute in its BGP best path selection process (see "BGP Best Path Selection Algorithm" section below).

Except in the case of BGP Confederations **[RFC5065]**, a BGP router must not include the Local Preference attribute in UPDATE messages it sends to eBGP peers. The Local Preference attribute is configured on an iBGP router and exchanged between iBGP peers; it is not passed to eBGP peers. If an eBGP router receives an UPDATE message containing the Local Preference attribute from an internal or external peer, then it must ignore this attribute, except in the case of BGP Confederations **[RFC5065]**.

3.8.7   COMMUNITIES ATTRIBUTE (COMMUNITIES)

The BGP Communities attribute was proposed in **[RFC1997]** for the grouping of network destinations so that routing decisions can be based on the identity of the group. The goal is to facilitate and simplify the control of routing information based on defined groups (or Communities). This is done to significantly simplify a BGP router's configuration and to control the distribution of routing information. Reference **[RFC1997]** defines a BGP Community as a logical group of network destinations based on the notion that the member destinations share some common property.

The administrator of each autonomous system has the flexibility to define which Communities a network destination is a member of. A Community is a logical grouping and has no physical boundaries with member networks being able to belong to any autonomous system (e.g., government, educational, or health care communities). It can be defined not to be restricted to a single network or autonomous system.

A network administrator can define Communities to simplify routing policies by identifying routes based on a logical property rather than on ASN and IP prefixes. BGP routers can use the defined BGP Communities attribute in conjunction with other BGP path attributes to control which routes to prefer, accept, and propagate to other BGP neighbors. BGP router can use the BGP Communities attribute to filter incoming or outgoing routes. A BGP router can tag routes with an indicator (a BGP Community) to allow other BGP routers to make routing decisions based on that tag.

3.8.7.1   BGP Communities Attribute

The BGP Communities path attribute defined in **[RFC1997]** is an optional transitive BGP path attribute of variable length. The Communities attribute consists of a set of four bytes (32 bits) values each one specifying a BGP Community. All BGP routes carrying this attribute belong to the Communities specified in the attribute.

The BGP Communities attribute has been assigned a Type Code of 8. BGP Communities are treated as 32-bit values; however, for administrative assignment, the following presumptions may be made:

* The BGP Communities attribute values in the range 0x0000000 to 0x0000FFFF and 0xFFFF0000 to 0xFFFFFFFF are reserved.
* The rest of the BGP Communities attribute values, 0x02B20000 through 0x02B2FFFF, are allowed for use. In common practice, the ASN of an autonomous system are encoded in the first two bytes (16 bits). The semantics of the last two bytes are defined by the autonomous system and is a value defined in relationship to that autonomous system. For example, a service provider (having AS 286) may define an educational and research Community using the BGP Communities attribute value 286:1 expressed in decimal notation. The 286 indicates that this particular service provider defined this BGP Communities attribute, while the 1 has special meaning to that service provider, which in this case represents the educational and research Community. The service provider uses this BGP Communities attribute for policy routing as defined by the policies of that autonomous system.

Reference **[RFC1997]** defines the following well-known BGP Communities which have global meaning, and their operations can be implemented in any BGP router that understands the BGP Communities attribute:

- **NO_EXPORT (0xFFFFFF01)**: All routes carrying this BGP Communities attribute value must not be advertised outside a BGP Confederation (see BGP Confederations below and in **[RFC5065]**). The route must not be advertised to any eBGP neighbors. However, routes with the NO_EXPORT attribute will be advertised to neighboring eBGP peers in autonomous systems within the same BGP Confederation (see Figure 3.28). A stand-alone autonomous system that is not part of a BGP Confederation is considered a BGP Confederation on its own.
- **NO_ADVERTISE (0xFFFFFF02)**: All routes carrying this BGP Communities attribute value must not be advertised to any BGP neighbors (iBGP or eBGP peers) as shown in Figure 3.29.
- **NO_EXPORT_SUBCONFED (0xFFFFFF03)**: All routes carrying this BGP Communities attribute value must not be advertised to eBGP neighbors including eBGP peers in autonomous systems within the same BGP Confederation (see Figure 3.30). This includes BGP peers belonging to other member autonomous systems (Member-AS) within a BGP Confederation.

In addition to the well-known BGP Communities attributes, private BGP Communities attributes can be defined for special use. Reference **[RFC1998]** describes methods

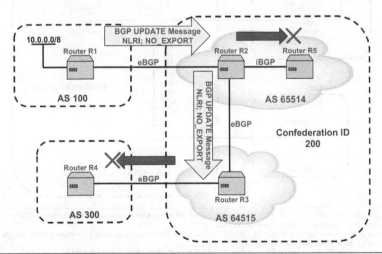

- Router R1 learns and advertises the network 10.0.0.0/8 with the NO_EXPORT attribute set.
- When Router R2 receives the network 10.0.0.0/8 with the NO_ADVERTISE attribute set, it will not advertise this route to any eBGP peers, except eBGP peers within the same BGP Confederation.
- Thus, Router R2 will advertise the network 1.0.0.0/8 to its eBGP confederation peer R3.
- Note that Router R3 will not advertise the network 1.0.0.0/8 to R4.

FIGURE 3.28　Using the BGP Communities Attribute NO_EXPORT

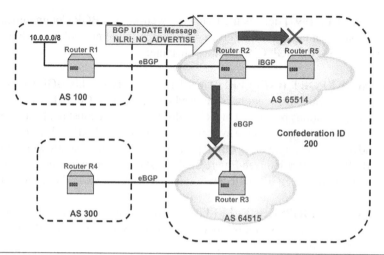

- When the NO_ADVERTISE attribute is set, Router R2 will not advertise the network 10.0.0.0/8 to any BGP peers (i.e., iBGP and eBGP peers).
- Router R1 may use a route map named, for example, NO_ADVERTISE -Community to add the NO_ADVERTISE community attribute to its route advertisements to R2.
- The route-map is applied outbound of R1. R1 is also configured to send BGP Communities attributes to R2.
- Router R2 will not advertise the network 1.0.0.0/8 to R3.

FIGURE 3.29   Using the BGP Communities Attribute NO_ADVERTISE

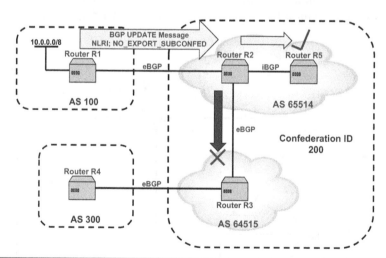

- When the NO_EXPORT_SUBCONFED attribute set, R2 will not advertise the network 10.0.0.0/8 to any eBGP neighbors including eBGP peers within the same BGP confederation.
- R1 may used a route-map to add the NO_EXPORT_SUBCONFED community attribute to its route advertisements to R2.
- The configuration is applied outbound of R1.
- However, Router R2 will advertised the network 10.0.0.0/8 to its iBGP neighbor R5 in AS 65514.
- The NO_EXPORT_SUBCONFED attribute prevents R2 from advertising the network 1.0.0.0/8 to the eBGP confederation peer R3 in AS 64515.

FIGURE 3.30   Using the BGP Communities Attribute NO_EXPORT_SUBCONFED

through which routers in service provider networks can use BGP Communities to manipulate BGP path selection.

A BGP router can use the BGP Communities attribute to control which routing information it prefers, accepts, and distributes to other BGP neighbors:

- A BGP router that receives a route that does not carry the BGP Communities attribute may append this attribute to the route when advertising the route to its BGP peers.
- A BGP route may carry one or more BGP Communities attributes. A BGP router that receives a route with multiple BGP Communities attributes may process one, some, or all of the received attributes. The router has the choice of adding or modifying the BGP Communities attributes before propagating the route to other internal and eBGP peers.
- A BGP router that receives a route carrying the BGP Communities attribute may modify the attribute according to the routing policies that are configured local. If a number of routes are aggregated and the resulting BGP Aggregates attribute (see section below) does not carry the BGP Atomic Aggregate attribute, then the resulting aggregate route should contain a BGP Communities attribute which contains all BGP Communities from all of the aggregated routes.

3.8.7.2   BGP Extended Communities Attribute

The BGP Extended Communities attribute defined in **[RFC4360]** describes extensions to the BGP Community Attribute in **[RFC1997]** by providing an extended range (ensuring that Communities can be assigned for a wide range of applications [without the possibility of overlaps]), and the addition of a Type field that provides structure for the Community space. The addition of structure allows routing policies to be used based on the application for which the BGP Communities value was defined.

For example, this structure allows all BGP Communities of a particular type to be filtered out, or allow only certain BGP Communities values for a particular type of BGP Community to be used. It also allows whether a particular BGP community can be specified as transitive or non-transitive across an autonomous system boundary. Without providing structure, accomplishing the above can only be done by explicitly enumerating all BGP Communities values that will be allowed or denied, and passed to BGP routers in neighboring autonomous systems based on the transitive property.

The Extended Communities Attribute is assigned the Type Code 16 and is a transitive optional BGP attribute. This attribute consists of a set of "extended BGP Communities". All routes carrying the BGP Extended Communities attribute belong to the Communities specified in the attribute. Each Extended Community is encoded as an 8-byte value and contains the following (see Figure 3.31):

- **Type Field**: 1 or 2 bytes

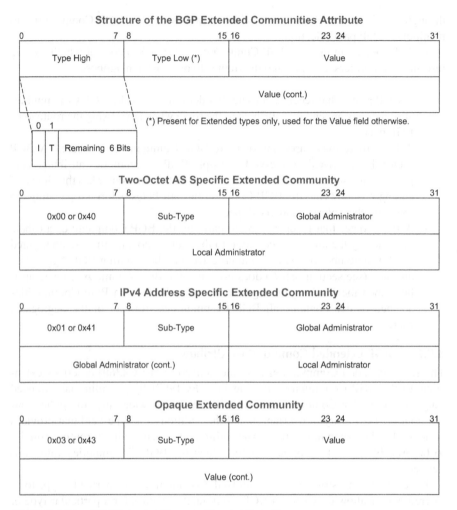

FIGURE 3.31 BGP Extended Communities Attribute Formats

- o Two classes of Type Field are defined in **[RFC4360]**, referred to as Regular type and Extended type. The size of Type Field for Regular types is 1 byte, while the size of the Type Field for Extended types is 2 bytes.
- o In the Type Field, the value of the high-order byte specifies if an Extended Community is a Regular type or an Extended type. The class of a type (Regular or Extended) is not encoded in the structure of the type itself but is specified in the (Internet Assigned Numbers Authority [IANA]) registry.
- o For the high-order byte of the Type Field, we have the following:
  - ■ First leftmost (or first high-order) bit is the I bit (IANA authority bit). A value of 0 means IANA-assignable type using the "First Come First Serve" policy. A value of 1 means part of this Type Field space is for types assignable by IANA using either the Standard Action or the Early

IANA Allocation policy. The rest of this Type Field space is for Experimental use.
- The second leftmost (or second high-order) bit is the T bit (Transitive bit). A value of 0 means the BGP Community is transitive across autonomous systems. A value of 1 means the BGP community is non-transitive across autonomous systems.
- The remaining 6 bits of the high-order byte indicates the structure of the Community.
• **Value Field**: These are the remaining bytes of the 8-byte Extended Community value
  o The value carried in the Value Field is determined by the "type" of the BGP Community as specified by the Type Field.

To specify any Community value, the two elements in the tuple <Type, Value> should be enumerated. The remaining bytes of the Community are interpreted based on the Type Field value. Two Extended Communities are considered equal only when all 8 bytes of the Extended Community are the same or equal.

Reference **[RFC4360]** describes a few Extended types and the format of the Value Field for those types. The high-order byte of the Extended Community Type field determines if the Community is a Regular type or an Extended type, and the lower-order byte (subtype) indicates a particular type of Extended Community.

• **Two-Octet AS Specific Extended Community** (see Figure 3.31): This is a BGP Extended Community type with Type Field consisting of 2 bytes and Value Field of 6 bytes.
  o The value of the high-order byte of this Extended Community type is either 0x00 or 0x40. The low-order byte is used to indicate subtypes.
  o The Value Field consists of two subfields:
    ▪ A 2-byte Global Administrator subfield which contains an ASN assigned by IANA.
    ▪ A 4-byte Local Administrator subfield in which the organization identified by ASN in the Global Administrator subfield can encode any information. The format and meaning of the value encoded in this subfield is defined by the subtype of the Community.
• **IPv4 Address Specific Extended Community** (see Figure 3.31): This is an Extended Community type with Type Field consisting of 2 bytes and Value Field of 6 bytes.
  o The value of the high-order byte of this Extended Community type is either 0x01 or 0x41. The low-order byte is used to indicate subtypes.
  o The Value field consists of two subfields:
    ▪ A 4-byte Global Administrator subfield which contains an IPv4 unicast address assigned by one of the Internet Registries.
    ▪ A 2-byte Local Administrator subfield in which the organization that has been assigned the IPv4 address written in the Global Administrator subfield can encode any information. The format and meaning of this value encoded in this subfield is defined by the subtype of the community.

- **Opaque Extended Community** (see Figure 3.31): This is an Extended Community type with Type Field consisting of 2 bytes and Value Field of 6 bytes.
  - o The value of the high-order byte of this Extended Community type is either 0x03 or 0x43. The low-order byte is used to indicate subtypes. This is a generic Community of Extended type. The IANA is to assign the value of the subtype that should define the Value Field.
- **Route Target Community**: The Route Target Community is transitive across the Autonomous System boundary, and identifies one or more routers that may receive a set of routes carried by BGP (that contain this Community).
  - o The Route Target Community is of an Extended Community type. The high-order byte of the Type field can have a value equal to 0x00, 0x01, or 0x02. The value of the low-order byte is 0x02.
  - o When the high-order byte of the Type field has value equal to 0x00 or 0x02, the Local Administrator subfield carries a number assigned from a numbering space that is administered by the organization to which the ASN written in the Global Administrator subfield has been assigned.
  - o When the high-order byte of the Type field has a value equal to 0x01, the Local Administrator subfield carries a number assigned from a numbering space that is administered by the organization to which the IP address written in the Global Administrator subfield has been assigned.

  Reference **[RFC4364]** specifies one possible use of the Route Target Community.
- **Route Origin Community**: The Route Origin Community is transitive across the Autonomous System boundary and identifies one or more routers that inject a set of routes into BGP (that carry this Community).
  - o The Route Origin Community is of an Extended Community type. The high-order byte of the Type field can have a value equal to 0x00, 0x01, or 0x02. The value of the low-order byte is 0x03.
  - o When the high-order byte of the Type field has a value equal to 0x00 or 0x02, the Local Administrator subfield carries a number assigned from a numbering space that is administered by the organization to which the ASN written in the Global Administrator subfield has been assigned.
  - o When the high-order byte of the Type field has a value equal to 0x01, the Local Administrator subfield carries a number assigned from a numbering space that is administered by the organization to which the IP address written in the Global Administrator subfield has been assigned.

Reference **[RFC4364]** specifies one possible use of the Route Origin Community.

3.8.7.2.1 Use of the BGP Extended Communities Attribute

The Extended Communities attribute may be used by a BGP router to control which routing information it is willing to accept or distribute to its BGP peers:

- BGP routers must not use the BGP Extended Community attribute to modify the BGP best path selection algorithm in a way that leads to forwarding loops.

- When A BGP router receives a route that does not contain the BGP Extended Communities attribute, it may append this attribute to the route when propagating it to its BGP peers.
- When a BGP router receives a route that carries the BGP Extended Communities attribute, it may modify this attribute according to the local routing policies.
- By default, if a BGP router is to aggregate a range of routes and the resulting aggregates path attributes do not carry the Atomic Aggregate attribute, then the resulting aggregate route should have a BGP Extended Communities attribute that contains the set union of all the BGP Extended Communities from all of the individual routes being aggregated. It is possible to override this default behavior through local configuration, which means the handling of the BGP Extended Communities attribute when performing route aggregation becomes a matter of the local policy of the BGP router doing the aggregation.
- If a route has a non-transitive Extended Community, then before a BGP router advertises the route across the Autonomous System boundary, it should remove the Community from the route. However, the Community should be removed when the BGP router advertises the route across the BGP Confederation boundary.
- A route may carry both the **[RFC1997]** defined BGP Communities attribute and the **[RFC4360]** defined Extended BGP Communities attribute. In this case, the BGP Communities attribute is handled according to **[RFC1997]**, and the Extended BGP Communities attribute is handled as specified in **[RFC4360]**.

3.8.8 ATOMIC AGGREGATE ATTRIBUTE (ATOMIC_AGGREGATE)

The Atomic Aggregate Attribute (Type Code 6) is a well-known discretionary BGP path attribute of length 0. A BGP router uses this attribute to inform its peers that it is advertising a less-specific route (i.e., an aggregate route) that includes more-specific routes. The attribute informs the BGP peers that the received route may not necessarily be the most complete routing information available. The Atomic Aggregate attribute is set to either "True" or "False" with former alerting the BGP peers that multiple more-specific routes have been summarized into a single aggregate route.

When a BGP router aggregates several routes to be advertised to a particular peer, the router normally includes an AS_SET that is formed from the set of autonomous systems from which the aggregate was formed in the AS-Path attribute of the aggregate route. In many cases, the network administrator has to determine if the BGP router can safely advertise the aggregate route without the original AS_SET, without creating routing loops.

If a BGP router creates an aggregate route that excludes at least some of the ASNs present in the AS-Path attribute of the individual more-specific routes as a result of dropping the AS_SET, the aggregate route should include the Atomic Aggregate attribute when advertising it to its BGP peer (Figure 3.32). This means if a BGP router elects to aggregate routes, then it should either include all autonomous systems used to form the aggregate in an AS_SET, or add the Atomic Aggregate attribute to the aggregate route. The receiving router performs the following:

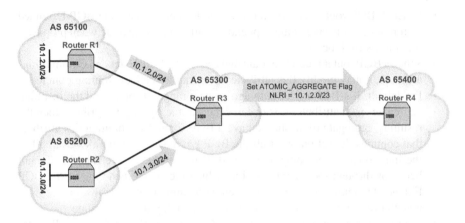

FIGURE 3.32 Illustrating the Use of the BGP Atomic Aggregate Attribute (ATOMIC_AGGREGATE)

- A BGP router that receives a routing advertisement for a route that contains the Atomic Aggregate attribute should not remove the attribute when advertising the route to other BGP routers.
- A BGP router that receives a routing advertisement for a route containing the Atomic Aggregate attribute must not make any route in the NLRI associated with that route more specific when propagating this route to other BGP routers.

A BGP router that receives a routing advertisement for a route containing the Atomic Aggregate attribute needs to know that the actual path to the network destinations specified in the UPDATE message's NLRI field of the route (while providing the loop-free routing) may not be the path specified in the AS-Path attribute of the route.

3.8.9 AGGREGATOR ATTRIBUTE (AGGREGATOR)

The Aggregator Attribute (Type Code 7) is an optional transitive BGP path attribute of length 6. This path attribute contains the last ASN that formed the aggregate (or summary) route (encoded as a 2-byte ASN value), followed by the IP address of the BGP router that created the aggregate route (encoded as a 4-byte value) (Figure 3.33). This IP address should be the same IP address used as the BGP Identifier of that BGP router.

The UPDATE messages a BGP router sends to its peers, carrying network prefixes formed by route aggregation, may include the Aggregator attribute. After the BGP router has performed route aggregation, it may include the Aggregator attribute in the UPDATE messages containing its own (local) ASN and IP address as shown in Figure 3.33. The IP address in the Aggregator attribute is the same as the BGP Identifier of the local BGP router.

Including the local ASN and the BGP Identifier in the routing update for the aggregate route enables a network administrator to determine which BGP router created a particular instance of an aggregate route. Tracing an aggregate route to its creator may be necessary for troubleshooting purposes.

3.8.10 WEIGHT ATTRIBUTE (WEIGHT)

Weight is a Cisco defined BGP path attribute and has only local significance in the Cisco router on which it is configured. It is the first path attribute a Cisco BGP router checks when determining the best path using the BGP path selection algorithm. When multiple paths exist to the same destination, BGP prefers the path with the highest Weight value (Figure 3.34). Configuring a Weight of 1,000 on Router R4 for the route through R2 to network 10.2.2.0/24 forces R4 to prefer this route over the route through R3 to 10.2.2.0/24 which has a Weight of 200.

The Weight attribute is a locally configured attribute and is not advertised to other routers through BGP UPDATE messages – it is not exchanged between BGP routers (not even routers within the same autonomous system). The major difference between the Local Preference (LOCAL_PREF) and the Weight (WEIGHT) attributes is that,

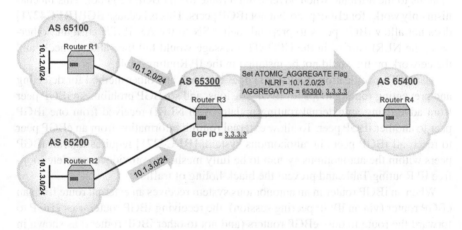

FIGURE 3.33 Illustrating the Use of the BGP Aggregator Attribute (AGGREGATOR)

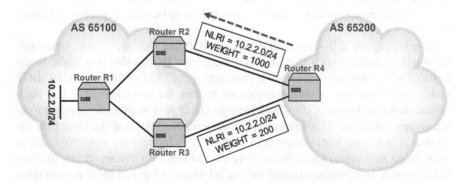

FIGURE 3.34 Illustrating the Use of the BGP Weight Attribute (WEIGHT)

when the Local Preference attribute is changed, that change is propagated throughout the local autonomous system. The new Local Preference attribute value will be advertised to all other BGP routers in the autonomous system, unlike the Weight attribute, which has local significance only on the BGP router on which it is configured.

This parameter can be set locally in a BGP router either through the BGP `neighbor weight` command [CISCBGPCOMD19] [CISCMASCOMD14], via an AS-Path access list, or via route maps. The Weight attribute can take any value from 0 to 65,535. Routes that a router originates are assigned (by default) a Weight value of 32,768. Routes that originate externally to the router (not originated by the router itself) are assigned a Weight value of 0.

3.9   UNDERSTANDING THE IBGP FULL MESH REQUIREMENT

BGP uses the AS_PATH path attribute as a mechanism for detecting and preventing routing loops by allowing BGP routers to prepend the ASN of their local autonomous systems to the attribute when advertising a route to an eBGP neighbor. This mechanism only works for eBGP peers but not iBGP peers. This is because BGP **[RFC4271]** does not allow iBGP peers to prepend their ASN to the AS_PATH attribute; otherwise, the NLRI carried in the UPDATE message would fail the validity check, and the network prefix would not be installed in the IP Routing Table.

The BGP specification **[RFC4271]** does not have any other method for detecting and preventing routing loops for iBGP sessions, and so, BGP prohibits an iBGP peer from advertising an external routing update (with NLRI) received from one iBGP peer to another iBGP peer. To allow external routing information from an eBGP peer to reach all iBGP peers in autonomous system, **[RFC4271]** requires that all iBGP peers within the autonomous system to be fully meshed to provide a complete loop-free IP Routing Table and prevent the black-holing of traffic.

When an iBGP router in an autonomous system receives an external route from an eBGP router (via an iBGP peering session), the receiving iBGP router uses eBGP to forward the route to only eBGP routers (and not to other iBGP routers) as shown in Figure 3.35. In simple words, an iBGP peer cannot advertise an external route learned from one iBGP peer to another iBGP peer. This behavior of iBGP is why it is necessary for iBGP routers within an autonomous system to be fully meshed as illustrated in Figure 3.35.

Considering the above observation about iBGP and loop prevention, other than the use of techniques such as BGP Route Reflection **[RFC4456]** (see "BGP Route Reflection" section below), network engineers would fully interconnect all iBGP router within the same single autonomous system, the so-called iBGP full mesh requirement (see Figure 3.35). This allows an eBGP peer to advertise external routing updates to all the iBGP routers in the autonomous system at the same time (as illustrated in Figure 3.35). However, a number of techniques are available (such as BGP Route Reflection) for circumventing the full mesh iBGP requirement to allow iBGP peers to re-advertise external routing information from an iBGP peer to other iBGP peers within the same autonomous system.

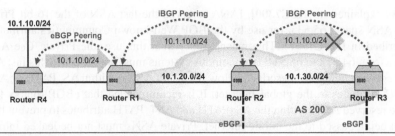

Without iBGP full mesh connectivity:
- Let us assume Routers R1, R2, and R3 are all within a single autonomous system AS 200.
- Router R1 forms an iBGP session with R2, and R2 forms an iBGP session with R3.
- Router R1 advertises the external network 10.1.10.0/24 to R2, which processes and installs it in its (R2's) BGP Table.
- Router R2 does not advertise the 10.1.1.0/24 NLRI to R3 because the prefix was received from an iBGP peer.
- eBGP routes learned via R2 are passed to both R1 and R3, and eBGP routes learned via R3 are passed only to R2. However, eBGP routes learned via R3 will not be passed by R2 to R1.

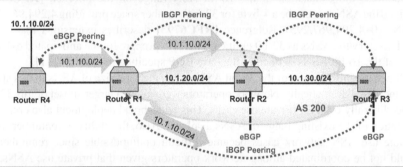

With iBGP full mesh connectivity:
- To resolve the issue in the first diagram, Router R1 must form, in addition, a multihop iBGP session with R3 so that it can receive the network prefix 10.1.1.0/24 directly from R1.
- Router R1 connects to R3's IP address (10.1.30.33), and R3 connects to R1's IP address (10.1.10.11).
- Routers R1 and R3 need a static route or IGP route to connect each other.

FIGURE 3.35   iBGP Full Mesh Requirement

3.10   AUTONOMOUS SYSTEM NUMBERS (ASNS)

Each autonomous system that communicates on the public Internet is assigned a globally unique ASN. The autonomous system uses this globally unique identifier in both the exchange of routing information with other neighboring autonomous systems, and as an identifier for itself. The autonomous system, however, will normally use one or more IGPs for exchanging reachability information within itself.

3.10.1   ORIGINAL 16-BIT ASN SPACE

The base BGP specification in **[RFC4271]** specifies the ASN as a two-byte entity (see "My Autonomous System" field in the BGP OPEN message format in Figure 3.13). This 16-bit ASN field limits the number of unique ASNs to 65,536 (0–65,535). Within this 16-bit limit, the IANA reserved the block of 1,023 ASNs from 64,512 to 65,535 for private use, that is, as "Private Use ASNs" that are not to be advertised on the global Internet **[RFC1930]**.

As explained in **[RFC7300]**, IANA reserved the last ASN of the 16-bit Private Use ASN space, 65,535, for use by the BGP Well-known Communities attribute as described in **[RFC1997]**. This means, other than this use, the Private Use ASNs range from 64,512 to 65,534. Autonomous systems must not advertise this last ASN (the corresponding one described below for 32-bit ASNs) within AS_PATH or AS4_ PATH attributes to the global Internet. It is recommended that eBGP routers filter these reserved ASNs within the AS_PATH and AS4_PATH attributes to prevent them from being passed to the public Internet. Private ASNs must not be leaked into the public Internet because they are not globally unique.

3.10.2 NEWER 32-BIT ASN SPACE

To address the exhaustion of the 16-bit ASN range on the Internet, **[RFC6793]** defined the ASN space to be a 4-byte (or 32-bit) number space providing 4,294,967,296 ASNs (0–4,294,967,295). Reference **[RFC6793]** describes extensions that allow BGP to carry the ASNs as 32-bit entities. This is to prepare for the anticipated exhaustion of the 16-bit ASNs defined in the base BGP specification **[RFC4271]**.

Since the introduction of the 32-bit ASN space, the total size of ASNs in this space has increased dramatically. Network operators were given a larger subset of this 32-bit space liberally even for private use cases. This widespread deployment also created a depletion of the existing range of ASNs available. Also, the ability to renumber these private use ASN network cases became difficult or impossible since renumbering could not be coordinated among network operators given that private use ASNs, by definition, are not IANA registered. To address this, **[RFC6996]** defined a second larger range of Private Use ASNs out of the 32-bit ASN space defined in **[RFC6793]**.

In addition to the contiguous block of 1,023 Private Use ASNs from the original 16-bit ASNs (namely 64,512–65,534), the IANA also reserved, for private use, the contiguous block of 94,967,295 ASNs from the 32-bit ASNs (namely 4,200,000,000– 4,294,967,294). Similar to the last ASN of the 16-bit ASN space, 65,535, reference **[RFC7300]**, also explains that the last ASN of the 32-bit Private Use ASN space, 4,294,967,295, could find similar use, and so is better reserved.

If an autonomous system uses Private Use ASNs and originates network prefixes carrying these ASNs, it must remove these ASNs from AS-Path attributes before advertising routes to the global Internet (as well as the AS4_PATH if the 32-bit ASN space is utilized). The eBGP routers of an autonomous system may also use normal AS-Path filtering methods to prevent network prefixes originating from Private Use ASNs within the autonomous system from being advertised to the global Internet.

3.10.2.1 Extensions for Carrying 32-bit ASNs in BGP

Reference **[RFC6793]** defines a BGP Capability Code that BGP routers can use to indicate their support for 32-bit ASNs. This document introduces two BGP path attributes, AS4_PATH and AS4_AGGREGATOR, that routers can use to propagate 32-bit based AS-Path information to other BGP routers that do not support 32-bit ASNs. The document also describes mechanisms for constructing the AS-Path information from these two new BGP path attributes. The extensions specified in **[RFC6793]** allow the gradual transition from 16-bit ASNs to 32-bit ASNs.

In BGP, the ASNs are carried in the "My Autonomous System" field of the OPEN message (Figure 3.13), and in the AS_PATH, AGGREGATOR, and BGP Communities attributes when carried in an UPDATE message. A 32-bit ASN BGP router uses the BGP Capabilities Advertisements defined in **[RFC5492]** to advertise to its internal or external neighbors that it supports 32-bit ASN extensions. The 32-bit ASN capability the BGP router advertises to its peers also carries the ASN (encoded as a 32-bit entity) in the Capability Value field of the advertisement. The sending router sets the Capability Length field of the advertisement to 4.

The AS-Path information that any two 32-bit ASN BGP routers exchange is carried in the existing AS_PATH attribute (as defined in **[RFC4271]**), except that each ASN in the AS_PATH attribute is encoded as a 32-bit entity (instead of a 16-bit entity). This also applies when the AGGREGATOR attribute is used between any two 32-bit ASN BGP routers – the ASN carried in this attribute is also encoded as a 32-bit entity. However, the AS_PATH attribute and the AGGREGATOR attribute carried between a 32-bit ASN BGP router and a 16-bit ASN BGP router have the ASNs encoded as 16-bit entities. Some of the protocol extensions in **[RFC6793]** for the use of 32-bit ASNs are as follows:

- To allow the AS-Path information with 32-bit ASNs to be preserved across 16-bit ASN BGP routers, reference **[RFC6793]** defines a new BGP path attribute, AS4_PATH. This is defined as an optional transitive BGP path attribute that contains the AS-Path encoded with 32-bit ASNs. Other than being an "optional transitive" attribute and carrying 32-bit ASNs, the new AS4_PATH attribute has the same semantics and encoding as the older AS_PATH attribute. The AS4_PATH attribute is carried across a series of 16-bit ASN BGP routers without modification, which also helps to preserve the non-mappable 32-bit ASNs in the AS-Path information.
- A BGP UPDATE message that is exchanged between any two 32-bit ASN BGP routers must not carry the new BGP attributes, AS4_PATH and AS4_ AGGREGATOR. A 32-bit ASN BGP router that receives an UPDATE message from another 32-bit ASN BGP router with the AS4_PATH or AS4_ AGGREGATOR path attribute, must discard the path attribute, and continue processing the message.
- To prevent BGP Confederation-related path segments from being propagated outside of a BGP Confederation, the AS_CONFED_SEQUENCE and AS_CONFED_SET path segment types (defined in **[RFC5065]**) are declared invalid for the AS4_PATH attribute, and BGP routers must not include these path segment types in the AS4_PATH attribute of an UPDATE message.
- Reference **[RFC6793]** also defines a new optional transitive BGP path attribute, AS4_AGGREGATOR, which has the same semantics and encoding as the older AGGREGATOR attribute (in **[RFC4271]**), except that it carries a 32-bit ASN.
- Currently assigned 16-bit ASNs are converted into 32-bit ASNs by setting the two high-order bytes (i.e., high-order 16 bits) of the 32-bit field to zero. Such a 32-bit ASN is said to be mappable to a 16-bit ASN.

Reference **[RFC6793]** provides further details on the interaction between 16-bit and 32-bit ASN BGP routers (with respect to BGP peering, generating BGP updates, and processing of received updates), BGP Communities handling, BGP error handling, and transitional issues when going from 16-bit ASNs to 32-bit ASNs.

3.11   BGP PATH ATTRIBUTES AND ROUTE AGGREGATION

Route aggregation is the process by which a router combines the characteristics of several different routes into a single route to be advertised to other routers. A BGP router can perform route aggregation as part of the best path selection process (see Decision Process below) to reduce the amount of routing information that will be placed in the Adj-RIBs-Out database. BGP route aggregation only applies to routes that are in the BGP Routing Table (see "Understanding the Role of the BGP Routing Table and IP Routing Table in BGP" below for details on the BGP Routing Table).

3.11.1   ROUTE AGGREGATION RULES

Using route aggregation, a BGP router can reduce the amount of routing information it stores locally and exchanges with other BGP routers. A BGP router can aggregate routes by applying the following procedure, separately, to BGP path attributes of the same type and to the NLRI **[RFC4271]**:

- Routes that have different MED attributes are not to be aggregated.
- If the first element in the AS_PATH attribute of the aggregated route is an AS_SET, then the router that originates the route is not to advertise the MED attribute with this route.
- Routes with BGP path attributes that have different Attribute Type codes cannot be aggregated together. The router may aggregate routes with path attributes of the same Attributes Type code according to the following rules:
  o **NEXT_HOP Attribute**: When routes that have different NEXT_HOP attributes are being aggregated, the NEXT_HOP attribute of the resulting aggregate route has to identify an interface on the BGP router that performed the route aggregation.
  o **ORIGIN Attribute**: If at least one route among the routes to be aggregated has ORIGIN attribute with the type code value of 2 (meaning INCOMPLETE), then the resulting aggregate route must also carry the ORIGIN attribute with the type code INCOMPLETE. Otherwise, if at least one route among the routes to be aggregated has ORIGIN attribute with the type code value of 1 (i.e., EGP), then the aggregate route must also carry the ORIGIN attribute with the type code EGP. In all other cases, the type code of the ORIGIN attribute of the aggregated route is 0 (i.e., IGP).
  o **AS_PATH Attribute**: If routes to be aggregated have identical AS_PATH attributes, then the aggregate route carries the same AS_PATH attribute identical to the attribute of each individual route.

o **ATOMIC_AGGREGATE Attribute**: If at least one of the routes to be aggregated has ATOMIC_AGGREGATE path attribute set, then the aggregate route is to have this attribute set as well.

o **AGGREGATOR Attribute**: Any AGGREGATOR attributes that is present in the individual routes to be aggregated must not be included in the resulting aggregate route. The BGP router performing the route aggregation, if necessary, may attach a new AGGREGATOR attribute to the aggregate route.

3.11.2 AS_PATH ATTRIBUTE AND ROUTE AGGREGATION

When a router performs route aggregation it summarizes a range of IP address prefixes into one or more aggregate routes to minimize the number of routes maintained in the IP Routing Table. However, route aggregation has the drawback that it results in the loss of the route granularity that exists in the individual routes that form the aggregate route. For example, the AS_PATH attribute information that exists in the individual routes will be lost when the router summarizes the multiple routes into a single advertisement. This loss of granularity could lead to routing loop being created because a route that was propagated in the autonomous system might get accepted by the same autonomous system as a new route.

In order to prevent this problem from happening, the AS_PATH object AS_SET (in which the autonomous systems are listed as an unordered set) is used. Since this object includes the autonomous systems that a route has passed through, aggregate routes carrying an AS_SET would list the collective set of path attributes that exist in the individual routes that the router has summarized (see example in Figure 3.36). The autonomous system that advertises an aggregate route becomes the originator of that route, irrespective of where the individual routes originated.

If AS 65,300 advertises the aggregate route 10.1.2.0/23 with AS_PATH = {65300}, the AS_PATH information would be just {65300}. This would cause routing loops because the originators of the member routes AS 65100 and AS 65200 are not listed in the AS_PATH. If some other autonomous system somehow advertises the aggregate route back to AS 65100 and AS 65200, both would accept the route leading to routing loops. However, advertising the aggregate route 10.1.2.0/23 with AS_PATH = [65300 {65100, 65200}] will not create routing loops since AS 65100 and AS 65200 will both reject the route if is somehow advertised back to them by some autonomous system.

Similarly, AS 65300 will reject the route if it receives such an advertisement. This preserves the integrity of BGP's loop prevention mechanism (by default), since all three autonomous systems would not accept a route with an AS_PATH listing their ASNs. In Cisco IOS, to include an AS_SET in an aggregate route, append the `as-set` keyword to the `aggregate-address` command [CISCBGPCOMD19] [CISCMASCOMD14]. The `aggregate-address` command is used to summarize a set of network addresses into a single prefix. Now we make the following note about AS_PATHs and route aggregation:

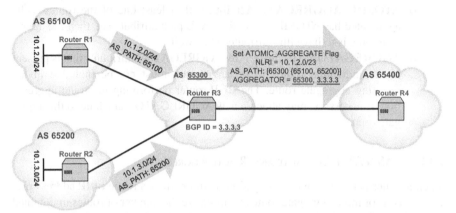

- Routers R1 and R2 are advertising 10.1.2.0/24 and 10.1.3.0/24, respectively, to Router R3, which is in AS 65300.
- Router R3 aggregates these two advertisements to wards Router R4, advertising the single prefix 10.1.2.0/23.
- Here, weex plain why and how Router R3 builds the AS-Path in the route it advertises to router R4:
  - Router R3 cannot operate or act as if AS 65100 and AS 65200 do not exist, since that would break the inherent loop detection capabilities of the BGP AS Path attribute, and this would also break any routing policies base don the AS-Path containing AS 65100 or AS 65200 in AS 65300 (or down stream).
  - R3 also cannot include both of these autonomous systems in the BGP AS-Path sequentially, since that would imply that the path to reach either of these networks (10.1.2.0/24 or 10.1.3.0/24) passes through both AS 65100 and AS 65200.
- To solve this problem, R3 includes both originating autonomous systems (AS 65100 and AS 65200) as an *AS_SET*, that is, {65100,65200}. An AS_SET includes an un ordered set of autonomous systems possibly included in the path to reach a given destination.
- So, when advertising the aggregate route 10.1.2.0/23, Router R3 would advertise (65300{65100, 65300}) in the AS-Path, that is, group AS 65100 and AS 65200 into an AS_SET, and prepending the local AS Number, AS 65300. So, the AS-Pathis composed of both an AS_SEQUENCE (65300) and an AS_SET {65100, 65200}.
- Router R3 includes the Atomic Aggregate attribute in the advertisement to inform Router R4 that while the AS-Path, as presented, is loop free, the prefix has some longer prefix length components that are reachable through an autonomous system notlisted in the AS-Path.

FIGURE 3.36   Impact of Aggregation on BGP Route Attributes

- When a BGP router aggregates multiple routes, it will select the longest, most common shared chain of AS_SEQUENCE tuples and place this in the AS_PATH for the aggregate route (in Figure 3.36, this is just AS 65300). Any autonomous system that is not common to each route is placed in the AS_SET. The end result is that, any BGP router will be able to detect a routing loop because each autonomous system that the aggregate route has traversed, will either be in the AS_PATH or AS_SET. If a BGP router sees its own ASN in any of these sets, it definitely means a routing loop has occurred.

On Router R3, the most common AS_SEQUENCE toward networks to 10.1.2.0/24 and 10.1.3.0/24 is just AS 65300. So, this is placed in the AS_PATH. The uncommon ASNs to both of these networks are AS 65100 and AS65200. These ASNs are placed in the AS_SET. This provides routing loop detection even with an aggregate route.

3.12   CONSIDERING IGP COST WHEN DECIDING THE BEST EXIT POINT OF AN AUTONOMOUS SYSTEM

The BGP best path selection algorithm discussed below may need to consider the IGP cost of paths in an autonomous system when determining the best path to a given

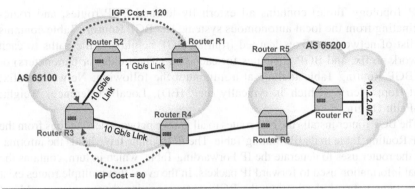

- Router R3 receives via iBGP two paths from Routers R1 and R4 that have been advertised from AS 65200 about network 10.2.2.0/24.
- Router R3 sees these two paths from the exit points R1 and R4 to be the same from a usability perspective.
- From R3's perspective, the cost of sending traffic from either exit point to the destination network is the same, so it has to decide which exit point has the shortest internal path to it so that it can use that exit point to send traffic to AS 65200.
- In this situation, the most suitable way to deter mine the appropriate exit point with the shortest path to AS 65200 is to use IGP metrics to compare the internal paths to the two exit points.
- In this diagram, R4 provides the shortest path since its IGP cost is 80.

FIGURE 3.37   Including IGP Cost in BGP Metrics

network destination (Figure 3.37). During best path selection, BGP also considers the IGP cost of reaching the best exit point from within a given autonomous system. In the algorithm, the lowest IGP metric check (or best exit from the autonomous system) assumes that the same IGP is running throughout the autonomous system. If a single autonomous system is running several different IGPs, then their metrics are not comparable, and the results of this check can actually produce suboptimal routing to the exit points of the autonomous system.

The discussion in Chapter 1 of Volume 1 of this two-part book shows that IGP metrics are typically derived from a number of factors. For example, the bandwidth or delay on a particular link may be considered. Other factors could include the bandwidth of a link (e.g., a 10 Gb/s Ethernet link would be more preferable than a 1 Gb/s Ethernet link. Also, the higher link bandwidth may reflect the reliability of the Physical Layer medium. As seen in Chapter 1 of Volume 1 of this two-part book, the actual method used to derive IGP metrics varies and is influenced by many other factors as well (e.g., the traffic load, reliability, hop count).

3.13   UNDERSTANDING THE ROLE OF THE BGP ROUTING TABLE AND IP ROUTING TABLE IN BGP

A BGP router receives UPDATE messages from its peers, runs some routing policies which may include performing filtering on the UPDATE messages, and then propagate the routes to other BGP peers. For practical purposes, and in order to facilitate the processing and maintenance of routing information, a BGP router stores all BGP routing updates (routes) it receives in a single BGP Routing Table, which is maintained separately from the router's IP Routing Table. The BGP Routing Table (also known as the BGP Table, BGP Topology Database, or

BGP Topology Table) contains all externally learned BGP routes, and routes originating from the local autonomous system. The BGP Routing Table contains the list of network prefixes learned from all BGP neighbors, the paths to each network prefix, and BGP attributes for each path. The fields (or parameters) of the BGP Routing Table include at a minimum the following: Network prefix; Next-Hop; Metric (which is typically the MED); Local Preference; Weight; AS-Path; Origin.

The BGP router installs the best routes to all known network destinations from the BGP Routing Table in its IP Routing Table. The IP Routing Table holds the information the router uses to generate the IP Forwarding Table, which in turn, contains the actual information used to forward IP packets. In the event that multiple routes exist to the same network destination, the BGP router receiving these routes would not advertise all of them to its peers; instead, it will determine the best route and advertises that one.

In addition to propagating external routes from eBGP peers, the BGP router can originate routes to internal networks within its own autonomous system. The BGP router then installs the local routes originated in the autonomous system and the best routes learned from the eBGP peers in its IP Routing Table. The IP Routing Table holds the final information the BGP router uses to generate its IP Forwarding Table.

3.14   ALTERNATIVES TO THE IBGP FULL MESH REQUIREMENT

As discussed earlier, BGP requires that all BGP router within the same single autonomous system be fully interconnected, that is, fully meshed (Figure 3.38). All iBGP routers within an autonomous system need to be fully meshed, because generally iBGP routers do not re-advertise routing updates to other iBGP routers (see Figure 3.39). The full mesh requirement is to allow any external routing information to be re-distributed to all iBGP routers within that autonomous system. However, maintaining a full mesh between iBGP routers in the autonomous system does not scale well in large networks.

Other than in autonomous systems with a small number of BGP routers, the full mesh iBGP requirement poses serious scalability problems in large systems. For N BGP routers within an autonomous system, $N(N-1)/2$ unique iBGP sessions are required. This "full mesh" requirement makes it difficult to scale an autonomous system when there are a large number of iBGP routers within it, as is common in many networks today. An autonomous system with 10 iBGP routers will require 45 iBGP peering sessions if all these routers are to be fully meshed.

The discussion in the following three sections presents extensions to BGP (BGP Confederations [RFC5065], Route Reflection [RFC4456], and BGP Route Servers [RFC7947]) that can be used to design autonomous systems that operate in a manner that does not require "full mesh" iBGP peering. These extensions allow network operators to design fully operational autonomous systems without incurring the high policy administration and management complexity of large autonomous systems.

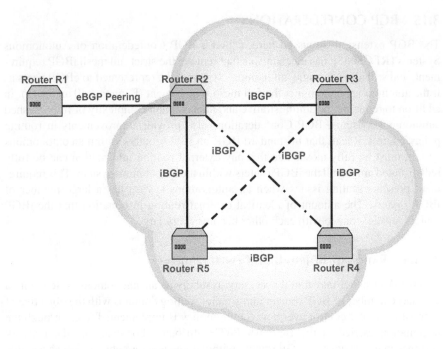

FIGURE 3.38 Fully Meshed BGP Configuration

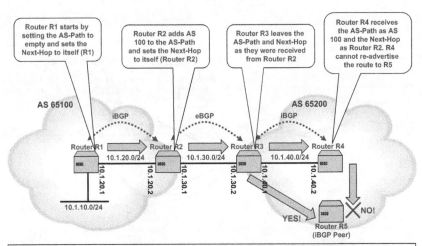

- Router R1 originates the prefix 10.1.10.0/24 with an empty AS-Path list and a BGP Next-Hop of Router R1.
- Router R1 advertises this prefix to Router R2. Router R2, when advertising this route to Router R3, adds AS 65100 to the AS-Path list and sets the BGP Next-Hop to 10.1.30.1, because Router R3 is an exterior peer (a BGP peer outside the autonomous system).
- Router R3 then advertises the prefix 10.1.10.0/24 to Router R4 without changing the AS-Path or the BGP Next-Hop, since Router R4 is an interior peer (a BGP peer within the same autonomous system).
- Router R4 will need a path to Router R2 in order to consider this prefix reachable; generally, the BGP Next-Hop reachability information is provided by advertising the link between R2 and R3 through an IGP, or through iBGP originating the link as a prefix from Router R3 into AS 65100.
- Note: Router R4 cannot re-advertise the route received from R3 to R5 because R5 is also an iBGP peer. This means all iBGP peers in AS 65200 need connectivity to R3 so that they can receive external routing information.

FIGURE 3.39 iBGP and eBGP Peering Example

3.15   BGP CONFEDERATIONS

The BGP extension discussed here, called a BGP Confederation of Autonomous Systems [RFC5065], has mechanisms that remove the strict full mesh iBGP requirement, and still allow a single autonomous system to be represented to eBGP peers as if the autonomous system met the full mesh requirement (Figures 3.40 and 3.41). In addition to reducing the administrative and management complexity of fully meshed autonomous systems, BGP Confederations also provide improvements in routing policy control. Recall that in standard BGP, all iBGP routers within an autonomous system must be fully meshed so that any external routing information can be fully redistributed among all the iBGP routers within the autonomous system. This requirement presents scaling issues when an autonomous system has a large number of iBGP routers. The amount of identical external routing information that the iBGP routers must exchange with each other becomes very large.

3.15.1   RATIONALE BEHIND BGP CONFEDERATIONS

A network designer may find it necessary to subdivide an autonomous system with a very large number of BGP routers into smaller routing domains with the objective of exercising greater control over how routing policy is implemented (e.g., through the information carried in the BGP AS_PATH attribute). For example, the network designer may consider all BGP routers within a given geographic region as a single

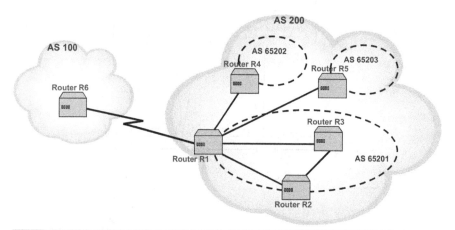

• There are multiple BGP routers in AS 200 and to reduce the number of IBGP connections, AS 200 is divided into three sub-autonomous systems: AS 65201, AS 65202, and AS 65203.
• AS 65201 has established fully meshed iBGP connections between the three routers R1, R2 and R3.
• BGP routers outside the BGP confederation AS 200 such as Router R6 in AS 100, do not know about the existence of the sub-autonomous systems (AS 65201, AS 65202, and AS 65203) in the BGP Confederation.
• The AS Confederation ID is the Autonomous System Number that isused to identify the entire BGP Confederation. For BGP Confederation AS 200, the AS Confederation ID is 200.
• eBGP peers communicating with the BGP Confederation have no knowledge that they are peering with a Confederation, and they reference the AS Confederation ID in their configuration.

FIGURE 3.40   BGP Confederation Topology Example 1

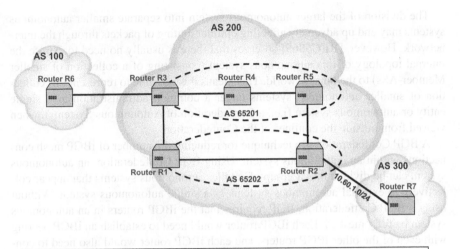

AS 200

AS 100

Router R6   Router R3   Router R4   Router R5

AS 65201

AS 65202

Router R1   Router R2

AS 300

Router R7

10.60.1.0/24

- This figure shows a BGP Confederation with the AS Confederation ID of 200 (which is the AS Number of AS 200).
- The Member-ASs are AS 65201 and AS 65202. We assume R3 provides Route Reflection in Member-AS 65201.
- R6 resides in AS 100 and does not see any of the BGP Sub-Confederation information. R1 is not aware that AS 200 is subdivided into a BGP Confederation. R3's BGP Table participates in the Member-AS 65201.
- The next-hop addressis not modified for the 10.60.1.0/24 (network between R2 and R7) when R2 sends an update to R7 even though AS 65202 is aMember-ASofAS200.
    – R2 will lists in the AS_CONFED_SEQUENCE segment type an ordered set of Member-AS Numbers the update has passed through in the AS 200 Confederation. The AS_CONFED_SEQUENCE segment will include the Sub-AS 65201 the update has just passed through.

FIGURE 3.41   BGP Confederation Topology Example 2

entity (autonomous system) that can be divided into smaller routing domains (or autonomous systems).

Although potential improvements in routing policy control may be gained, if additional mechanisms are not used and smaller routing domains are created, BGP still requires all BGP routers within the same autonomous system to establish full mesh TCP connections among themselves to allow exterior routing information to be fully exchanged. Particularly in large autonomous systems, the number of intra-domain TCP connections that need to be established by each BGP router can be very high.

Dividing a large autonomous system into smaller domains (or autonomous systems) can produce a significant reduction in the total number of intra-domain BGP connections, since the BGP connectivity requirements simplify or reduce to the particular model used for inter-domain BGP connections. Unfortunately, subdividing the autonomous system may end up resulting in an increase in the complexity of the routing policy based on AS_PATH information for all members of the internetwork. Additionally, this subdivision of the autonomous system increases the administrative and maintenance overhead of coordinating eBGP peering when the internal topology of this collection of smaller autonomous systems is modified. Each (smaller) autonomous system still has to pass its routing information to all other autonomous systems in the internetwork. Route changes and changes in reachability information in any one autonomous system still have to be passed to other autonomous systems.

The division of the larger autonomous system into separate smaller autonomous systems may end up adversely affecting optimal routing of packets through the inter-network. However, **[RFC5065]** observes that there is usually no need to expose the internal topology of this autonomous system (consisting of a collection of smaller Member-ASs) to the outside world. This means it is possible to represent this collection of smaller autonomous systems under a common administration as a single entity or autonomous system (i.e., a Confederation of Autonomous Systems), when viewed from outside the confines of the Confederation.

A BGP Confederation is a technique for reducing the number of iBGP mesh connections within an autonomous system. Using BGP Confederation, an autonomous system can be divided into a group of smaller autonomous systems that appear collectively to external autonomous systems as a single autonomous system. Without using a BGP Confederation, BGP requires that the iBGP routers in an autonomous system be fully meshed. Each iBGP router would need to establish an iBGP peering with each of the other iBGP routers, and each iBGP router would also need to connect to an external autonomous system via eBGP peering.

Using a BGP Confederation reduces the number of iBGP peers within the autonomous system. The network operator can divide the autonomous system into multiple mini-autonomous systems and assign these to a BGP Confederation. The iBGP member routers inside each mini-autonomous system would still be fully meshed, and each mini-autonomous system would have a connection to the other mini-autonomous systems within the BGP Confederation. Within each mini-autonomous system, the standard iBGP route advertisement rules still apply. Because each mini-autonomous system is assigned a unique ASN, eBGP must be used between them.

Even though the mini-autonomous systems would have eBGP peers connecting to other mini-autonomous systems within the BGP Confederation, they would exchange BGP routing updates as if they were using iBGP. Routing inside the BGP Confederation behaves like iBGP routing in a single standard autonomous system **[RFC4271]**. That is, BGP path attributes such as NEXT_HOP, MULTI_EXIT_DISC, and LOCAL_PREF would be preserved when crossing mini-autonomous system boundaries. These BGP path attributes, which normally are restricted to a single RFC 4271 standard autonomous system, are allowed to be propagated to all the member mini-autonomous systems of the BGP Confederation.

As discussed below, a BGP Confederation is identified by its own ANS, the AS Confederation Identifier. To the outside world, the BGP Confederation appears like a single standard autonomous system. The internal topology of the BGP Confederation is hidden from external autonomous systems. Because a BGP Confederation is treated as if it were a single autonomous system, the same routing policy can be applied to all the member mini-autonomous systems of the BGP Confederation.

3.15.2   BGP CONFEDERATION PARAMETERS

Terminology:

- **A BGP Autonomous System Confederation**: This is a collection of autonomous systems that are represented and advertised as a single ASN to BGP routers that are not members of that BGP Confederation.

- **The AS Confederation Identifier**: This is an externally visible ASN that identifies a particular BGP Confederation as a whole.
- **A Member Autonomous System (Member-AS)**: This is an autonomous system that is contained in, that is, belongs to a given BGP Confederation.
- **Member-AS Number**: This is an ASN identifier used to represent a specific Member-AS within a particular BGP Confederation and is visible only within that Confederation.

Figures 3.40 and 3.41 illustrate the concept of BGP Confederations **[RFC5065]** as an alternative solution to iBGP full mesh scalability issues discussed earlier. A Confederation consists of subautonomous systems each referred as a Member-AS that combine to form a larger autonomous system known as a BGP Confederation or Autonomous System Confederation. Each subautonomous system must be uniquely identified within the BGP Confederation by a Member-AS Number. A Member-AS normally uses a Member-AS Number taken from the private ASN range (64,512–65,535). eBGP peers interacting with the BGP Confederation have no knowledge that they are peering with a Confederation, and they reference the AS Confederation Identifier in their transactions.

Within any given Member-AS, the same iBGP full mesh requirement exists. Connections from Member-AS to other BGP Confederations are made with standard eBGP, and BGP peers outside the Member-AS are treated as external. Connections between a BGP Confederation and other BGP autonomous systems is done via eBGP. To prevent routing loops, a Member-AS uses Confederation lists (see AS_CONFED_SEQUENCE and AS_CONFED_SET below), which operate like AS_SEQUENCE and AS_SET in a standard autonomous system but use only the privately assigned Member-AS Number.

These special lists are loop prevention mechanisms used to detect routing updates leaving one mini-autonomous system and attempting to reenter the same mini-autonomous system. A routing update that tries to reenter a mini-autonomous system that originated it will be detected because the mini-autonomous system will see its own Member-AS Number listed in the routing update's Confederation list.

The BGP Confederation appears as a complete autonomous system to other BGP Confederations. The AS_PATH received by other autonomous system carry only the globally assigned ASN and does not include the Confederation sequences (AS_CONFED_SEQUENCE and AS_CONFED_SET), or the privately assigned Member-AS Numbers. The Member-AS Numbers are removed when the route is advertised out of the BGP Confederation.

We discussed earlier above that the AS_PATH attribute is a well-known mandatory BGP path attribute that consists of a sequence of AS path segments with each segment represented by a triple <Path Segment Type, Path Segment Length, Path Segment Value>. Reference **[RFC4271]**, as discussed above, defines the Path Segment Type as a 1-byte field with the following two values defined: Value = 1 corresponds to Segment Type AS_SET, and Value = 2 corresponds to AS_SEQUENCE. Reference **[RFC5065]** specifies two additional Segment Types:

- Value = 3: Segment Type is AS_CONFED_SEQUENCE, which is ordered set of Member-ASs the UPDATE message has traversed in the local BGP Confederation.

- Value = 4: Segment Type is AS_CONFED_SET, which is an unordered set of Member-ASs the UPDATE message has traversed in the local BGP Confederation.

3.15.3 Operation

Reference **[RFC5065]** defines the following BGP Confederation operation rules:

- A BGP router that is a member of a particular BGP Confederation must use the associated AS Confederation Identifier in all of its transactions with BGP peers that are not members of that Confederation. This is because the assigned AS Confederation Identifier is the "externally visible" ASN to outside BGP routers. The member BGP router uses this number in BGP OPEN messages and is the ASN advertised in the AS_PATH attribute.
- A BGP router that is a member of a particular BGP Confederation must use its Member-AS Number in all of its transactions with BGP peers that are members of the same BGP Confederation.
- A BGP router that receives an AS_PATH attribute carrying an ASN matching its own AS Confederation Identifier should treat the path as if it had received a path containing its own ASN.
- A BGP router that receives an AS_PATH attribute carrying an AS_CONFED_SEQUENCE or AS_CONFED_SET that contains its own Member-AS Number should treat the path as if it had received a path containing its own ASN.

3.15.4 AS_PATH Modification Rules

In BGP Confederations, BGP routers use the rules described below and specified in **[RFC5065]**, instead of the AS_PATH rules specified in **[RFC4271]** discussed earlier. The AS_PATH which is a well-known mandatory BGP path attribute identifies the autonomous systems through which the routing information carried in the BGP UPDATE message has traversed. The AS_PATH can carry AS_SETs, AS_SEQUENCEs, AS_CONFED_SETs, or AS_CONFED_SEQUENCES.

3.15.4.1 Originating a BGP Route

When a BGP router originates a route, it follows these steps:

1. The originating BGP router includes its own AS Confederation Identifier in a path segment, of type AS_SEQUENCE, in the AS_PATH attribute of all UPDATE messages it sends to BGP routers located in neighboring autonomous systems that are not members of its BGP Confederation. In this case, the originating router's own AS Confederation Identifier will be the only entry of the path segment, and this path segment will be the only segment in the AS_PATH attribute.
2. The originating BGP router includes its own Member-AS Number in a path segment, of type AS_CONFED_SEQUENCE, in the AS_PATH attribute of all UPDATE messages it sends to BGP routers located in neighboring

Member-ASs that are members of its BGP Confederation. In this case, the originating router's own Member-AS Number will be the only entry of the path segment, and this path segment will be the only segment in the AS_PATH attribute.

3. The originating BGP router includes an empty AS_PATH attribute in all UPDATE messages it sends to BGP router located within the same Member-AS. An empty AS_PATH attribute is one with Length field containing the value zero.

3.15.4.2 Propagating a BGP Route

When a BGP router propagates a route that it has learned from the UPDATE message sent by another BGP router, it modifies the AS_PATH attribute of the route based on the location of the BGP router to which the route will be sent:

1. When the BGP router advertises the route to another BGP router located in its own Member-AS, the router does not modify the AS_PATH attribute associated with the route.

2. When the BGP router advertises the route to a BGP router located in a neighboring autonomous system that is a member of the same BGP Confederation, the router updates the AS_PATH attribute as follows:

   a. If the first path segment of the AS_PATH is of type AS_CONFED_SEQUENCE, the BGP router prepends its own Member-AS number as the last element of the sequence (placed in the leftmost position with respect to the position of bytes in the BGP message). If prepending will cause an overflow in the AS_PATH segment (i.e., exceed 255 autonomous systems), the router prepends a new segment of type AS_CONFED_SEQUENCE, and then prepends its own ASN to this new segment.

   b. If the first path segment of the AS_PATH is not of type AS_CONFED_SEQUENCE, the BGP router prepends a new path segment of type AS_CONFED_SEQUENCE to the AS_PATH, including its own Member-AS Number in that segment.

   c. If the AS_PATH is empty, the BGP router constructs a path segment of type AS_CONFED_SEQUENCE, then writes its own Member-AS Number into that segment, and places that segment into the AS_PATH.

3. When the BGP router advertises the route to a BGP router located in a neighboring autonomous system that is not a member of its BGP Confederation, the router updates the AS_PATH attribute as follows:

   a. If any path segments of the AS_PATH are of the type AS_CONFED_SEQUENCE or AS_CONFED_SET, the router removes those segments from the AS_PATH attribute, leaving the cleaned up AS_PATH attribute to be processed by steps 2, 3 or 4.

   b. If the first path segment of the remaining AS_PATH is of type AS_SEQUENCE, the BGP router prepends its own AS Confederation Identifier as the last element of the sequence. If prepending will cause an overflow in the AS_PATH segment, the router prepends a new segment of type AS_SEQUENCE, and then prepend its own ASN to this new segment.

    c. If the first path segment of the remaining AS_PATH is of type AS_SET, the BGP router prepends a new path segment of type AS_SEQUENCE to the AS_PATH, including its own AS Confederation Identifier in that segment.

    d. If the remaining AS_PATH is empty, the BGP router constructs a path segment of type AS_SEQUENCE, then writes its own AS Confederation Identifier into that segment, and places that segment into the AS_PATH.

Whenever the modification of the AS_PATH attribute calls for the local BGP router to include or prepend its own AS Confederation Identifier or Member-AS Number, the router may include or prepend more than one instance of that value in the AS_PATH attribute. This behavior is one that is controlled via local configuration of the router.

Reference **[RFC5065]** discusses other BGP Confederations issues such as BGP UPDATE message error handling, common administrative issues related to running the same or different IGPs in Member-ASs, BGP MED and LOCAL_PREF attribute handling, BGP AS_PATH and path selection, and compatibility and deployment considerations.

3.16 BGP ROUTE REFLECTION

This section describes a method known as "Route Reflection" which is another method used to address the need for full mesh iBGP connections in an autonomous system **[RFC4456]**. Route Reflection also includes its own routing loop prevention mechanism through the addition of two new optional non-transitive BGP attributes. Route Reflection breaks one of the fundamental rules of normal or standard iBGP operation by allowing routers called Route Reflectors (RRs) to re-advertise routes learned from an iBGP peer to other iBGP peers and external peers, and vice versa.

3.16.1 CONCEPT OF ROUTE REFLECTION

Let us start by looking at the basic idea of Route Reflection by considering the networks shown in Figure 3.42. In the first figure, there are three iBGP routers (Routers R1, R2, and R3). With the standard BGP model as described in **[RFC4271]**, when R1 receives an eBGP route and it is selected as the best path, it must re-advertise the learned external route to both R2 and R3. However, Routers R2 and R3 (both being iBGP routers) must not re-advertise this learned route to other iBGP peers.

However, by relaxing this standard BGP rule and allowing Router R3 to advertise routes learned from an iBGP peer to other iBGP peers, then R3 could re-advertise (or reflect) the routes learned from the iBGP peer R1 to R2 and vice versa. With this, there would be no need for the iBGP session between R1 and R2 and so, it can be eliminate as shown in the network on the right-hand side of Figure 3.42. This basic idea forms the basis of the Route Reflection technique is described in **[RFC4456]**.

The term "Route Reflection" is used to describe the operation of a BGP router that advertises a route learned from an iBGP peer to another iBGP peer. Reference **[RFC4456]** calls such a BGP router a "Route Reflector", and such a route, a "reflected route". In an autonomous system with N iBGP routers, instead of having $N(N-1)/2$

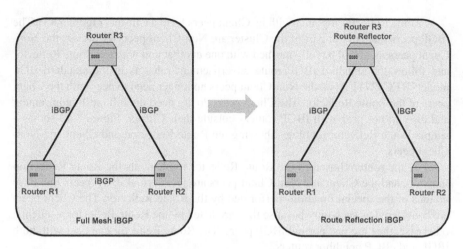

FIGURE 3.42   Illustrating Full Mesh iBGP and Route Reflection iBGP

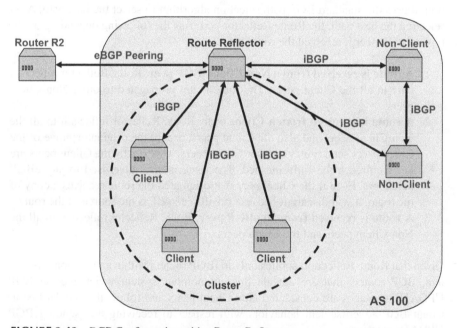

FIGURE 3.43   BGP Configuration with a Route Reflector

unique iBGP peering sessions if all of the N routers are to be fully meshed, using a Route Reflector will require only $(N-1)$ iBGP sessions. Having ten iBGP routers in an autonomous system will require only nine iBGP sessions if one of them is designated the Route Reflector. All the nine iBGP routers will peer only with the Route Reflector.

The internal peers of a Route Reflector are categorized into two groups: Client peers and Non-Client peers (see Figure 3.43). The Route Reflector reflects routes between Client and Non-Clients, and may reflect routes among Client peers.

A Route Reflector together with its Client peers form a Cluster (Figure 3.43). The iBGP peers that are not part of the Cluster are Non-Client peers. However, the Non-Client peers must still be fully meshed with one another and with the Route Reflector and follow the standard iBGP route advertisement rules as in the standard BGP model **[RFC4271]**. Also, the Non-Client peers no longer need to peer with the Client peers of the Route Reflector. The Client peers do not need the full mesh requirement and also do not peer with iBGP routers outside their Cluster. Figure 3.43 shows a simple Route Reflector topology showing the Route Reflector, and Client and Non-Client peers.

Only the router chosen as the Route Reflector implements the Route Reflection function, and the Client and Non-Client peers are just normal iBGP peers that have no idea of the special functions performed by the Route Reflector. The Client peers are considered as such only because they are listed by the Route Reflector as clients; otherwise, they are just normal iBGP peers. The Route Reflector can peer with both iBGP and eBGP neighbor routers.

When a Route Reflector receives multiple routes for the same network destination, it uses the standard BGP path selection algorithm to select the best path. After selecting the best path, the Route Reflector performs the following depending on the type of BGP peer it received the route from:

1. **A route is received from a Non-Client iBGP peer**: Route Reflector reflects it only to all the Client peers. The route is not propagated to other Non-Client peers.
2. **A route is received from a Client peer**: Route Reflector reflects it to all the Non-Client peers and also the Client peers, except the originator; none of the Client peers send routes to other Client peers. This is why the Client peers are not required to be fully meshed. The route is also advertised to any eBGP neighbors. Even if the Client peer that originated the route receives a copy of the route, it will discard it because it will see itself as the source of the route.
3. **A route is received from an eBGP peer**: Route Reflector reflects it to all the Non-Client peers and the Client peers.

Given that Route Reflection applies only to iBGP routers within an autonomous system, BGP routers that are outside this autonomous system which receive BGP UPDATE messages are considered Non-Client peers and follow the standard Non-Client BGP advertisement behaviors with respect to receiving and sending BGP UPDATE messages.

An autonomous system may support multiple Route Reflectors. In such a case, each Route Reflector treats other Route Reflectors just like any other iBGP router. In some situations, it may be desirable to configure a Route Reflector to have other Route Reflectors operating in a Client group or Non-client group. In a simple configuration, the network designer may divide the network backbone into many smaller Clusters.

Each Cluster could then be assigned a Route Reflector. Each Cluster Route Reflector could then be configured to operate with other Cluster Route Reflectors as Non-Client peers. This means all the Cluster Route Reflectors will be fully meshed.

The Clients in each Cluster will be configured to maintain iBGP session only with their Cluster Route Reflector. Using Route Reflection, all the iBGP routers will receive reflected routing information from their Cluster Route Reflector; the iBGP routers within a Cluster do not have to be fully meshed. This Route Reflector configuration is discussed in detail below.

An autonomous system may also have conventional BGP routers that do not understand the concept of Route Reflection. However, Route Reflection allows such conventional BGP router to coexist with Route Reflectors. The conventional BGP routers could be members of either a Client group or a Non-Client group. This arrangement allows the current standard iBGP model to be easily and gradually migrated to the Route Reflection model. The network designer may start by creating Clusters, and then configuring a single router as the designated Route Reflector for each Cluster. The network designer configures other Route Reflectors and their Clients as normal iBGP peers. Additional Clusters with their designated Route Reflectors can then be created gradually as the need arises.

Thus, an autonomous system can support more than one Route Reflector. When an autonomous system has multiple Route Reflectors, each Route Reflector treats the other Route Reflectors as normal iBGP peers. Also, a Cluster can have more than one Route Reflector, and an autonomous system can have more than one Cluster.

By default, a Route Reflector does not change the BGP Next-Hop attribute (Next-Hop IP address) when it reflects a route. However, the **neighbor next-hop-self** command [CISCBGPCOMD19] can be used to change the Next-Hop address for routes reflected from an eBGP peer to any Route Reflector client.

3.16.2 ROUTE REFLECTION WITH REDUNDANCY

In a simple configuration, a Cluster of Client peers will have a single designated Route Reflector. In this simple case, the BGP Identifier of the Route Reflector can serve as the identifier of the Cluster. However, this simple configuration has a weakness because the single Route Reflector represents a single point of failure, and the Cluster identifier is tied to the Route Reflector. When a Route Reflector failure occurs, the Cluster also loses its identifier.

Furthermore, with the lack of full mesh iBGP peering within a Cluster with one Route Reflector, redundancy and reliability becomes important issues. If the Route Reflector fails, then the Client peers become isolated and cannot participate in the BGP route advertisement process. The presence of multiple Route Reflectors in a (single) Cluster provides redundancy and allows the Client peers to simultaneously peer with the multiple Route Reflectors. In this scenario, when one Route Reflector fails, the other Route Reflector(s) are still available to provide Route Reflection functions.

So, in many cases, it might be important and desirable to provide Route Reflector redundancy by configuring multiple Route Reflectors for a Cluster (Figure 3.44). When two or more Route Reflectors are defined in a Cluster for redundancy, each client must have a physical connection to each of these Route Reflectors and must peer with each one of them. If one Route Reflector fails, the Client peers in the Cluster will still have connectivity to the other Route Reflectors, and Route Reflection

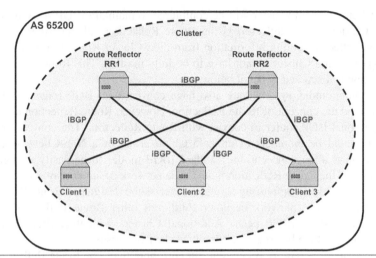

Route Reflectors RR1 and RR2 are in the same Cluster (same CLUSTER_ID). RR1 and RR2 establish an iBGP connection so that each Route Reflector is a **Non-Client** of the other Route Reflector.

– If Client 1 receives an updated about a route from an external BGP peer, it will advertise that route to RR1 and RR2 through iBGP.

– RR1 after receiving the route update, will reflect the route to other Clients (Client 2 and Client 3) and the Non-Client (RR2). RR1 will add the local CLUSTER_ID to the head of the CLUSTER_LIST.

– RR2 after receiving the reflected route, will check the CLUSTER_LIST and will find that its CLUSTER_ID is listed in the CLUSTER_LIST received; therefore, it will discard the updated route.

If RR1 and RR2 are configured with different CLUSTER_IDs, each Route Reflector will receive both the route from Client 1 and the updated route reflected from the other Route Reflectors. This means, configuring the same CLUSTER_ID for RR1 and RR2 reduces the number of routes that each RR receives as well as memory

FIGURE 3.44   Configuring a Backup Route Reflector

will not be lost. To have multiple Route Reflectors in the same Cluster, all the Route Reflectors in that Cluster can be assigned a 4-byte CLUSTER_ID so that any one of the multiple Route Reflectors can discard routes sent from the other Route Reflectors in the Cluster.

3.16.3   PREVENTING ROUTING INFORMATION LOOPS DURING ROUTE REFLECTION

BGP uses the information in the BGP AS_PATH attribute to detect routing loops. A BGP UPDATE message that attempts to reenter the autonomous system that originated it will be discarded by the border router of that autonomous system. Without additional mechanisms, the introduction of Route Reflectors can potentially create routing loops within the autonomous system. There is the possibility that a BGP UPDATE message that leaves a Cluster can reenter that Cluster creating a routing loop. Recall that routing loops inside an autonomous system cannot be detected using the standard BGP AS_PATH approach **[RFC4271]** (one of the reasons for the BGP full mesh requirements). Thus, the use of Route Reflectors requires extra mechanisms for loop prevention within an autonomous system.

It is possible through misconfiguration that route redistribution loops could be formed when a route is being reflected. Reference **[RFC4456]** defines the following attributes for detecting and avoiding routing information loops during Route Reflection:

- **ORIGINATOR_ID**: This is a 4-byte, optional non-transitive BGP attribute with Type Code of 9. This attribute carries the Router ID of the originator of a route in the local autonomous system. The first Route Reflector that receives the route creates the attribute and sets the attribute's value to the Router ID of the router that advertised/injected the route into the autonomous system. The Route Reflector then adds this attribute to the UPDATE message it sends.

  If the ORIGINATOR_ID is already listed in an NLRI in the UPDATE message, the Route Reflector does not overwrite it. If a router receives an UPDATE message with its Router ID already listed in the ORIGINATOR_ID attribute, it discards the message. If the UPDATE message comes back to the route's originator because of network misconfiguration, the originator should discard the message.

- **CLUSTER_LIST**: This is an optional, non-transitive BGP attribute with Type Code of 10. It carries a sequence of CLUSTER_ID values of the Clusters along the reflection path through which the UPDATE message has traversed. The Route Reflector appends (not overwrites) its CLUSTER_ID to this attribute. If a Route Reflector receives an UPDATE message that already contains its CLUSTER_ID in the CLUSTER_LIST attribute, it discards the message.

When a Route Reflector reflects a route, it must prepend the CLUSTER_ID of the local Cluster (the Cluster in which it is a member of) to the CLUSTER_LIST. If the CLUSTER_LIST is empty, the Route Reflector creates a new one listing its CLUSTER_ID. A Route Reflector uses this attribute to determine if the routing information has looped back to the same Cluster due to network misconfiguration. If the Route Reflector finds that the local CLUSTER_ID is already listed in the CLUSTER_LIST, it ignores the advertisement containing that information. The CLUSTER_LIST provides loop detection inside an autonomous system while the AS_PATH information provides loop detection when BGP UPDATE messages traverse multiple external autonomous systems.

In addition, a Route Reflector is not allowed to modify the BGP path attributes of reflected routes; the Route Reflector does not change the way iBGP handles BGP path attributes. When a Route Reflector reflects a route, the following BGP path attributes should not be modified: NEXT_HOP, AS_PATH, LOCAL_PREF, and MED. Otherwise modifying any of them could potentially cause routing loops.

Table 3.2 summarizes the rules that Route Reflectors (RRs) use when filling in the CLUSTER_LIST of a reflected route with CLUSTER_IDs [JUNBGPGUIDE20].

3.16.4 ROUTE REFLECTION WITH MULTIPLE CLUSTERS

To address the iBGP full-mesh requirement, most networks todays use Route Reflectors to simplify configuration. Since the full-mesh iBGP model does not scale well, using a Route Reflector, BGP routers can be grouped into Clusters and identified by CLUSTER_IDs (which are identifiers unique to the autonomous system).

TABLE 3.2
Rules for Route Reflectors

Route Reflection Scenario	Transaction Flow	Route Reflector Action
Reflecting a route from one of the Clients to a Non-Client router.	Client -> RR -> Non-Client	The RR fills the CLUSTER_ID associated with that Client in the CLUSTER_LIST of the reflected route.
Reflecting a route from a Non-Client router to a Client router.	Non-Client -> RR -> Client	The RR fills the CLUSTER_ID associated with that Client in the CLUSTER_LIST of the reflected route.
Reflecting a route from a Client router to another Client router that is in a different Cluster.	Client1 -> RR -> Client2 (different Cluster)	The RR fills the CLUSTER_ID associated with Client1 in the CLUSTER_LIST before reflecting the CLUSTER_ID to Client2. The CLUSTER_ID associated with Client 2 is not added.
Reflecting a route from a Client router to a Non-Client router that is in a different autonomous system.	Client -> RR -> Non-Client (different AS)	The RR does not fill the CLUSTER_LIST with the CLUSTER_ID before reflecting the route to the Non-Client device because the CLUSTER_ID is specific to one autonomous system.

Within the Cluster, a BGP session is configured from a single router (the Route Reflector) to each iBGP Client peer. This configuration allows the iBGP full-mesh requirement to be met somehow without actually implementing one.

To use Route Reflection in an autonomous system, one or more routers are designated as a Route Reflector, typically, one Route Reflector per point-of-presence (POP). A Route Reflector works by re-advertising routes learned from an iBGP Client peer to other internal Client peers as discussed earlier. So, instead of configuring all iBGP Client peers to create a full mesh between them, Route Reflection requires only the Route Reflector to be fully connected with each iBGP Client peer.

Figure 3.45 shows a Route Reflector in a simple one Cluster topology. In this figure, all of the iBGP Client peers and the Route Reflector form a Cluster. Multiple Clusters topologies can also be configured and then linked together by configuring a full mesh of Route Reflectors as shown in Figure 3.46. This figure is one example of a multiple Cluster topology with Route Reflection. In Figure 3.46, when a router in any given Cluster advertises a route to its Route Reflector, that Route Reflector re-advertises the route to the other Route Reflectors. These Route Reflectors, in turn, re-advertise that route to their Client peers and remaining routers within the same autonomous system.

However, as the number of Clusters in the autonomous system become large, creating a full mesh with Route Reflectors becomes very difficult to implement. The difficulties in creating a full mesh between Route Reflectors also becomes very significant. To help address this problem, Clusters of iBGP routers can be grouped together into Clusters of iBGP Clusters and to implement hierarchical Route Reflection as illustrated in Figures 3.47 and 3.48.

Figures 3.47 and 3.48 show that an autonomous system can have multiple Clusters and where a Route Reflector in one of the Clusters can also be a client in another

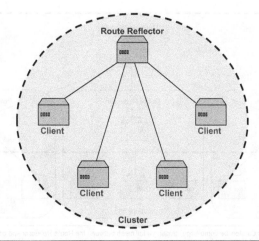

One router is configured as the Route Reflector for the Cluster. The other routers are designated internal Client peers within the Cluster. BGP routes are advertised to the Route Reflector by any of the internal Client peers. The Route Reflector then readvertises these routes to all other Client peers within the Cluster.

FIGURE 3.45   Example Simple One Cluster Topology with a Single Route Reflector

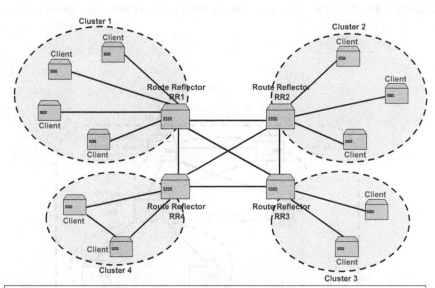

- Route Reflectors RR1, RR2, RR3, and RR4 as fully meshed internal BGP peers.
- For example, when a router advertises a route to RR1, RR1 will read vertise that route to the other Route Reflectors, which, inturn, will read vertise the route to the remaining routers within the autonomous system.
- Route Reflection allows the route to be propagated through out the autonomous system without the scaling problems associated with, otherwise, the requirements of full mesh BGP connections.
- However, it must be noted that, a Route Reflector that supports multiple Clusters does not accept a route with the same CLUSTER_ID from a Non-Client router. This means, in this scenario, a different CLUSTER_ID must be configured for a redundant Route Reflector to reflect that route to other Clusters.

FIGURE 3.46   An Autonomous System Configured with Multiple Clusters

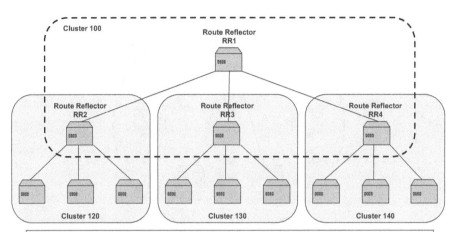

- When the number of Clusters be come large, creating a full mesh between the Route Reflect or sbe comes difficult which also creates scal ability problems.
- One way to over come this problem, is to group Cluster so frouters to gether into a Clusters of Clusters to allow hierarchical Route Reflection.
- Route Reflectors RR2, RR3, and RR4 act as Route Reflectors for Clusters 120, 130, and 140, respectively.
- Rather than creating a fully mesh between these Route Reflectors, the Route Reflectors are configured as part of another Cluster (Cluster 100) for which RR1 is the Route Reflector.
- For example, when a router advertises a route to RR2, RR2 read vertises the route to all the routers with in its own Cluster, and then read vertises the route to RR1.
- Route Reflect or RR1 read vertises that route to the routers in its Cluster, and these routers (Route Reflectors) in turn propagate the route down through their Clusters.

FIGURE 3.47   Example 1: Hierarchical Route Reflectors in a Clusters of Clusters Topology

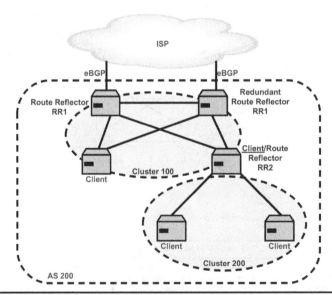

- ISP provides Internet routes for AS 200 and two eBGP connections are established between the ISP and AS 200.
- AS 200 is divided into two Clusters. The four routers in Cluster 100 are core routers.
- Route Reflector RR2 is deployed in Cluster 2. and is a Client of RR-1.

FIGURE 3.48   Example 2: Hierarchical Route Reflectors in an Autonomous System

Cluster in the same autonomous system. Since a Route Reflector can peer with another iBGP router outside its Cluster, this means the Route Reflector can be a client in another Cluster within the same autonomous system. These examples illustrate that Route Reflection allows routes to be propagated throughout an autonomous system without the scaling problems created by the full mesh iBGP requirement.

3.17 BGP ROUTE SERVER

In the typical Internet Exchange (IX) architecture, the edge eBGP routers of service providers share a common subnet in order to provide full mesh connectivity between them (Figure 3.49). Within the IX, the eBGP router of each service provider maintains a connection to the common subnet or switching infrastructure, as shown in Figure 3.49. Even though the eBGP peers share a common subnet, peering between the different service providers often demands the establishment of full mesh BGP connectivity between them, creating scalability problems as discussed earlier above (and as illustrated in Figure 3.50).

BGP sessions must still be configured and maintained individually between each of the service provider eBGP routers, for every service provider with which a given service provider wants to establish a peering relationship. Assuming each service provider wants to establish peering with every other service provider in the IX, the resulting full mesh of BGP sessions established can be significantly complex (Figure 3.50). Peering is the connection of two different service providers for the purpose of exchanging routing information and end-user traffic.

The BGP Route Server is a feature mostly used in the IX environment where each eBGP peer needs only to set up a BGP connection to the BGP Route Server in order

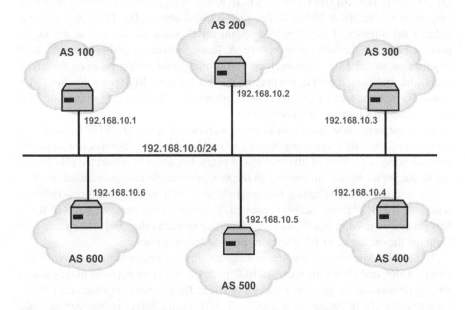

FIGURE 3.49 Internet Exchange Shared Switching Infrastructure

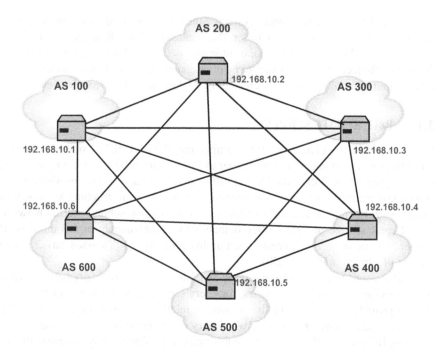

FIGURE 3.50   Internet Exchange eBGP Full Mesh

to receive BGP route advertisements sent by all other eBGP peers having a BGP connection with the BGP Route Server as illustrated in Figure 3.51 **[CISCROUTESERV]**
[JUNBGPGUIDE20] [RFC7947]. A BGP Route Server can be viewed as the eBGP
equivalent of the iBGP Route Reflector discussed above. The BGP Route Server
reduces the number of direct point-to-point eBGP sessions required in the IX for
peering between the service providers. A Route Server simplifies the full mesh
requirement in the IX for the exchange of routing information between the eBGP
peers, and speeds up the set of the peering process for new eBGP peers. Without BGP
Route Servers, the service providers need full mesh of BGP peering to provide BGP
prefix exchanges between the providers.

Route Servers allow the IX to keep the number of necessary BGP sessions low
because only one BGP session is needed to each of the Route Servers to receive the
network address prefixes of all other eBGP peers. For a newly connected eBGP peer,
the Route Server speeds up peering as the new peer can receive all network prefixes
of all other peers without having to setup individual BGP peering sessions – only one
session to the BGP Route Server is enough. To provide redundancy, two BGP Route
Servers can be used in the IX that are redundant to each other and at the same time
facilitate the exchange of BGP announcements between peers.

The BGP Route Server provides customized routing policy support for each service provider, and allows the standard BGP path selection process to be overridden by
routing policies set per particular service provider. Each service provider sets policies
for particular eBGP neighbors manually. A BGP Route Server is transparent to the
user traffic path but allows a service provider to override the normal BGP best path

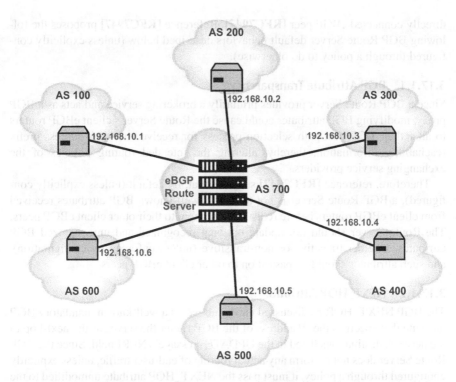

FIGURE 3.51   Internet Exchange with eBGP Route Server

with an alternative path based on the service provider's policy, or suppress all BGP routes for a particular network prefix and therefore not advertise that network prefix.

The Route Server operates within its own autonomous system (with an ASN) but is transparent to the AS_PATH, which means that it does not include its ASN in the AS_PATH while advertising routes to eBGP peers. A Route Server forwards BGP announcements among the eBGP peers according to rules set by the service providers themselves, without altering the BGP AS_PATH or the NEXT_HOP. The BGP Route Server provides AS_PATH, NEXT_HOP, and MED (MULTI_EXIT_DISC) transparency so that peering service providers at the IX will still appear to be directly connected. The BGP Route Server mediates the peering between service providers, but the peering relationship is not visible outside of the IX.

BGP Communities can be used to instruct a BGP Route Server to know which eBGP peer to announce to (or not announce to) network address prefixes. Incoming filter lists can be used on an eBGP peer to control what network prefixes it can accept (or not accept) when the Route Server announces prefixes to the peer.

3.17.1   BGP ROUTE SERVER BEHAVIORS

A BGP Route Server is transparent to BGP path attribute propagation allowing a BGP route received from a Route Server to carry the same set of BGP attributes (AS_PATH, NEXT_HOP, MED, and Communities) if the BGP route is sent from a directly connected eBGP peer **[RFC7947]**. Reference **[RFC7947]** proposes the following BGP Route Server default behaviors described below (unless explicitly configured through a policy to do otherwise).

3.17.1.1   BGP Attribute Transparency

Since a BGP Route Server provides primarily a brokering service and acts as a BGP proxy, modifying BGP attributes could cause the Route Server's client eBGP routers to alter their BGP best path selection process for received network address prefix reachability information, thereby, altering the intended routing policies of the exchanging service providers.

Therefore, reference **[RFC7947]** suggests that, by default (unless explicitly configured), a BGP Route Server is not to update well-known BGP attributes received from client eBGP routers before redistributing them to their other client eBGP peers. The Route Server should not update optional recognized and unrecognized BGP attributes, whether transitive or non-transitive (unless enforced by configuration), and such attributes should be passed on to other client eBGP peers.

3.17.1.2   NEXT_HOP Attribute

The BGP NEXT_HOP, as discussed earlier above, is a well-known mandatory BGP attribute that specifies the IP address of the BGP router the serves as the next-hop to the network destinations listed in the UPDATE message's NLRI field. Since the BGP Route Server does not perform any actual routing of end-user traffic, unless explicitly configured through a policy, it must pass the NEXT_HOP attribute unmodified to the client eBGP routers, similar to the "third-party" Next-Hop feature described earlier and also in **[RFC4271]**.

3.17.1.3   AS_PATH Attribute

The BGP AS_PATH is a well-known mandatory BGP path attribute that identifies the autonomous systems through which the routing information contained in an UPDATE message has traversed. Because a BGP Route Server does not participate in the process of forwarding packets between client eBGP routers, modifying the AS_PATH attribute could affect the BGP best path selection process of the client eBGP peers. So, the BGP Route Server is not to prepend its own ASN to the AS_PATH nor modify the AS_PATH in any other way.

Reference **[RFC7947]** recommends that the client eBGP routers of the BGP Route Server be able to accept BGP UPDATE messages that have the leftmost ASN in the AS_PATH attribute not equal to the ASN of the Route Server that sent the UPDATE message. If the client eBGP router implements this capability, it must allow this check to be disabled on a per-peer basis.

3.17.1.4   MED (MULTI_EXIT_DISC) Attribute

The MED is an optional non-transitive BGP attribute that is used on external (inter-autonomous system) links to allow BGP routers to discriminate among multiple exit or entry points to the same neighboring autonomous system. Reference **[RFC7947]** recommends that, if the MED attribute applies to an NLRI UPDATE sent to a Route

Server, the Route Server should propagate it to other client eBGP routers, and the value of the attribute should not be modified. The Route Server must propagate the MED attribute (optional, non-transitive) as received.

3.17.1.5 BGP Communities Attributes

The BGP Route Server may receive transitive and non-transitive BGP Communities attributes (the BGP Communities **[RFC1997]** and Extended Communities **[RFC4360]** attributes) carried in BGP UPDATE messages from the client eBGP routers. When these apply to an NLRI in an UPDATE message sent to the Route Server, these attributes should not be modified, processed, or removed, except when defined by local policy.

The Route Server must propagate all BGP Communities attributes, including NO_ADVERTISE, NO_EXPORT, and non-transitive Extended BGP Communities, as received. If a particular BGP Communities attribute is intended for processing by the BGP Route Server itself, as determined by local policy, the router server may modify or remove it.

3.18   BGP BEST PATH SELECTION PROCESS

A BGP router first applies the input or import policies in its local Policy Information Base (PIB) to the routes stored in its Adj-RIBs-In database, and then uses its BGP path selection process (called the Decision Process in **[RFC4271]**) to select routes for subsequent advertisement to its BGP peers. The output of the BGP path selection process is the set of routes that the BGP router will advertise to its peers. The BGP router stores these selected routes in its Adj-RIBs-Out database, after applying the output or export routing policies configured on it. The Adj-RIBs-Out database contains the routes that the router itself will advertise to its peers via BGP UPDATE messages.

The BGP router stores in the Adj-RIBs-In database unprocessed routing information that have been advertised to it by its peers. The Loc-RIB database contains the routes that the router has selected using its BGP path selection process (algorithm) in addition to all routes that are originated locally by the router (e.g., directly connected networks and static routes). The Adj-RIBs-Out database contains routes in the Loc-RIB database that the router has selected to be advertised to its peers after applying the output policies. Routes that is not in the Loc-RIB will not be placed in the Adj-RIBs-Out and will not be advertised to BGP peers.

Figure 3.52 summarizes the BGP route processing steps in a typical BGP router. Typically, a BGP router supports a process that is responsible for processing routes contained in the Adj-RIBs-In database, and this process is responsible for the following:

- Selecting routes to be used locally by the BGP router itself.
- Selecting routes that the BGP router will advertise to other BGP peers.
- Performing route aggregation aimed at reducing routing information during route advertisement.

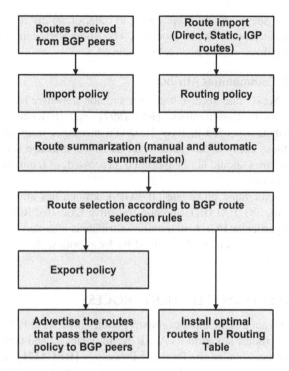

FIGURE 3.52   BGP Route Processing

Reference **[RFC4271]** divides the BGP route processing steps into the following three distinct phases, each triggered by a different system event:

- **Phase 1**: In this phase, the BGP router calculates the degree of preference for each route it has received from a BGP peer.
- **Phase 2**: A BGP router enters this phase upon completing Phase 1. Here, the router chooses the best route out of all the routes available to each distinct network destination, and installs each selected route in its Loc-RIB.
- **Phase 3**: The BGP router enters this phase after it has modified/updated its Loc-RIB. In this phase, the router disseminates routes in the Loc-RIB to each BGP peer, according to the policies contained in its PIB. Optionally, the router can perform route aggregation to reduce routing information in this phase.

The above BGP route processing phases are only conceptual, and a BGP implementation does not have to implement these phases precisely, as long as the implementation supports the intended functionality and exhibits the same externally visible behavior to other BGP routers.

3.18.1   PHASE 1: CALCULATING THE DEGREE OF ROUTE PREFERENCE

The BGP router enters Phase 1 whenever it receives an UPDATE message from a BGP peer that is advertising a new route to a specific destination, a replacement route, or withdrawn routes. In this Phase 1, the BGP router locks its Adj-RIBs-In database prior to processing any route stored in it, and unlocks it after processing all new or unfeasible routes in it.

For each new or replacement feasible route the BGP router receives, it determines a degree of route preference as follows:

- If the BGP router learned the route from an iBGP peer, it considers either the value of the LOCAL_PREF attribute as the degree of route preference, or it calculates the degree of route preference based on preconfigured local routing policy information. However, when using the latter, the router's administrator has to be aware that this may result in the formation of persistent routing loops.
- If the BGP router learned the route from an eBGP peer, then it calculates the degree of route preference based on preconfigured local policy information. If the result of the calculation indicates the route to be ineligible, the router may not use the route as input to the next phase of route selection process. Otherwise (if the route is eligible), the router must use the resulting value as the LOCAL_ PREF value in any iBGP re-advertisement it sends. In both here and above, the routing policy information used, and the nature of the computation carried out, are left to local implementation (vendor-dependent).

3.18.2 Phase 2: Best Path Selection

In Phase 2, the BGP router's best path selection process examines the Adj-RIBs-In database and considers all routes that are eligible in it. The router blocks the Phase 2 from running while the Phase 3 process is being run. The router locks the Adj-RIBs-In database before commencing the Phase 2 process, and unlocks it once the process completes. A few factors to consider in Phase 2 are the following:

- If the BGP router determines that the NEXT_HOP attribute of a BGP route is not resolvable, or if it would become unresolvable if it was installed in the IP Routing Table, that route must be excluded from the BGP best path decision process.
- If the router determines that the AS_PATH attribute of a BGP route contains routing information that can cause an autonomous system loop, that route is excluded from the best path selection process. The router detects and prevents an autonomous system loop by examining the full autonomous system path information (as specified in the AS_PATH attribute), and making sure that the ASN of its autonomous system does not appear in the AS_PATH attribute.
- It is critical that all BGP routers within an autonomous system make clear and consistent decisions that do not conflict with each other regarding route selection since doing otherwise would cause forwarding loops to occur.

For each set of network destinations for which a feasible route exists in the Adj-RIBs-In database, the BGP router identifies and selects the route that satisfies at least one of the following conditions:

a. The route is the only route to that destination, or
b. The route has the highest degree of route preference among all the routes that exist to the same set of destinations, or
c. The route is selected as a result of the Phase 2 best path selection process' tie breaking rules specified below.

After identifying the route as described above, the BGP router then installs that route in its Loc-RIB, replacing any route to the same network destination that is currently stored in the Loc-RIB. When the BGP installs the new route in its IP Routing Table, it must ensure that existing routes to the same destination that are now considered invalid are deleted from the IP Routing Table. Determining whether the new route entered in the IP Routing Table will replace an existing non-BGP route depends on the type of routing policies configured on the BGP router.

The BGP router must examine the NEXT_HOP attribute of the selected route and determine the immediate next-hop IP address of that route. If the BGP router determines that, either the immediate next-hop of the route or the IGP cost to the next-hop (i.e., where the next-hop is resolved through an IGP route) has changed, the router must perform Phase 2 best path selection again.

Even though the BGP router does not have to install routes in the IP Routing Table with the immediate next-hop(s), BGP implementations must ensure that, before the router forwards any packets along a given BGP route, its associated next-hop IP address is resolved to the immediate (or directly connected) next-hop IP address, and that, this directly connected IP address (or multiple IP addresses) is the address(es) used for actual packet forwarding. The BGP router has to remove unresolvable routes from its Loc-RIB database and the IP Routing Table. However, the router has to keep the corresponding unresolvable routes in the Adj-RIBs-In database (in case these routes become resolvable).

3.18.2.1 BGP Route Resolvability Condition

A BGP router excludes unresolvable routes from the best path selection process which ensures that only valid routes are installed in the Loc-RIB database and the IP Routing Table. Reference **[RFC4271]** defines the route resolvability condition as follows:

1. A BGP route X that references only the intermediate IP network address, is considered resolvable if there exists in the IP Routing Table at least one resolvable route Y that matches the intermediate IP network address of X and is not recursively resolved (directly or indirectly) through X. If there exist multiple matching routes, only the longest matching prefix route is considered.
2. A route that references a router interface (with or without an intermediate IP address) is considered resolvable if the state of the interface being referenced is up (i.e., operational) and if IP processing is enabled on that interface.

BGP routes (each expressed as an n-tuple <IP Address Prefix, Next Hop, AS-Path, [...other BGP attributes...]>) do not refer to routing device interfaces, but they can be resolved through the routes maintained in the BGP router's IP Routing Table. The routes in the IP Routing Table consist of routes that specify (outbound) interfaces and those that may not. Routes to directly connected networks and IGP routes are expected to specify their associated outbound interfaces. Static routes can specify their outbound interfaces, intermediate IP addresses, or both.

A BGP route is considered unresolvable when there exists no route in the BGP router's IP Routing Table matching the BGP route's NEXT_HOP. A route that resolves to itself (i.e., a mutually recursive route) also fails the BGP route resolvability check.

BGP implementations typically ignore any feasible route that would become unresolvable if it was installed in the IP Routing Table, even if the route's NEXT_HOP is resolvable using the current contents of the IP Routing Table. Mutually recursive routes are examples of such routes. The route resolvability check ensures that a BGP router will not install routes in its IP Routing Table that will later be removed and not used. This check helps to maintain local IP Routing Table stability, as well as improve the behavior of BGP in the network. Whenever the BGP router identifies a route that fails the BGP route resolvability check because it is a mutual recursive route, it will log an error message.

3.18.2.2   A Note on BGP Recursive Route Lookup

In BGP, neighbors do not necessarily have to be directly connected and may be located several hops away from each other. Without the use of recursive lookups, BGP routers may not be able to locate their next-hops since recursive routing forms the basis of the entire BGP concept. Recall that an iBGP peer does not modify the Next-Hop in a received routing advertisement (leaving it as it was originally received). Therefore, when a BGP router receives an IP packet for forwarding and performs a route recursive lookup for the next-hop address associated with a specific network address prefix, this can fail if there is no IGP route to the next-hop address advertised with the network address prefix. Figure 3.53 presents one example that explains the concept of BGP recursive lookup when a BGP packet receives an IP packet to be forwarded.

Entries in the BGP router's IP Routing Table are built from the Loc-RIB database. BGP never associates a BGP route with an outgoing router interface. Therefore, a given BGP router performs a recursive lookup to forward an IP packet toward an external destination (a BGP prefix). The following takes place when a BGP router receives an IP packet for forwarding:

- The BGP router first performs a lookup of the BGP route (prefix) in its IP Routing Table to determine the BGP next-hop needed to reach the destination in the remote autonomous system.
- Then, the router performs another lookup in the IP Routing Table using the BGP next-hop (found above) to determine a route (learned by an IGP) to reach the indicated BGP next-hop.

BGP recursive route lookup allows BGP use the BGP Next-Hop attribute in UPDATE messages to determine a path to a given network over an IGP route (which has an

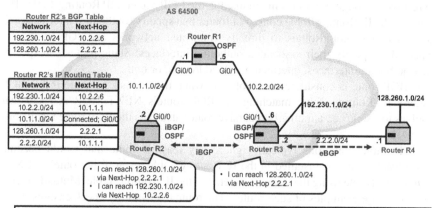

FIGURE 3.53   Illustrating BGP Recursive Lookup

associated outgoing interface, and since BGP routes are not associated with interfaces).

For a given BGP route (prefix), the BGP router extracts the associated next-hop IP address from the Next-Hop attribute in the UPDATE messages carrying the route. Since the BGP route has no outgoing interface associated with it in the IP Routing Table but instead a BGP next-hop, the BGP router uses an IGP to determine the outgoing interface associated with this BGP next-hop.

So, when the BGP router receives an IP packet with a destination in network "BGP route (prefix)", it searches its IP Routing Table and finds the installed BGP route (prefix). The router then takes the indicated (BGP) next-hop IP address, and performs another search in the IP Routing Table. This time it now finds a match with the IGP (e.g., OSPF) route to this next-hop IP address.

3.18.2.3   Deep Dive into BGP Recursive Lookup

In Figure 3.53, Router R3 is running an eBGP session with R4 and an iBGP session with R2. Router R3 learns BGP route 128.260.1.0/24 from R4. R3 also injects the local route 192.230.1.0/24 into BGP. R2 learns the route 192.230.1.0/24 via 10.2.2.6 which is the IP address of the iBGP peer (R3) advertising the route. Thus, 10.2.2.6 is the next-hop for R2 to reach 192.230.1.0/24. Similarly, R3 learns 128.260.1.0/24 from R4 via next-hop 2.2.2.1. When R3 passes this route to R2 via iBGP, it includes the next-hop information (2.2.2.1) unmodified. So, R2 would receive the BGP update about 128.260.1.0/24 with the next-hop 2.2.2.1. This example explains how the eBGP next-hop gets passed into iBGP.

As seen in Figure 3.53, the next-hop 2.2.2.1 is not necessarily reachable (from R2) via a direct connection. Router R2 sees the next-hop for 128.260.1.0/24 as 2.2.2.1, but reaching it requires going through 10.1.1.1. Thus, determining the next-hop mandates the Router R2 to perform a recursive lookup to know where to forward a packet to a destination such as 128.260.1.0/24 or 192.230.1.0/24. To reach the next-hop 2.2.2.1, R2 will have to perform a recursive lookup in its IP Routing Table to determine if and how 2.2.2.1 can be reached. The router performs this recursive search until it associates destination 2.2.2.1 with an outgoing interface. R2 performs the same recursive lookup to find how to reach 10.2.2.6. If the BGP router finds that the next-hop cannot be reached, it considers the route inaccessible.

In Figure 3.53, R2 determines from it BGP Table that 128.260.1.0/24 can be reached via next-hop 2.2.2.1. R2 performs a lookup in its IP Routing Table and determines that 2.2.2.0/24 can be reached via next-hop 10.1.1.1. R2 performs another recursive lookup in the IP Routing Table and finds that network 10.1.1.0/24 is directly connected via interface Gi0/0. This indicates that traffic destined toward 2.2.2.1 should be sent out interface Gi0/0 which also applies to traffic going to the next-hop 10.2.2.6. Thus, the IGP route from R2 has an outgoing interface set to Gi0/0 and a next-hop set to 10.1.1.1, meaning IP packets that are destined for network 128.260.1.0/24 should be forwarded via 10.1.1.1, which is directly reachable from Router R2 over interface Gi0/0.

The above discussion shows that, to successfully forward a packet to a destination in a BGP network, it must be ensured that the BGP NEXT_HOP is provided via an IGP or static routing. If the BGP NEXT_HOP is not reachable, the BGP route will be considered inaccessible.

3.18.2.3.1 Traditional non-BGP Recursive Route Lookup

BGP recursive lookup is different from the recursive lookup a non-BGP router typically performs using directly its IP Routing Table (instead of the more optimized IP Forwarding Table). Traditional router architectures (mostly those based on centralized CPU based forwarding) use the suboptimally structured (software) IP Routing Table for route lookups which may involve performing recursive lookups.

In these architectures, the router's CPU uses software-based lookups and is involved with every IP packet forwarding decision. For each IP packet, the CPU searches the IP Routing Table again and again (i.e., recursively) until it finds the outgoing interface for the packet, a process which can slow down the router considerably. Figure 3.54 explains the process of recursive lookup in an IP Routing Table in a traditional software-based non-BGP router.

As explained in Chapter 3 of Volume 1 of this two-part book, modern IP router architectures (especially high-performance routers) maintain an IP Forwarding Table (also called the Forwarding Information Base [FIB]) that contains mainly the information required for packet forwarding lookups (IP address prefix, next-hop IP address, next-hop Mac address, and the outgoing interface).

Network IP Address	Next-Hop IP Address	Outgoing interface
172.16.1.0/24	172.16.2.126	-
172.16.2.0/24	172.16.3.126	-
172.16.3.0/24	10.1.0.2	-
10.1.0.0/30	-	Gi0/1

Network 10.1.0.0/30 is directly connected through interface Gi0/1

- The router performs a lookup of the IP Routing Table in order to forward a packet to the destination IP address of 172.16.1.222.
- The route 172.16.1.0/24 is the best-match with the next-hop IP address 172.16.2.126.
- The router performs another look up in the IP Routing Table for 172.16.2.126 and the route 172.16.2.0/24 is the best-match with the next-hop IP address 172.16.3.126.
- Again, the IP Routing Table is searched to find the best match for the next-hop IP address 172.16.3.126. The route 172.16.3.0/24 is the best-match with the next-hop IP 10.1.0.2.
- Finally, the next-hop IP address 10.1.0.2 matches the route 10.1.0.0/30 in the IP Routing Table and the packet is forwarded over the outgoing interface Gi0/towards the destination 172.16.1.222.

FIGURE 3.54   Traditional non-BGP Recursive Route Lookup in an IP Routing Table

The IP Routing Table contains a lot of information that is not directly relevant for IP packet forwarding. The extra information is mainly used by the routing protocols and not for packet forwarding. Thus, in most high-end routers, the most relevant information for actual packet forwarding is distilled from the IP Routing Table and stored in the IP Forwarding Table. Any time the IP Routing Table is updated, the Forwarding Table is also updated to reflect the changes.

In this architecture, entries that involve recursive lookups are resolved before they are entered into the IP Forwarding Table. With this, even if the underlying IP Routing Table contains routing entries that are recursively chained, lookups in the Forwarding Table are not recursive.

3.18.2.4   BGP Best Path Selection Algorithm (with Tie Breakers)

A BGP router may have several routes to the same network destination in its Adj-RIBs-In database that have the same degree of route preference. The router can select only one of these routes to be installed in the associated Loc-RIB database. The router considers all routes with the same degree of route preference from both of its iBGP peers and external peers. We discuss in this section, the typical best path selection algorithm used in real-world BGP routers [CISCID13753] [JUNIL2VPNGU18] [RFC4271].

The tie-breaking rules described here assume that, for each candidate route (to a network destination), all the BGP routers within an autonomous system follow a consistent procedure to determine the path cost (interior distance) to the network address carried in the NEXT_HOP attribute of the route, and follow the same best path selection algorithm. The BGP router starts the algorithm by examining all equally preferable routes to the same network destination, and then selects the routes that should be removed from consideration. The router terminates the selection algorithm as soon as only one route remains in consideration.

The BGP router bases its best path section process on the BGP path attributes. When multiple paths (routes) to the same network destination exits, the router selects the best path for routing traffic to that destination. Some BGP implementations include other path parameters (dependent on the networking scenario) that are not included in the steps below (see, for example, **[CISCID13753] [JUNIL2VPNGU18] [RFC4456] [RFC5065]**). Typically, the BGP router applies the path selection criteria in the following order.

1. **Resolvable Next-Hop**: The router verifies that the next-hop is resolvable. If the (route to the) next-hop is not accessible, the router ignores that router. This is why having an IGP route to the next-hop is important.
2. **Best Route Preference Value**:
   a. **Largest Weight**: The router prefers the route with the largest WEIGHT attribute value. Weight is a Cisco specific path parameter, making this step only applicable to routers that understand this attribute **[CISCID13753]**.
   b. **Lowest Administrative Distance**: In non-Cisco routers, the route with the lowest Administrative Distance (or Route Preference) value is selected **[JUNIL2VPNGU18]**. For example, if a BGP and an OSPF route exist to a particular network destination, the OSPF route is preferred because the OSPF route has a default Administrative Distance of 110, while the BGP route has a default Administrative Distance of 170.
3. **Largest Local Preference Value**: If the Weights or Administrative Distances are the same, the router prefers the route with the largest LOCAL_PREF value.
4. **Shortest AS-Path Length**: If there are no locally originated routes and the routes have the same Local Preference value, the router prefers the route with the shortest AS_PATH value. The router prefers routes with the smallest number of ASNs (i.e., autonomous system hops) in their AS_PATH attributes. The router prefers routes with shorter AS-Paths over routes with longer AS-Paths.
5. **Lowest Origin Value**: If the routes have the same AS_PATH value, the router prefers the route that has the lowest Origin type code. Routes learned by an IGP have a lower Origin code (Type Code = 0) than routes learned by an EGP (Type Code = 1). Both routes have lower Origin codes than routes whose Origin is unknown (i.e., INCOMPLETE routes [Type Code = 2]).
6. **Lowest MED Value**: If the routes have the same Origin type, the router prefers the route with the lowest MED value if the routes were received from the same autonomous system. The router interprets the absence of a MED value as a MED of 0.
7. **Prefer eBGP over iBGP Routes**: If the routes have the same MED value, the router prefers eBGP over iBGP routes. The router prefers routes learned from an eBGP peer over those learned from an iBGP peer.
8. **Lowest IGP Metric (or Best Exit from the Autonomous System)**: If the routes considered up to this point are identical, the router prefers the route with the lowest IGP metric to the BGP next-hop. The router prefers the shortest internal path inside the autonomous system to reach the network destination – that is, the shortest path to the next-hop. The router skips this step if the Next-Hop for a route is reachable, but no IGP cost can be determined (i.e., all routes are considered to have equal costs).

9. **Lowest BGP Router Identifier**: If the internal paths considered above are the same, the router uses the Router ID as the tiebreaker. In this case, the router prefers the route provided by the BGP router having the lowest Router ID. Router ID determination is vendor-specific. Cisco routers use the loopback IP address if one is configured; otherwise, they use the highest IP address on the router. The Router ID can also be manually configured.
10. **Lowest BGP Peer IP Address**: At this point, the router prefers the route from the BGP peer with the lowest IP address. This IP address corresponds to the IP address the remote peer uses in its TCP connection to the local router.

BGP routers typically receive multiple routes to the same network destination. For each network address prefix, the BGP router selects a single best path. After the best path is selected, the router installs this in its IP Routing Table to be used for packet forwarding..

The route selection algorithm discussed above is only meant to show the range of parameters and their order of preference in a typical BGP router. Readers can refer to documents such as **[CISCID13753] [JUNIL2VPNGU18]** for example implementations of path selection algorithms (which are mostly vendor-dependent). For example, when considering scenarios with multipath routing, BGP Confederations, or Route Reflection Clusters, the path selection algorithm tend to have vendor-specific details or include other parameters as described in **[CISCID13753] [JUNIL2VPNGU18] [RFC4456] [RFC5065]**. Reference **[RFC5065]** (Section 5.3), for example, specifies additional rules that apply when performing path selection in the presence of BGP Confederations, while **[RFC4456]** (Section 9) specifies additional path selection rules when considering BGP Route Reflectors.

3.18.3 PHASE 3: BGP ROUTE DISSEMINATION

The BGP router invokes the route dissemination function upon completion of the best path selection process, or when any of the following events occur:

a. When routes in the Loc-RIB database of the router to local network destinations have changed
b. When routes that have been locally generated but learned by non-BGP means have changed
c. When a new connection to a BGP peer has been established

The router blocks the Phase 3 route dissemination function from running while the Phase 2 best path selection function is in progress. The router allows the Phase 3 function (which is a separate decision process) to run to completion when it has no further work to do.

The router processes all routes in its Loc-RIB into Adj-RIBs-Out database according to configured output or export routing policy. The routing policies may involve excluding a particular route in the Loc-RIB from being installed in a particular

Adj-RIB-Out. The router does not install a route in the Adj-Rib-Out database unless the network destination, and NEXT_HOP described by the route is able to be used by the IP Routing Table for packet forward. If the router excludes a route in the Loc-RIB from being installed in a particular Adj-RIB-Out database, then the router must withdraw from service this previously advertised route (if one exists in that Adj-RIB-Out) by sending of an UPDATE message. The router may also optionally apply route aggregation and information reduction techniques if necessary.

When the BGP router completes updating the Adj-RIBs-Out and the IP Routing Table, it runs the BGP Update-Send process described below.

3.18.3.1   BGP Update-Send Process

The Update-Send process in the BGP router is the process via which the router advertises UPDATE messages to all of its peers. For example, this process in the router is responsible for distributing the routes selected by the BGP best path selection process to other BGP routers located in either the same autonomous system or neighboring autonomous systems.

When a BGP router receives an UPDATE message from an iBGP peer, it should not re-distribute the routing information carried in this UPDATE message to other iBGP peers (unless the router is acting as a BGP Route Reflector [**RFC4456**] which means it can pass the message to internal peers).

The BGP router updates its Adj-RIBs-Out database as part of Phase 3 process. The router advertises to its peers by means of an UPDATE message all newly installed routes, and all newly unfeasible routes for which there is no replacement route. A router does not advertise a given feasible BGP route from its Adj-RIB-Out database if doing so would result in sending an UPDATE message containing the same BGP route as was previously advertised.

The BGP router removes any routes in the Loc-RIB database that are marked as unfeasible. The router also advertises in an UPDATE message changes to the reachable destinations within its own autonomous system. If a single route does not fit into an UPDATE message due to the limits imposed by the maximum size of an UPDATE message (see Figure 3.14), the BGP router must not advertise this route to its BGP peers and it may choose to log an error locally.

3.18.3.2   Originating BGP Routes

A BGP router may originate BGP routes by advertising routing information supplied by some other routing methods (e.g., routes acquired via an IGP). A BGP router that originates BGP routes assigns a degree of route preference (e.g., determined according to local configured policies) to these routes by passing them through the best path selection process (i.e., the Decision Process).

The router may also distribute these routes to other BGP routers within the local autonomous system as part of the BGP Update-Send Process (described above). The decision governing whether a BGP router should distribute non-BGP acquired routes within an autonomous system depends on the routing conditions within the autonomous system (e.g., the type of IGP used), and is an issue that is often controlled via local configuration in the autonomous system.

3.19 BGP SESSION SECURITY

The BGP authentication mechanisms are used by BGP to protect itself against the injection of spoofed TCP segments into the BGP connection. Many BGP implementations still use the TCP MD5 Signature Option [RFC2385] even though it has been obsoleted by the TCP Authentication Option (TCP-AO) [CISCPICONGUI19] [RFC5925]. In this section, we describe the two authentication options for the sake of completeness.

3.19.1 TCP MD5 Signature Option

Reference [RFC2385] describes an extension to TCP that can be used for BGP session authentication. BGP uses TCP as its Transport Layer protocol. So, a new TCP Option ([TCP Option Kind 19]) was defined in [RFC2385] to allow an MD5 digest to be carried in a TCP segment which may be carrying BGP data. The MD5 digest serves as a signature for the TCP segment, possibly, carry BGP routing information that needs to be authenticated by the receiving end. This TCP Option allows BGP to significantly reduce the danger from certain security attacks. With the TCP MD5 signature, the receiver can authenticate each TCP segment, including the TCP IPv4 pseudo-header, TCP header, and any TCP data carried.

Using the method in [RFC2385], every TCP segment sent on a TCP connection carries a 16-byte MD5 digest generated by applying the MD5 algorithm to the following items (and processed in the order listed):

1. TCP pseudo-header which consists of the 32-bit source IP address, 32-bit destination IP address, zero-padded 8-bit IP header protocol number field, and 16-bit TCP segment length.
2. TCP header, excluding all TCP Options, and assuming the 16-bit TCP checksum is all zeros.
3. TCP segment data (if any).
4. An independently specified secret key or password, and presumably connection-specific, that is known to both TCP end-points.

The TCP header and the TCP pseudo-header are arranged in network byte order where the bytes are ordered from most significant byte to least significant byte. The MD5 digest is carried in a 16-byte field in the TCP segment, and the TCP Option 19 appears in every TCP segment sent over the TCP connection (Figure 3.55). This means each TCP segment sent on the TCP connection between the endpoints is verified. The secret key or password is never sent over the connection and known only by the TCP endpoints. The length of the TCP MD5 Option is fixed, and the MD5 digest is carried in a 16-byte (128-bit) field following the 1-byte TCP Option Kind and 1-byte Length fields as shown in Figure 3.55.

The TCP receiver upon receiving a signed TCP segment, validates it by calculating its own MD5 digest from the same data listed above (using its own secret key). The receiver then compares the received MD5 digest (carried in the TCP segment) with the locally computed digest. A mismatch in the two digest results in the received

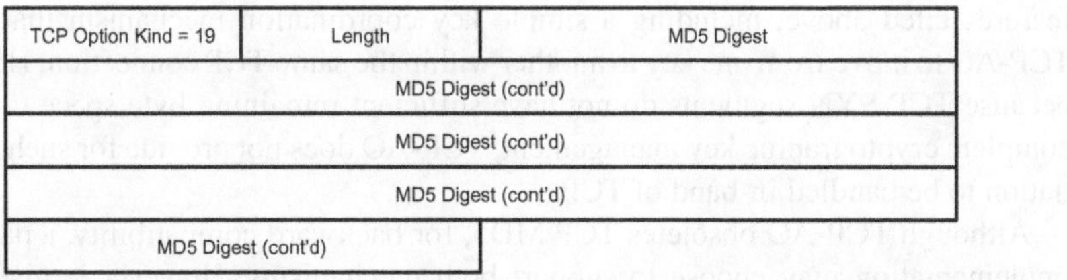

TCP Option Kind = 19	Length	MD5 Digest
MD5 Digest (cont'd)		
MD5 Digest (cont'd)		
MD5 Digest (cont'd)		
MD5 Digest (cont'd)		

Field	Meaning
TCP Option Kind (1 Byte)	This field carries a value of 19 indicating the TCP MD5 Option.
Length (1 byte)	This field indicates the length of the TCP Option in bytes including the Option Kind, Length, and MD5 digest fields.
MD5 Digest (16 Bytes)	This field carries the MD5 digest calculated the sender.

FIGURE 3.55   TCP MD5 Option Format

TCP segment being discarded, and the receiver does not send any response back to the sender. It is recommended that the receiver logs the failure in a local log. If the two digests are the same, the receiver assume the information it is receiving is from a trusted source.

3.19.2   TCP AUTHENTICATION OPTION (TCP-AO)

The TCP-AO is specified in **[RFC5925]** as a replacement for the TCP MD5 Signature Option (TCP MD5) **[RFC2385]** which is now considered obsolete. TCP-AO allows for the use of stronger Message Authentication Codes (MACs), specifies more details than TCP MD5 on the security association of TCP connections, and guards against replays even for long-lived TCP connections (using TCP Sequence Number Extensions [SNEs]). TCP-AO is designed to use either an external, out-of-band Master Key Tuple (MKT) management mechanism, or a static MKT configuration.

Using either options, TCP-AO also allows the protection of TCP connections when the same MKT is used across repeated instances of a TCP connection, by utilizing Traffic Keys generated from the MKT, and the coordination of MKT changes between TCP endpoints. The development of TCP-AO is intended to allow current infrastructure that use TCP MD5 to be supported, such as the protection of long-lived TCP connections (as used, for example, in BGP and Label Distribution Protocol [LDP]).

TCP-AO also supports a larger set of MACs without causing significant system and operational changes. TCP-AO is defined to use a different TCP Option identifier (TCP Option Kind 29) than TCP MD5 (TCP Option Kind 19), even though both mechanisms are never permitted to be used simultaneously. TCP-AO is designed to also support IPv6, and is fully compatible with the requirements proposed for replacing TCP MD5.

The development of TCP-AO stems from the need for a stronger and robust authentication method than TCP MD5 which is less resistant to attacks because of its use of a simple keyed hash for authentication. TCP MD5 also does not support both algorithm agility and key-management. TCP-AO adds the latter, as well as the other features cited above, including a simple key coordination mechanism that allows TCP-AO to move from one key to another within the same TCP connection. However, because TCP SYN segments do not have sufficient remaining byte space to handle complete cryptographic key management, TCP-AO does not provide for such a negotiation to be handled in-band of TCP.

Although TCP-AO obsoletes TCP MD5, for backward compatibility, a particular implementation may choose to support both mechanisms. However, a given TCP connection can only run one mechanism at any given time. Also, a TCP connection that uses TCP MD5 cannot be migrated to TCP-AO because TCP MD5 does not support any mechanism that allows changes to the security algorithm of a TCP connection once established. TCP-AO does not support data confidentiality (encryption). TCP-AO also does not support dynamic negotiation of parameters.

3.19.2.1   Format of the TCP Authentication Option

TCP-AO contains an Option Kind field, a KeyID field, an RNextKeyID field and a variable 12 to 16-byte field that contains a MAC as shown in Figure 3.56. Comparing Figures 3.55 and 3.56, it can be seen that TCP-AO has 2 bytes less than TCP MD5, that is, TCP-AO has 16 bytes overall (rather than the 18 bytes in TCP MD5) in the initially specified default case (using a 96-bit [12-byte] MAC) in **[RFC5925]**.

If a manual mechanism or an out-of-band protocol is available to provide new keys, TCP-AO is able to support rekeying during a TCP connection. As shown in

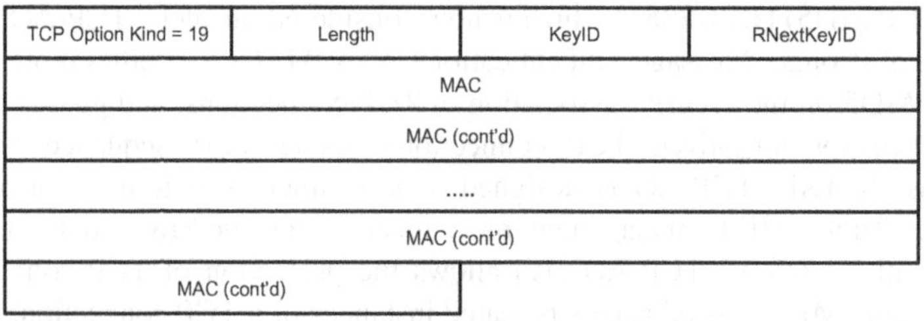

TCP Option Kind = 19	Length	KeyID	RNextKeyID
MAC			
MAC (cont'd)			
.....			
MAC (cont'd)			
MAC (cont'd)			

Field	Meaning
TCP Option (1 Byte)	This field carries a value of 29 indicating the TCP Authentication Option.
Length (1 Byte)	• This field indicates the length of the TCP option in bytes including the Option Kind, Length, KeyID, RNextKeyID, and MAC fields. • The Length value must be greater than or equal to 4. A TCP segment with a value less than 4,must be discarded. • The sum of the sizes of all TCP options, when added to the size of the basic TCP header (20 bytes), matches the TCP Offset field exactly.
KeyID (1 Byte)	• This field specifies the Master Key Tuple (MKT) used to generate the traffic keys that were used to generate the MAC that authenticates this TCP segment. • The KeyID supports efficient key changes during a TCP connection and/or to help with key coordination during connection establishment. • Note that the KeyID has no cryptographic properties and it need not be random.
RNextKeyID (1 Byte)	This field specifies the MKT that is ready at the sender to be used to authenticate received segments, i.e., the desired 'receive next' key ID. The RNextKeyID has no cryptographic properties and it need not be random
MAC (Variable Length)	The contents of this field are determined by the particulars of the security association.Typical MACs are 96-128 bits (12-16 bytes),but any length that fits in the header of the TCP segment being authenticated is allowed.

FIGURE 3.56   TCP Authentication Option Format

Figure 3.56, TCP-AO includes a KeyID field containing a "key ID" value that allows the efficient and concurrent use of multiple keys, plus a key coordination mechanism that uses the RNextKeyID field (containing the "receive next key ID") to manages the key change within a TCP connection.

TCP-AO fields do not explicitly or implicitly indicate the MAC algorithm used (the latter as in TCP MD5). The particular MAC algorithm used in TCP-AO is considered to be part of the configuration of the TCP connection security parameters and is a process that is managed separately.

3.19.2.2 TCP-AO Keys and Their Properties

At each endpoint, TCP-AO uses two sets of keys (MKTs and Traffic Keys) to authenticate incoming and outgoing TCP segments. TCP-AO uses the MKTs to derive unique Traffic Keys, and include the data used to generate the Traffic Keys, in addition to the parameters associated to the use of the Traffic Keys. These parameters include whether TCP Options are authenticated, and indicators of the algorithms TCP-AO uses for Traffic Key derivation and MAC computation. The Traffic Keys serve as the keying data TCP-AO uses to calculate the MAC of individual TCP segments.

TCP-AO includes the socket pairs of TCP (i.e., source IP address, destination IP address, source TCP port number, destination TCP port number) as a security parameter index (SPI) (together with the KeyID value), instead of specifying a separate field as the parameter index (such as the SPI in IPsec).

3.19.2.2.1 Master Key Tuple (MKT)

TCP-AO uses the MKT to describe the properties it associates with one or more TCP connections. The MKT comprises the following items:

- **TCP Connection Identifier**: This is a TCP socket pair consisting of a local IP address, a remote IP address, a local TCP port number, and a remote TCP port number.
- **TCP Option Flag**: This flag indicates whether the TCP sender has included TCP Options in a TCP segment apart from the TCP-AO in the MAC computation. When other TCP Options are included in the MAC computation, the TCP sender includes the content of all the TCP Options (in the order present in the TCP segment), with MAC field zeroed out. When the TCP Options are not included, the TCP sender excludes all TCP Options other than TCP-AO from all MAC computations (i.e., they are skipped over and not zeroed). Note that, regardless of the setting of this flag, TCP-AO always includes its MAC field while zeroed out, in the MAC computation. This option flag when set, applies to TCP Options carried in both outgoing and incoming TCP segments (both directions).
- **IDs**: These represent the values used in the KeyID or RNextKeyID fields of the TCP-AO. The TCP senders uses the KeyID to differentiate MKTs in concurrent use (during a TCP connection), and the RNextKeyID to indicate when MKTs are ready for use to authenticate TCP segments sent in the opposite direction (i.e., coming from the TCP receiver).

RNextKeyID specifies the preferred MKT information the TCP sender wants to use for subsequent TCP segments it receives from the remote end ('receive next'). It is a way for the TCP sender to indicate to the remote end the MKT that is ready for use for future incoming TCP segments. This is to allow the TCP receiver to know when to switch to another MKT (and thus use its KeyIDs and associated Traffic Keys).

Each MKT uses two types of IDs, a *SendID* which the TCP sender inserts in the KeyID field of the TCP-AO option of outgoing TCP segments, and a *RecvID* which the sender uses to match against the TCP-AO KeyID field of incoming TCP segments. The RNextKeyID indicates the desired MKT's *RecvID* to be used for incoming TCP segments.

- **Master Key**: This is a secret byte sequence that TCP sender uses for generating Traffic Keys. The Master Key may be derived from a separate shared secret key over a separate channel using a protocol external to TCP-AO.
- **Key Derivation Function (KDF)**: This specifies the KDF and its parameters that the TCP sender uses to derive Traffic Keys from Master Keys.
- **MAC Algorithm**: This specifies the MAC algorithm and its parameters the TCP sender uses for a particular TCP connection.

The MKT components must not be changed during a TCP connection; instead, the set of MKTs used for a TCP connection may be changed during that TCP connection. Since MKT parameters are not changed, the TCP connection can install, instead, new MKTs to allow it to change which MKT it wants to use whenever it deems necessary. TCP does not have a handshake mechanism for modifying the TCP state after a connection has been established. This means the values of the MKT component cannot be changed during the TCP connection. The endpoints only coordinate TCP state during TCP connection establishment.

3.19.2.2.2   Traffic Keys

A Traffic Key is a key that a TCP endpoint derives from the MKT and the TCP socket pair (consisting of a local IP address, a remote IP address, a local TCP port number, and a remote TCP port number), and, for TCP connections that are already established, the TCP Initial Sequence Numbers (ISNs) in each direction. TCP segments that the TCP endpoints exchanged before a TCP connection is established use the same items for deriving the Traffic Keys, except zeros are substituted for unknown values (e.g., that ISNs not yet coordinated).

Using TCP's ISNs for differentiation, TCP-AO is able to ensure that per-TCP connection Traffic Keys are as unique as the TCP connection itself, even when TCP-AO uses static MKTs across repeated instances of TCP connections on a single socket pair.

A TCP endpoint can use a single MKT to derive any one of the following four Traffic Keys:

- *send_SYN_traffic_key*: This is the Traffic Key the TCP sender uses to authenticate outgoing TCP SYN packets (to the receiver). The source ISN is known

(the local ISN of the TCP connection), and the destination (or remote) ISN is
unknown (and so takes the value of 0).

- *receive_SYN_traffic_key*: This is the Traffic Key the TCP sender uses to
  authenticate incoming TCP SYN packets (from the receiver). The source ISN
  is known (the remote ISN of the TCP connection), and the destination (remote)
  ISN is unknown (and so takes a value of 0).
- *send_other_traffic_key*: This is the Traffic Key the TCP sender uses to authenticate all other (non-SYN) outgoing TCP segments.
- *receive_other_traffic_key*: This is the Traffic Key the TCP sender uses to
  authenticate all other (non-SYN) incoming TCP segments.

All four Traffic Keys are unidirectional, and typically, only three of these Traffic
Keys are used by a given TCP connection. This is because a connection typically
uses only one of the SYN keys. However, all four Traffic Keys are used only when a
TCP connection executes a "simultaneous open", also called "simultaneous active
open on both sides".

3.19.2.3   Per-Connection TCP-AO Parameters

TCP-AO associates a number of parameters with each TCP connection that uses
TCP-AO:

- *current_key*: This is the MKT TCP currently uses to authenticate outgoing
  TCP segments (i.e., currently active MKT for outgoing segments). The sender
  inserts the appropriate *SendID* for this MKT in the KeyID field of outgoing
  TCP segments. The sender authenticates incoming TCP segments using the
  MKT corresponding to the TCP segment and its TCP-AO KeyID when matched
  against the MKT TCP Connection Identifier and the *RecvID*. A TCP connection uses only one *current_key* at any given time. Also, at most one *current_key*
  must be specified for every TCP connection in a non-IDLE state.
- *rnext_key*: This is the MKT that the TCP sender currently prefers to use for
  incoming (received) TCP segments. The sender inserts the appropriate *RecvID*
  for this MKT in the RNextKeyID field of outgoing TCP segments. At most one
  rnext_key must be specified for each TCP connection in a non-IDLE state.
- **A Pair of SNEs**: A TCP endpoint uses the SNEs to protect TCP segments
  against replay attacks. The endpoint initializes each SNE to zero upon TCP
  connection establishment. The SNEs are used in the MAC calculation.
- **One or More MKTs**: These represent the MKTs that match the TCP socket
  pair of a particular TCP connection. The TCP endpoint uses the MKTs, together
  with other parameters of a TCP connection, to generate the Traffic Keys that
  are unique to each TCP connection.

The 32-bit TCP Sequence Number may, for long-lived TCP connections, roll over
and repeat. This could cause TCP segments within a TCP connection to be replayed.
So, to prevent replay attacks, TCP-AO adds a 32-bit SNE to transmitted and received
TCP segments.

The TCP Sequence Number is extended by the SNE so that TCP segments within a single TCP connection are always unique. When a TCP Sequence Number roll over occurs, it is possible that a TCP segment could be repeated in whole. The SNE differentiates between TCP segments even if identical TCP segments are sent with identical TCP Sequence Numbers at different times in a TCP connection.

TCP-AO uses the SNE to emulate a 64-bit Sequence Number space by inferring when to increment the SNE (representing the high-order 32-bit portion) based on evolution of the TCP Sequence Number (considered the low-order portion). Thus, in TCP-AO, the TCP sender maintains one SNE, SND.SNE, for transmitted TCP segments, and another SNE, RCV.SNE, for received TCP segments, both initialized to zero when a TCP connection begins. TCP-AO uses these two SNEs, together with the 32-bit TCP Sequence Numbers, to emulate a 64-bit overall Sequence Number space.

3.19.2.4 Cryptographic Algorithms

Each TCP endpoint uses a MAC algorithm to compute the TCP-AO MAC that is used to authenticate TCP segments and their headers. TCP endpoints also use KDFs, which are also cryptographic algorithms, to convert MKTs of TCP connections into unique Traffic Keys for each TCP connection. The MAC and KDFs are described in the following sections.

3.19.2.4.1 MAC Algorithms

A MAC algorithm takes specific data of variable length and a secret key as input, and generates an output that is a fixed-length number. A TCP endpoint uses this fixed-length number to determine whether the input variable length data comes from a source that possesses that same secret key, and whether the input data have been tampered with in transit.

TCP-AO MACs are computed according to the following:

$MAC = MAC\_alg(traffic\_key, message)$

INPUT: MAC_alg, $traffic\_key$, $message$

OUTPUT: MAC

where

- *MAC_alg*: This represents the specific MAC algorithm used for the MAC computation. The MAC algorithm specifies the length of the MAC, that is the length of the OUTPUT, which means the algorithm does not require a separate OUTPUT length parameter.
- *traffic_key*: This represents the Traffic Key used for the MAC computation. This Traffic Key is computed from the current MKT of the TCP connection.
- *message*: This represents the input variable length data over which the MAC is computed. In TCP-AO, this is the TCP segment prepended by the TCP IP4 (or IPv6) pseudo-header plus the TCP header options
- *MAC*: This is the output of the MAC algorithm which is of fixed-length, given the input parameters, *traffic_key*, and *message*.

The MAC algorithms defined for TCP-AO in **[RFC5925]** each truncate the OUTPUT to 96 bits, although each algorithm can output a larger MAC. It is observed in **[RFC5925]** that a 96-bit MAC provides a reasonable trade-off between security and TCP message size. As discussed above, the MKT of a TCP connection defines the MAC algorithm used for the MAC computation.

The input data (*message*) to the MAC algorithm are processed in the following sequence (interpreted in network-standard byte order): the SNE; the TCP IP4 (IPv6) pseudo-header; the TCP header (by default including TCP Options, and with the TCP Checksum and TCP-AO MAC fields all set to zero); and the TCP data (i.e., the TCP segment's payload).

The MKT's TCP Option flag discussed above indicates whether the TCP Options are included in the MAC computations. For the MAC computations, the MAC field is zeroed as discussed above. Note that the Traffic Key, which is derived from the current MKT, is not included as part of the TCP data passed to the MAC algorithm.

TCP-AO necessitates a change in the computation of MACs when a TCP connection restarts, even when TCP-AO is reusing a TCP socket pair (source/destination IP addresses, and source/destination TCP port numbers).

3.19.2.4.2 Traffic Key Derivation Functions (KDFs)

TCP-AO derives the Traffic Keys from the MKTs using KDFs. The Traffic Keys are computed using the KDFs according to the following:

$$traffic\_key = KDF\_alg(master\_key, context, output\_length)$$

INPUT: *KDF_alg, master_key, context, output_length*

OUTPUT: *traffic_key*

where

- **KDF_alg**: This is the specific KDF as indicated in the MKT that is used to derive the traffic key.
- **master_key**: This is the *master_key* string associated with the MKT.
- **context**: This is the context (data) used as input to the KDF used to derive the Traffic_Key.
- **output_length**: The represents the desired length of the KDF output, that is, the length to which the KDF will truncate its output.
- **traffic_key**: The represents the desired output of the KDF, with length *output_length*, which TCP-AO uses as input to the MAC algorithm.

The *context* is used as input to the KDF and consists of the combination of the TCP socket pair (local IP address, a remote IP address, a local TCP port number, and a remote TCP port number), the source ISN and the destination ISN of a TCP connection. The *context* (data) is unique to each TCP connection instance, which allows TCP-AO to derive unique Traffic Keys for that TCP connection, even when an MKT is used across repeated TCP connections that share a TCP socket pair, or used across many different TCP connections. TCP-AO allows unique Traffic Keys to be derived for each TCP connection even without relying on external key management methods.

The inclusion of both source and destination ISNs in the *context* for the computation of the Traffic Key ensures that TCP segments cannot be replayed across repeated TCP connections reusing the same TCP socket pair. The 32-bit space of the ISNs prevents their repeated use except when the system is rebooted, and reuse assumes both TCP source and destination endpoints repeat their (ISN) use on the same TCP connection. It is more beneficial for the TCP endpoints to select the ISNs pseudo-randomly.

3.19.2.5 Sending TCP Segments with TCP-AO

This section describes the steps involved in processing a TCP segment carrying TCP-AO at the sending TCP endpoint. TCP-AO is the last TCP Option processed when the TCP endpoint sends a TCP segment carrying other TCP Options and the TCP-AO, because the MAC computation may include the values of other TCP Options.

1. **The TCP Sending Endpoint Receives a TCP Segment with TCP-AO and Retrieves the Per-Connection TCP-AO Parameters for the Segment**:
   a. If the TCP segment is a SYN packet, then this obviously is the first segment of a new TCP connection. The endpoint then looks for the matching MKT for the segment based on the TCP socket pair of the TCP segment:
      i. If no matching MKT is found, the TCP endpoint omits the TCP-AO and proceeds with transmitting the TCP segment.
      ii. If a matching MKT is found, then the TCP endpoint sets the per-connection TCP-AO parameters as required and proceeds to Step 2.
   b. If the TCP segment is not a SYN packet, and the TCP endpoint determines that the TCP connection is using TCP-AO, it uses the MKT as indicated by the *current_key* value from the per-connection TCP-AO parameters and proceeds to Step 2.
2. **Using the Per-Connection TCP-AO Parameters**:

   a. The TCP endpoint augments the TCP header of the segment with TCP-AO, which involves inserting the appropriate values in the Length and KeyID fields based on the MKT indicated by *current_key* (using the *SendID* of the *current_key* MKT as the TCP-AO KeyID). The TCP endpoint then updates the TCP header length accordingly.
   b. The TCP endpoint proceeds to determine the SND.SNE.
   c. The TCP endpoint determines the appropriate Traffic Key, that is, as pointed to or indicated by the *current_key*. That is, the TCP endpoint uses the *send_SYN_traffic_key* for SYN packets and the *send_other_traffic_key* for other TCP segments.
   d. The TCP endpoint determines the RNextKeyID value as pointed to by the *rnext_key*, and inserts it in the RNextKeyID field of TCP-AO (using the *RecvID* of the *rnext_key* MKT as the TCP-AO KeyID).

    e. The TCP endpoint calculates the MAC using the MKT (and Traffic Key determined above) and data taken from the TCP segment.

    f. The TCP endpoint then inserts the MAC in the MAC field of TCP-AO (see MAC field in Figure 3.56).

    g. Finally, the TCP endpoint proceeds with the transmission of the TCP segment containing the TCP-AO.

3.19.2.6 Receiving TCP Segments with TCP-AO

This section describes the steps involved in processing a TCP segment carrying TCP-AO at the receiving TCP endpoint. The TCP-AO is the first TCP Option the TCP receiver must process on incoming TCP segments carrying other TCP Options and the TCP-AO, because the MAC computation may include the values of other TCP Options that could change during TCP Option processing. Also, by processing the TCP-AO first, the behavior of all other TCP Options is protected from the impact of spoofed TCP segments or modified TCP header information. The TCP endpoint must perform TCP-AO checks for all incoming SYN packets, where required, to avoid accepting SYNs that lack TCP-AO.

1. **The TCP Receiving Endpoint Receives an Arriving TCP Segment with TCP-AO from the TCP Connection and Retrieves the Per-Connection TCP-AO Parameters for the Segment**:

    a. If the arriving TCP segment is a SYN packet, then this means it is the first segment of a new TCP connection. The TCP endpoint then looks for the matching MKT for the TCP segment, using the TCP socket pair of the segment and the KeyID of the TCP-AO, and matches this against the MKT's TCP Connection Identifier and the *RecvID* of the MKT:

      i. If no matching MKT is found, the TCP endpoint removes TCP-AO from the TCP segment and proceeds with further TCP handling of the segment. If there is no matching MKT, the TCP segment is silently accepted.

      ii. If a matching MKT is found, then the TCP endpoint sets the per-connection TCP-AO parameters as required and proceeds to Step 2.

2. **Using the Per-Connection TCP-AO Parameters**:

    a. The TCP endpoint checks if the Length field value of the TCP-AO in the TCP segment matches the length indicated by the MKT:

      i. If the lengths are different, the TCP endpoint silently discards the TCP segment and logs and/or signals the event.

    b. The TCP endpoint determines the RCV.SNE of the TCP segment.

    c. The TCP endpoint determines the Traffic Key of the TCP segment from the MKT. That is, the TCP endpoint uses the *receive_SYN_traffic_key* for SYN packets and the *receive_other_traffic_key* for other TCP segments.

    d. The TCP endpoint calculates the MAC of the TCP segment using the MKT and data taken from the TCP segment:

      i. If the calculated MAC does not match the value in the MAC field of the TCP-AO, the TCP endpoint silently discards the TCP segment and logs and/or signals the event.

e. The TCP endpoint compares the received RNextKeyID field value in the TCP-AO to the currently active outgoing KeyID value (i.e., the *SendID* of the *current_key* MKT):

   i. If these two values match, the TCP endpoint performs no further action.

   ii. If these two values differ, the TCP endpoint determines whether the MKT corresponding to RNextKeyID field value is ready for use:

      1. If the MKT corresponding to the TCP socket pair of the segment and RNextKeyID field value is not available, the TCP endpoint performs no action (the RNextKeyID field value of a received TCP segment needs to match the *SendID* of the MKT).

      2. If the matching MKT corresponding to the TCP socket pair of the segment and RNextKeyID field value is available:

         a. The TCP endpoint sets the *current_key* to the MKT corresponding to the RNextKeyID field value.

f. The TCP endpoint then proceeds with the processing of the TCP segment.

3.20 FACTORS AFFECTING BGP DEVICE AND NETWORK CONVERGENCE

The routing process in a router is considered to have converged when a network topology or state change occurs and the router has performed all control plane actions needed to react to the change. For example, in regards to BGP convergence, the routing process is said to have converged when a network change results in BGP altering the best route for a single network prefix and has advertised this best route to its downstream peers. In contrast, in OSPF, the routing process is said to have converged when OSPF has completed its SPF (shorted path first) calculations and has advertised the required link-states to neighbor routers. The process of routing convergence, in general, can be divided into the following three distinct phases **[RFC4098]**:

- Routing convergence across the entire internetwork,
- Routing convergence within a single Autonomous System,
- Routing convergence within a single routing device.

With respect to a single routing device, the convergence can be split into the following two parts:

- Convergence with respect to the data plane forwarding process(es)
- Convergence with respects to the routing process(es) in the device.

Some of the factors that have a significant effect on BGP convergence are discussed in this section **[RFC4098]**.

3.20.1 NUMBER OF BGP PEERS

A BGP router tends to invoke its route selection algorithm increasingly as the number of BGP peers in the internetwork increases. Furthermore, as the number of peers grows, and depending on the amount of routing updates generated by each additional peer, the rate of routing updates received from the various peers tend have an increasing effect on the convergence process on the router. The processing workload on the router for TCP (since BGP operates over TCP) and BGP KEEPALIVE messages also increase as the number of peers increases.

3.20.2 NUMBER OF ROUTES PER BGP PEER

In addition to increasing the processing load on the BGP router, the number of routes generated per BGP peer also the affects the convergence process. The number and relative proportion of routes withdrawn or added by each BGP peer affects the convergence process, as well as, the IGP routes supplied to the router.

3.20.3 ROUTING POLICY PROCESSING/RECONFIGURATION

The number of routes sent by BGP peers and number of path attributes being processed and filtered in routing policies implemented by the router expressed as a fraction of the router's route table size are parameters that also affect the BGP convergence process. In a minimal routing policy where the router receives all routes and, sends all without filtering, surely imposes relatively less load and allows faster convergence. On the other hand, if the router implements the extreme case (an extensive routing policy) where up to 100% of the total routes received or sent have applicable policy, then this can have a marked effect on the convergence process.

3.20.4 INTERACTIONS WITH OTHER ROUTING PROTOCOLS

BGP interacts with other routing protocols by way of other factors such as the Administrative Distance (or Route Precedence) of routing protocols (see Chapter 2 of Volume 1 of this two-part book), route redistribution to/from other protocols, route filtering, route duplication, protocol timers, and the BGP route selection criteria. This means, to understand BGP convergence, an understanding of its interactions with both the IGPs running in the autonomous system, and the protocols associated with the underlying Physical Layer media (such as Ethernet, SONET/SDH, and DWDM), is worthwhile.

3.20.5 BGP ROUTE FLAP DAMPING

BGP routers continuously adjust their IP Routing Tables to reflect actual changes in the internetwork, such as links failures, link restoration, and routers becoming unavailable/unreachable and becoming available/reachable again. These are events that occur in all internetworks and it is not unusual for these to happen in BGP internetworks. However, the occurrence of such events is supposed to be relatively infrequent. If a router or link is faulty, mismanaged, or misconfigured, then some components of the network may get into a rapid cycle of alternating down and up states.

A route flap is a condition that describes the unstable change in the state of a route, and whose availability changes repeatedly over time (i.e., repeated withdrawal and re-advertisement). This pattern of repeated route withdrawal and re-advertisement is known as route flapping which can cause excessive activities in the routers in the internetwork that know about the affected route. The routers react by continuously injecting and withdrawing the same flapping route from their IP Routing Tables. When route flaps occur, BGP routers send an excessive number of BGP UPDATE messages to their BGP peers which in turn increases the routing traffic and processing load on the peers (because they are receiving and sending updates continuously). This situation may make it impossible for a BGP router to forward user traffic while routes are being updated. In the internetwork, the flapping route may cause outages in some parts for some period of time.

A mechanism known as BGP Route Flap Damping **[RFC2439]** was proposed as a way to mitigate the effects of route flapping. It was suggested at that time that without Route Flap Damping, the routers may experience excessive activities that can place heavy processing load on their control planes processors. This in turn can cause the routers to delay routing updates to other routes, thereby affecting overall routing stability.

BGP Route Flap Damping was meant to allow BGP routers to prevent or reduce the propagation of flapping routes to peers without affecting the convergence time of routes that are not experiencing flapping (i.e., stable routes). The goal was to decrease the routing traffic and processing load on BGP peers while maintaining the overall network stability, since stable routes (or network prefixes) are still advertised (by BGP peers) while the propagation/advertisement of flapping routes is suppressed until they are declared stable again.

With damping, a router decays a route's flapping exponentially. Upon first occurrence, when the router senses that a route is unavailable and then quickly reappears, it does not allow damping to take effect, which allows the normal fail-over times of BGP to be maintained. Upon second occurrence, the router ignores that network prefix for a certain period of time. However, upon subsequent occurrences, the router times out the route (network prefix) exponentially. After the route flapping ceases (and hopefully also the network abnormalities), and a suitable length of time has elapsed for the affected route, the router reinstates the network prefixes in its Routing Table and declares the route stable. Route Flap Damping was also proposed for mitigating denial of service attacks where damping timings require highly customizable timers.

BGP Route Flap Dampening **[RFC2439]** is now considered obsolete and is no longer considered good BGP practice. With routers increasingly supporting more powerful processors, the practice of using BGP dampening to prevent high processor utilization and resource consumption and depletion is now seen as unnecessary. Modern BGP routers have enough processing power to support a large number of routes and frequent BGP routing updates without significantly affecting performance. The RIPE Routing Working Group has published "Recommendations On Route-flap Damping" which lists a number of negative effects associated with the use of BGP Route Flap Damping in ISPs, and therefore does not recommend its use (see RIPE website, initially published May 10, 2006 **[RIPEBRFD06]**, and revised

January 7, 2013 **[RIPEBRFD13]**). For these reasons, many BGP routers disabled this feature.

Research has shown that BGP Route Flap Damping can instead prolong BGP convergence times in some cases, and can even cause links that are not experiencing flapping to have interruptions in connectivity **[ZHOUMSIG02] [ZHANICDCS05]**. Also, an ISP that implements flap damping can cause side effects in their customer networks thereby affecting the Internet users in those networks. It is argued that side effects caused by using flap damping could be worse than the impact caused by simply allowing flapping to occur and not using flap damping at all **[RIPEBRFD06] [RIPEBRFD13]**.

3.20.6   HANDLING BGP CONTROL PLANE TRAFFIC

The effect of the overall traffic arriving at the control plane of a router on overall convergence depends on the router architecture and the processing power available to that control traffic. In router architectures that use the same processor(s) for both control plane traffic and packet forwarding, receiving an unusually higher percentage of control traffic in addition to normal data traffic can prevent the router from processing and sending adequately routing updates leading to network instability (see Chapter 4 in **[AWEYA1BK18]**).

In such centralized processing architectures, route flaps can exacerbate the processing load on the single processor. Fortunately, current BGP routers (which are typically used for routing in internetworks like large enterprise networks, ISPs and the Internets), typically use powerful processors, and also do not use architectures that present such control plane processing bottlenecks. Also, to guard against a surge in control plane traffic that can overwhelm the route processor, modern day routers implement what is known as Control Plane Policing (CoPP) (or Traffic Rate Limiting) **[CISCCoPPIMPL] [CISCCoPPFG06]**. CoPP is a feature that allows a network administrator to configure filters in a router to manage the flow of control plane packets in order to protect the processing and memory resources of the control plane. CoPP is designed to allow a router to manage the flow of unnecessary control traffic, that, if left unmanaged, could overwhelm the control plane processor (i.e., the route processor). Such control traffic if unregulated could affect system performance and network stability.

Under normal conditions, BGP routers receive routing updates from peers that define the internetwork and these updates show the routers and network in a steady state of operation, only sending appropriately BGP KEEPALIVE messages. However, in reality, the state of internetwork always changes caused by both normal and pathological events. For example, normal network activities include a router interface going down and the associated network prefix being withdrawn from the IP Routing Table. This route withdrawal is considered a normal event, although it contributes to the router control plane churn.

Also, if the router receives a routing update indicating the withdrawal of a route it has already advertised, or an advertisement of a route it did not previously learn, it will re-advertise all this type of information. This example is also a normal event and constitutes part of control plane churn. Routine router operations range from sending single routing updates advertising new route additions or withdrawals, to advertising an entire IP Routing Table (mostly done as an initialization condition when a router first powers up). Normal operations include the sending of BGP KEEPAIVE messages and TCP processing for peering. However, BGP route flapping **[RFC2439]** as well as route oscillations **[RFC3345]** are considered pathological events and contributors to control plane churn.

3.20.7 HANDLING BGP DATA PLANE TRAFFIC

In the centralized processing architectures, receiving a relatively higher percentage of data plane traffic can overwhelm the control plane if the combined offered load is excessive. Particularly, when the data plane traffic is high and at the same time the control traffic increases as a result of route flaps, the router will update its Routing and Forwarding Tables more frequently, leading to degradation in the data plane packet forwarding performance. The decoupling of the control plane and the data plane operations as seen in high-performance distributed forwarding eliminates the processing bottlenecks present in the centralized forwarding architectures **[AWEYA1BK18] [AWEYA2BK19]**.

3.20.8 BGP TIMERS

The settings of the BGP timers (Hold Time and Keepalive Time) can affect BGP performance and convergence. A BGP router uses the BGP Hold Time and KEEPALIVE messages to make sure that neighbors are still alive and operating. The Hold Time specifies how long a BGP router has to wait for incoming BGP messages before declaring the neighbor as unavailable or unreachable. The Keepalive Time specifies the frequency (in seconds) at which a BGP router sends KEEPALIVE messages to its peer.

A BGP router periodically sends KEEPALIVE messages (as a way to keep the BGP session alive) when it has no UPDATE messages to send. The BGP specification **[RFC4271]** suggests the default value for the Hold Time as 90 seconds, and BGP KEEPALIVE messages to be sent at intervals of 30 seconds (i.e., the Keepalive Time) which is one-third the Hold Time. In Cisco routers, the default Hold Time is 180 seconds while the default Keepalive Time is 60 seconds.

A network administrator can change the default setting to suit their particular network conditions. The BGP timer settings can take values from the range 0–65,535 seconds. A BGP router starts the Hold Time timer at zero and lets it count up to the Hold Time value before declaring the peer as dead. For the *KeepaliveTimer*, the router initializes the timer at the Keepalive Time value and counts down to zero and then sends out another BGP KEEPALIVE message to the peer.

3.20.9 BGP AUTHENTICATION

BGP routers using authentication (TCP MD5 **[RFC2385]** or TCP-AO **[RFC5925]**) can suffer some performance penalties which can affect the convergence process. Particularly, in routers with a large number of BGP peers, the processing of the authentication data when a large amount of routing updates is received or have to be sent, can have an impact on the control plane of the router.

3.21 HIGH-LEVEL BGP ROUTER ARCHITECTURE, PROCESSES, AND DATABASES

The typical BGP router has a high-level architecture that can be split into modular components as illustrated in Figure 3.57. This architecture allows the various components to be appropriately distributed to the route processor (or routing engine) and the line cards. This architecture also allows a router to support multiple instances of BGP as well as other processes and databases as typically required for virtual routers and VPNs. We describe these elements in greater detail below.

When a BGP router establishes a connection with a peer, it sends all the routes in its local IP Routing Table to that peer using BGP UPDATE messages. The BGP peer extracts the contents of these messages to create new routes to be added to its own local IP Routing Table. If a BGP router learns multiple routes to the same network destination, it runs a route selection algorithm over the competing routes to decide which route is the most preferred one to that destination. The BGP router then installs the most preferred route in its local IP Routing Table, and also advertises that route to other BGP peers.

The BGP router also adds the routes learned from other routing protocols (e.g., via redistribution from RIP, EIGRP, or OSPF) to the (combined) IP Routing Table (Figure 3.57). The combined IP Routing Table contains all the routing information to

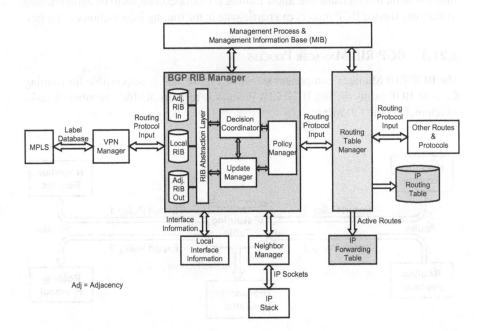

FIGURE 3.57 BGP Processes and Databases

network destinations the router knows about which includes the next-hop IP address and outgoing interface associated with each destination.

The IP Routing Table may be implemented as a set of distinct logical databases (not necessarily physical) that collectively contain all network destination address prefixes to which the router can forward packets, plus one or more next-hop addresses for these prefixes the are currently reachable. Routes included in the IP Routing Table may have been selected from several routing information sources, including static routes, Interior Gateway Protocols (RIP, EIGRP, OSPF), and Exterior Gateway Protocols (BGP). A router imposes additional criteria when installing routes in the IP Routing Table such as the Administrative Distance (or Route Preference) given to the different routing information sources (see Chapter 2 of Volume 1 of this two-part book).

The IP Forwarding Table, also called the FIB (Figures 3.57 and 3.58) contain the actual information necessary to forward IP packets to their destinations. At minimum, this table contains the outgoing router interface (identifier) and next-hop information (i.e., IP and Layer 2 addresses) for each reachable destination network prefix. The IP Forwarding Table is a router data plane construct used solely for the packet forwarding. It is generated from the IP Routing Table and it effectively holds a subset of the IP Routing Table information. The Forwarding Table is used by the forwarding plane (forwarding engine) to make per-packet forwarding decisions.

BGP, through the use of implementation-specific policies (Figure 3.58), is designed to allow routers to modify routes (via their BGP path attributes) before they are distributed to peers. BGP also uses timers to prevent a route that is rapidly changing from being continually advertised throughout the internetwork. BGP supports authentication mechanisms that allow routing protocol exchanges to be authenticated so that only trusted BGP routers can participate in the routing information exchanges.

3.21.1   BGP RIB MANAGER PROCESS

The BGP RIB Manager component shown in Figure 3.57 is responsible for running the core BGP protocol. The BGP RIB Process is responsible for a number of tasks which include the following:

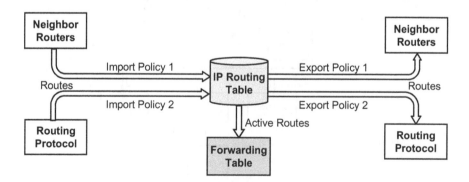

FIGURE 3.58   Import and Export Routing Policies in BGP

- **Exchanging Routing Updates with BGP peers through the Neighbor Manager Process**: BGP sends routing updates over TCP (specifically, using TCP port number 179). As discussed earlier, BGP peers or neighbors refers to any two BGP routers that have established a TCP connection to each other to allow them to exchange routing information. BGP peers start by exchanging their full IP Routing Tables with each and thereafter, they send incremental routing updates only when network changes occur.

  BGP peers of a particular (local) router that all share the same update policies can be configured as members of a BGP peer groups to help reduce the load and the amount of system CPU and memory resources used in generating routing updates, especially in a large network with many peers. In a BGP peer group, the local router checks its IP Routing Table only once, and then replicates routing updates to all peer group members instead of performing this process individually for each BGP peer. Depending on the number of members in the peer group, the number of IP address prefixes in the IP Routing Table, and the number of addresses prefixes advertised, the system load can be significantly reduced.

  Very often, many BGP neighbors/peers are configured with the same update policies (e.g., the same outbound update policies using route maps, filter lists, distribute lists, and update source) thereby making it beneficial to group these neighbors into BGP peer groups. Grouping BGP neighbors with the same update policies into peer groups helps to simplify system configuration, makes routing updates more efficient, and results in improved system performance. In the absence of the peer group, each peer is handled separately which can increase the load on the local router.

  Now, let us look at how BGP advertises networks. A network that resides within a BGP autonomous system is said to originate from that autonomous system. To inform other autonomous systems about networks that are within it, an autonomous system via BGP will advertise these networks. This allows an autonomous system to inform other systems about its internal networks. A BGP router advertises the BGP routes that it learns from both its internal and external neighbors. BGP always advertises routing information that it learns from one BGP peer to other peers. Additional, BGP routers provide techniques via a number of router commands for advertising routes that originate from an autonomous system (e.g., in Cisco routers the `redistribute` and `network` router configuration commands can be used advertise static and dynamic routes) **[CISCBGPCOMD19] [CISCFUNNETD] [CISCMASCOMD14] [CISCUSINGBGP]**. Chapter 7 of Volume 1 of this two-part book discusses in greater detail, various router configuration commands and path control tools.

- **Management of the BGP RIBs, the Loc-RIB, Adj-RIBs-In, and Adj-RIBs-Out [RFC4271]**: A BGP router stores BGP routes in a BGP RIB, also known as its BGP Routing Database, which conceptually consists of the following three logical subdatabases:

  o **Adj-RIBs-In**: The BGP router stores in the Adj-RIBs-In unprocessed routing information that has been advertised by its peers. The BGP router stores routing information that it has learned from inbound BGP UPDATE messages in the Adj-RIBs-In. This database contains the BGP NLRI in the original form before any input policies are applied. The contents of the Adj-RIBs-In database represent routes that are presented as input to the BGP Decision Coordinator in Figure 3.57 after being manipulated or filtered by the input policies associated with the peer. The input policies are processed by the Input Policy Engine (in the Policy Manager). The BGP Decision Coordinator is also referred to as the BGP Path Selection Process or Decision Process in **[RFC4271]**.

o **Loc-RIB**: The Loc-RIB contains only the preferred or best routes that the BGP router has selected by applying its BGP Path Selection Process to the routing information contained in its Adj-RIBs-In. This database also includes all routes that are originated locally by the router. The types of routes considered include iBGP routes, eBGP routes, IGP routes, interface attached routes, and static routes. The contents of the Loc-RIB are routes that are the output of the BGP Path Selection Process after the router has applied appropriate incoming local policies via the Input Policy Engine in the Policy Manager. After the routes in the Adj-RIBs-In have passed the validity and next-hop reachability check, and appropriate input policies have been applied, the BGP best path selection algorithm is applied to select the best routes for each destination. The routes in the Loc-RIB are used to populate the local router's IP Routing Table along with best routes learned by other routing protocols (Figure 3.57). The Loc-RIB contains the preferred/best routes to each available network destination as selected by the BGP router's BGP Path Selection Process. These are the routes that can also be advertised to BGP peers.

o **Adj-RIBs-Out**: The BGP router stores the routing information that it has selected for advertisement to its peers in the Adj-RIBs-Out. The BGP router writes the routing information stored in the Adj-RIBs-Out in BGP UPDATE messages and advertises these to its peers. The Adj-RIBs-Out contains routing information from the Loc-RIB that the BGP router has selected to be advertised to its peers (via UPDATE messages) after it has applied the associated output policies (via the Output Policy Engine in the Policy Manager). A route that is not installed in the Loc-RIB cannot be advertised to BGP peers.

Note that the Adj-RIBs-In and Adj-RIBs-Out are per-peer databases maintained in each BGP router. Although, the Adj-RIBs-In, Loc-RIB, and Adj-RIBs-Out are described as separate databases, there is no requirement in a real implementation for these to be kept separate. These three databases are simply logical subcomponents of the RIB that a BGP router may choose to implement as a single integrated database in order to preserve route processor memory. Also, a router may choose to implement the Loc-RIB and the IP Routing Table as a single integrated table since the Loc-RIB contains a subset of the information in the IP Routing Table.

Note that the BGP router maintains in the IP Routing Table the routing information it uses to construct the IP Forwarding Table used for forwarding packets. This IP Routing Table maintains routes to directly connected networks, static routes, routes learned by the IGPs, and routes learned by BGP. The BGP router also uses the IP Routing Table for resolving the next-hop addresses specified in BGP UPDATE messages.

- **Implementing Routing Policy to Control the Passing of Routes between the RIBs (see Figure 3.58)**: This is done by defining conditions for redistributing routes from one routing protocol to another (e.g., RIPv2 to OSPF) or controlling routing information when injected into and out of BGP. Import routing policies can be defined to control which routes a specific routing protocol places in its IP Routing Table, and export routing policies to control which routes the routing protocol advertises from its IP Routing Table to its neighbors.

  To define a routing policy, the network administrator creates a *term* that specifies *match conditions*, which are criteria that a route must match, and *actions*, which specify what the router should to do if a route matches the *conditions*. The *actions* can specify whether the router should accept or reject the route or whether it should evaluate the next term or routing policy. *Actions* can also specify how the router should manipulate the characteristics associated with a route to control which routes are selected as the active routes to reach network destinations (i.e., BGP path attributes used in the best path selection process). *Match conditions* and *actions* are defined within a *term*. After defining a routing policy, the network administrator then applies it to a routing protocol or to the Routing Table.

  A network administrator can specify a *match condition* for policies based on protocol by naming a protocol from which the route is learned or to which the route is to be advertised. For example, a routing policy can be created to inject or redistribute OSPF routes into the IP Routing Table.
- **Updating the IP Routing Table through the Routing Table Manager Process**: The routes learned by BGP are combined with routes learned from other routing protocols (e.g., OSPF) to generate the IP Routing Table for the router. This IP Routing Table contains all the destinations the router knows about, that is, each destination prefix associated with a next-hop IP address and outgoing interface.
- **Performs BGP Packet Verification and Authentication**: A BGP session can authenticate protocol exchanges to ensure that only trusted routers participate in the routing information exchange. Many routers still use TCP MD5 authentication **[RFC2385]** even though it has been superseded (or obsoleted by) the TCP-AO **[RFC5925]**. A session that uses MD5 authentication between two BGP peers, allows each TCP segment sent on the TCP connection between the peers to be verified using the techniques in **[RFC2385]**. The MD5 authentication algorithm creates an encoded message digest using a secret authentication key or password that is added to the transmitted protocol packet. The receiving BGP peer uses the same secret authentication key (password) to verify the MD5 message digest carried in the received protocol packet.

  When MD5 authentication is used, the same secret password must be configured on both BGP peers; otherwise, the two cannot establish the connection between themselves. Configuring authentication causes the BGP peers to generate and check the MD5 digest of every segment they exchange on the TCP connection.

- **Updating Other Routing Tables, for example, through the VPN Process (see Figure 3.57)**: A BGP/MPLS Layer 3 VPN **[CISCMPLSVPN18] [JUNIL3VPNGU19] [RFC4364] [RFC4659]** consists of a collection of customer sites that are interconnected using an MPLS provider core network. Each customer site supports one or more CE (Customer Edge) routers that are connected directly to one or more PE (Provider Edge) routers. A PE router attaches a VPN label to incoming packets from the CE routers based on the subinterface or interface on which they are received. The PE router also attaches the correct MPLS core labels. P (Provider) routers in the core of the provider network run MPLS but do not attach VPN labels to routed packets. The VPN labels are used (by the PE routers) to direct data packets to the correct CE router or private network.

To create a conventional VPN, a full mesh of tunnels or PVCs (Permanent Virtual Circuits) must be configured to all sites in the VPN. This makes conventional VPN not scalable or easy to maintain, because adding a new site requires configuring again each edge device in the VPN. MPLS-based VPNs, on the other hand, are created at Layer 3 (IP Layer) and are based on what is referred to as the "peer model". The peer model allows the customer and the service provider to exchange IP routing information. The service provider is responsible for transporting traffic between the customer sites without any customer involvement. MPLS VPNs are more scalable, and easy to manage than conventional VPNs. Adding a new site to an MPLS VPN requires only the PE router (of the service provider that provides services via the CE routers to the customer site) to be updated.

The tasks that the PE router performs include, among others, the following: exchanging routing updates with the CE routers; translating the routing information provided by the CE routers into VPN routes; exchanging VPN routes with other PE routers through the MP-BGP (Multiprotocol Border Gateway Protocol); maintaining VRF (Virtual Routing and Forwarding or VPN Routing and Forwarding) tables; distribution of VPN routing information; distribution of VPN routing information using BGP; MPLS forwarding.

Each VPN created in the network is associated with one or more VRF instances. A VRF defines the VPN to which a customer site is a member of on the attached PE router. A VRF consists of the following components (collectively called a VRF instance): An IP unicast Routing Table and its associated IP Forwarding Table; the set of interfaces (on the PE router) that use the IP Forwarding Table; the set of routing protocol parameters and rules that control the routing information that is added to the IP Routing Table.

The PE router stores packet forwarding information in the IP Routing Table and the Forwarding Table for each VRF. The PE router maintains a separate set of Routing Table and Forwarding Table is for each VRF which prevents routing information for a particular VPN from being forwarded outside it, and also prevents packets that are flowing outside that VPN from being forwarded to a router that is within the VPN. Details of BGP/MPLS Layer 3 VPN are given in the following example references: **[CISCMPLSVPN18] [JUNIL3VPNGU19] [RFC4364] [RFC4659]**.

3.21.2 Neighbor Manager Process

The Neighbor Manager Process component in Figure 3.57 manages the connections of the BGP router to its peers. Its responsibilities include the following:

- **Negotiating Capabilities between BGP peers (see RFC 5492, Capabilities Advertisement with BGP-4 [RFC5492])**: A BGP router that supports Capabilities Advertisement can send a BGP OPEN message to its BGP peer, with the message including an optional parameter, called Capabilities. This parameter lists the capabilities supported by the sending BGP router. By examining the list of capabilities listed in the optional parameter Capabilities (carried in the BGP OPEN message that the router receives from its peer), a BGP router can determine the capabilities supported by the peer. If two BGP routers are configured to support a particular capability, any one of them may use this optional parameter to determine if the peer supports this capability.
- **Exchanging Routing Updates on Behalf of the BGP RIB Process**: When a BGP router detects changes to the IP Routing Table, its Neighbor Manager Process sends to its neighbors only those routes that have changed.
- **Running the Keepalive Protocol**: As discussed earlier, BGP peers exchange BGP KEEPALIVE messages to ensure that the BGP session connection between them is up. A router sends BGP NOTIFICATION messages when a peer fails to send a BGP KEEPALIVE or UPDATE message, receives an unsupported BGP option in an OPEN message, and, in response to errors or when something goes wrong with the BGP session. A *KeepaliveTimer* controls the interval at which BGP KEEPALIVE messages are sent. A Hold Time timer controls how long BGP must wait for a KEEPALIVE message before declaring a peer as unreachable or unavailable.

  BGP negotiates the Hold Time with each neighbor when establishing a BGP connection. The peers use the lower of the two configured Hold Times they have announced. Each BGP peer sets its *KeepaliveTimer* based on the negotiated Hold Time and the configured Keepalive Time.

3.21.3 Policy Manager Process

A BGP router receives external and/or internal routes (plus their BGP attributes) from peers via BGP UPDATE messages. The configured input policies (in the Policy Manager) of the BGP router determines which routes will eventually end up in the Loc-RIB of the router. A BGP router may implement a Routing PIB RPIB) that contains the set of incoming and outgoing policies it wants to apply to routes. Incoming policies are applied to routes in the Adj-RIBs-In as they are passed to the BGP route selection process. Outgoing Policies are applied to routes in the Loc-RIB (network

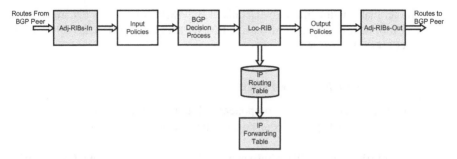

FIGURE 3.59 Relationship between the Different Databases in a BGP Router

prefix and path attribute tuple) before being placed in the Adj-RIBs-Out and advertised to a specific BGP peer. The relationship between the different databases in a
BGP router are depicted in Figure 3.59.

The policies in the RPIB are defined to have matching and action conditions.
Some of the common parameters a route must match include route (address) prefixes,
BGP AS-Paths, and BGP Communities. The action to be taken upon a match may
include dropping the routing update and not passing it to the Loc-RIB, or modifying
the BGP path attributes in the routing update in some form, such as changing the
BGP Local Preference attribute (on input) or MED Attribute (on output), adding or
deleting BGP Communities attribute, and prepending the current ASN in the BGP
AS-Path attribute.

The Policy Manager has two main parts: an Input Policy Engine and an Output
Policy Engine **[CISCHALABS00]**, which are described here.

3.21.3.1 Input Policy Engine

The Input Policy Engine, which is configurable by the network administrator, is
responsible for input route filtering and the manipulation of the BGP attributes of the
associated routes. Route filtering can be performed based on parameters such as IP
prefixes, BGP AS-Path attribute, and other BGP attributes. The Input Policy Engine
also provides a BGP router with a tool to manipulate the BGP path attributes in order
to influence how its local BGP Path Selection Process determines preferred/best
routes, and hence determine which routes will actually be used to reach available
network destinations. For example, if BGP is configured to give a specific IP prefix
from a peer a better BGP LOCAL_PREF attribute value, this indicates to the Input
Policy Engine that BGP prefers that specific IP prefix from that peer over a similar IP
prefix from other peers.

3.21.3.2 Output Policy Engine

The Output Policy Engine operates at the output side of BGP. This Engine receives
the best routes used by the BGP router itself plus routes that it has generated locally
for processing. The Engines determines if it has to apply certain route filters and
possibly, change some of the BGP path attributes (such as the BGP AS-PATH attribute or metric) before sending a BGP UPDATE message to a peer. The Output
Policy Engine must be designed to differentiate between external and iBGP peers

– routes learned by one iBGP peer cannot be passed onto another iBGP peer (except when the local router provides Route Reflection functions). The internal and external routes that have successfully passed through the Output Policy Engine are those advertised to BGP peers (and stored in the Adj-RIBs-Out database).

3.21.4 Routing Table Manager Process

The Routing Table Manager is described in detail in Chapter 5 of Volume 1 of this two-part book. Its main functions are summarized as:

- This provides a common IP Routing Table interface functions for all the IP routing protocols running on the router including OSPF, IS-IS, and BGP.
- This handles and processes policies for route import/export between the IP routing protocols in the router.

Multiple instances of the Routing Table Manager can be distributed to the line cards or multiple route processors to spread the routing processing load across multiple system processors.

3.21.5 BGP Sessions and Operational Events

The BGP process(es) in a BGP router might restart because of power failure caused a complete shutdown of the system, or upon operator intervention. When events such as these occur, the system needs a hard reset. For example, a BGP session with a peer could be lost because of a dropped TCP session or some action on the part of the peer. With a hard reset, the router can reestablish a BGP session with its peers which enables it re-advertise all relevant routes. However, if a routing policy change occurs (due to a configuration change) or a BGP router loses its peer, but the BGP process(es) running on the router have not failed, BGP can perform instead a "soft reset". The two reset modes can be described as follows:

- **Hard Reset**: This is an event that triggers the BGP routes on one or more BGP sessions to be completely torn down and re-initialized, resulting in the reestablishment of BGP sessions with peers, and the exchange of the IP Routing Table with them.
- **Soft Reset**: A BGP router performs a soft reset on a per-neighbor basis and the BGP sessions are not cleared while the router is re-establishing its peering relation. Also, the router does not stop the flow of traffic during soft reset. With soft reset, BGP router uses the BGP prefix information it has stored to reconfigure and activate its BGP Routing Tables without having to tear down existing BGP sessions with peers. The stored routing update information allows the router to apply a new BGP policy without having to tear down BGP sessions with peers. A soft reset can be performed using two methods:
  - o Graceful restart **[RFC4724]**, where the BGP router loses a peer but continues to forward traffic for a period of time before tearing down the peer's routes.

  o Soft refresh using the Route Refresh Capability **[RFC2918]**, also called
   dynamic inbound soft reset, where the BGP router can request the Adj-
   RIBs-Out database of a peer. This capability allows a BGP router to reset its
   inbound routing databases (the Adj-RIBs-In) dynamically through the
   exchange of ROUTE-REFRESH messages with BGP peers supporting this
   capability. Using the Route Refresh Capability, the BGP router does not
   have to store routing update information locally for routing policy changes
   making BGP operations non-disruptive. The BGP router relies, instead, on
   the dynamic exchange of ROUTE-REFRESH requests with its peers. The
   BGP peers must first advertise the Route Refresh Capability **[RFC2918]**
   through BGP Capability Advertisement **[RFC5492]** as described
   earlier above.

3.21.6 Processing of BGP UPDATE Messages: Sending and Receiving

BGP routers exchange routes to network destinations using UPDATE messages. A router sends UPDATE messages to inform peers about new routes, modified routes to a particular destination, and withdrawn routes to a destination. A BGP peer receives the routes contained in the UPDATE messages and does not immediately install them in the IP Routing Table until it runs the BGP best path selection process which yields the routes to be entered in the IP Routing Table. The IP Routing Table contains the final information the router needs to forward packets to their destinations.

  A BGP router may receive multiple routes to the same network destination from multiple peers. In this case, the router selects the best route to each destination and sends only these best routes to the other peers. Each BGP router runs the best path selection algorithm to determine the best routes to be installed in its IP Routing Table. This means each router installs in its IP Routing Table only the best routes learned from BGP peers plus its locally originated routes.

  A BGP router stores in its Adj-RIBs-In database, the routes it has received in UPDATE messages. When a router receives a route from its BGP peer, and if the Next-Hop IP address specified in the UPDATE message is not reachable, then that route will not be selected to be included in the Loc-RIB database. The Loc-RIB database holds all routes the router uses locally and the next-hop for each of these routes must also be present in the router's IP Routing Table (as well as the IP Forwarding Table). The router's best path selection process selects the best routes by applying locally configured routing policies to the routes stored in the Adj-RIBs-In database. The router stores these best routes in the Loc-RIB database and (after applying appropriate output policies) also places them in the Adj-RIBs-Out for advertisement to its BGP peers.

  When the router receives the routes from the BGP peers, it installs them in the Adj-RIBs-In database. These routes are subject to input routing policies which includes route filtering based on parameters such as the BGP AS-Path attribute and other BGP path attributes. Once the router has applied the input policies, it runs the best path selection algorithm to select the best routes to be placed in the Loc-RIB database and also in its IP Routing Table. These routes then become candidate routes that the router advertises to other BGP routers.

The router also places in the IP Routing Table routes that it has originated itself after they have gone through the input policy checks. These routes also become candidate routes to be advertised to its peers. Before advertisement, the router applies the output policy filters to the routes in the Loc-RIB and may change some BGP path attributes such as AS-Path before sending the routes in UPDATE messages to the peers. All the routes the router advertises to internal or eBGP peers pass through the configured output policies. Note that if for some reason a route is not placed in the IP Routing Table, then the router will not advertise that route to any BGP peer.

3.22 BGP LINK-STATE DISTRIBUTION

Some networks are designed to require an external entity that can be called upon to perform path computations on behalf of clients (e.g., routers and switches) based on collected network topology information (that can include traffic engineering [TE] information) and the current state of the connections within the network. Typically, an IGP is used to distribute this information within the network.

BGP-LS (BGP Link-State) is a mechanism that can be used to collect link-state and traffic engineering information from networks, and share this with external components using BGP [CISCBGPLINKS19] [RFC7752] [RFC8571]. BGP-LS defines a new BGP NLRI encoding format and attribute (BGP-LS Attribute) that can carry traffic engineering metric extensions defined for IGPs such as OSPF and IS-IS. BGP-LS can be applied to physical and virtual links running any of these IGPs. BGP-LS defines new Link Attribute TLVs that can be used to carry the traffic engineering metric extensions for OSPF [RFC3630] [RFC7471] and IS-IS [RFC5305] [RFC8570]. BGP-LS contains the following two parts for carrying link-state information:

- A new BGP NLRI that can be included in BGP update message to describe IGP link-state information such as links, nodes, and network address prefixes. This extends BGP with a new NLRI to allow all IGP information to be carried over BGP.
- A new BGP-LS optional non-transitive attribute that can be included in BGP update messages to carry link, node, and network address prefix properties and attributes, such as link and network prefix metric, or the auxiliary Router IDs of routers.

Two BGP routers must use BGP Capabilities Advertisement [RFC5492] in order for them to exchange Link-State NLRI and to ensure that they are both capable of properly processing BGP-LS NLRI.

IGPs like OSPF(-TE) and IS-IS(-TE) are great routing protocols for the distribution of link-state information (including traffic engineering information) within a routing domain or an autonomous system but not for link-state distribution across routing domains (or autonomous system), or to an external path computing entity such as a Application-Layer Traffic Optimization (ALTO) server [RFC5693] or a Path Computation Element (PCE) server [RFC4655]. These entities perform path computations on behalf of clients based on the network topology and state of links within the network, subject to traffic engineering constraints.

When distributing link-state information across routing domains or autonomous systems, an EGP such as BGP is the most appropriate choice. The extensibility of BGP makes it more suitable for communicating link-state and traffic engineering information across routing domains or autonomous systems to a centralized ALTO or PCE server. BGP can be used to distribute IGP information beyond the normal routing domain of the IGP. BGP can be used to collect link-state and traffic engineering information from multiple IGP routing domains for computing paths inter-domain MPLS Label Switched Paths (LSPs), or share with external path computing components. BGP-LS has high scalability and provides the capability to carry the link-state information from IGPs as part of BGP messages and subject to policy control. Applications of BGP-LS include ALTO and PCE servers as described in **[RFC7752]**.

3.22.1 NEED FOR LINK-STATE DISTRIBUTION AND TRAFFIC ENGINEERING ACROSS INTERNETWORKS

The information contained in a Link-State Database (LSDB) or the Traffic Engineering Database (TED) of IGPs such as OSPF and IS-IS describe only the nodes and link-states within an IGP routing domain or area. Deploying applications such as end-to-end traffic engineering in an internetwork of routing domains demands that routing updates carrying traffic engineering information be visible outside one area or autonomous system to enable the various network nodes make better routing and packet forwarding decisions. However, the LSDB or TED of OSPF or IS-IS has the scope of a single IGP routing domain, thereby, limiting its applicability for end-to-end traffic engineering across multiple routing domains.

If a router is required to compute a traffic engineering path across a number of IGP areas, its local IGP TED most likely will lack visibility of the complete end-to-end network topology. Lacking end-to-end topology information makes the router not being able to determine the end-to-end path, and even select the right exit router at each IGP area (i.e., ABR [Area Border Router]) for an optimal path. This becomes an issue in large-scale networks that need to be segmented into distinct IGP areas but still require paths that satisfy traffic engineering requirements.

The PCE defined in **[RFC4655]** is a mechanism that is used for computing end-to-end traffic engineering paths cross internetwork of routing domains that have multiple TEDs, or that require coordinated or CPU-intensive path computations. A PCE as a computation server that can be made to have visibility into multiple IGP areas or autonomous system, or may cooperate or interact with other PCE servers to perform distributed path computation.

BGP, which is a proven routing protocol and has better scaling properties, is the standard EGP used for exchanging routing and reachability information between autonomous systems. BGP distributes link-state information across multiple autonomous systems and allows the information to be control via routing policies that reflect the interests of the administrator of each autonomous system. Routing policies control how an autonomous system receives routing information as well as how it propagates information. BGP with its capability of supporting different routing policies serves well for inter-domain route distribution.

A network operator can explicitly configure which BGP neighbors to peer with and which routes are explicitly accepted into BGP. Furthermore, routing policies can be configured to filter and modify routing information, and provide administrative control over the Routing Tables in the network. The BGP extensions provided by BGP-LS, allows BGP to be more efficient for link-state distribution and traffic engineering across multiple routing domains or autonomous systems.

One or more routers in each IGP area in an autonomous system will be configured with BGP-LS to obtain link-state information from all IGP areas (and from other autonomous systems via eBGP peers). These BGP routers can collect TED information from the IGP areas within the network, filter it according to configurable policy (e.g., filters), and then distribute it to the PCE as necessary. This architecture allows the PCE to have access to the TED for all the IGP areas it serves. The PCE can collect traffic engineering information across the IGP areas or autonomous system, and construct the end-to-end traffic engineering paths. The PCE can update the network topology graphs dynamically as topology changes occur, for example, if a node or link goes down.

Similarly, the ALTO server defined in **[RFC5693]** is an entity that collects network information and generates an abstracted topology of the network to be provided to network-aware applications via a web-service-based API (Application Programming Interface). Reference **[RFC5693]** describes how an ALTO and a PCE server gather network topology information and capabilities of the network in order to generate the information needed by their clients.

An OSPF or IS-IS router may maintain one or more databases containing link-state information describing nodes and links in a given area in a routing domain. These databases may contain link attributes that include: local and remote IP addresses, local and remote interface identifiers, link metrics and traffic engineering metrics, link bandwidth, reservable bandwidth, preemption, per Class-of-Service (CoS) reservation state, and Shared Risk Link Groups (SRLGs).

The BGP process in a router can extract the network topology information from these databases to be distributed to client devices/applications, either directly or via a BGP peer router (typically a dedicated Route Reflector), using the BGP-LS encoding defined in **[RFC7752]** and **[RFC8571]**. A BGP router may apply configurable policy to the information extracted from these databases before distributing it. Also, the BGP router may apply a policy to determine when the client device/application is sent information updates in order to reduce information flow from the network to the client devices/applications.

3.22.2 EXAMPLE DEPLOYMENT SCENARIO OF BGP-LS

One of the limitations of OSPF and IS-IS is their inability to extend the distribution of link-state information outside a single area or autonomous system. The use of BGP-LS becomes more important when LSPs have to cross multiple routing domains, or when routing information learned by IGPs is required by external entities such as PCE or ALTO servers for optimized path computation. In both scenarios, IGPs are not suitable for distributing the routing information (which may include traffic engineering information) to the appropriate nodes.

Another application where link-state information acquired by an IGP needs to be distributed across multiple routing domains is when there is the need to compute a path for an MPLS LSP that traverses multiple routing domains, for example, an inter-area traffic engineering LSP. Recent adoption of service provider SDNs (Software Defined Networks) has fueled the deployment of BGP-LS. There are scenarios where BGP-LS implementations must interoperate or co-exist with other service provider SDN protocols such as Path Computation Element Protocol (PCEP) **[RFC5440]** and Segment Routing **[RFC8402]**.

A PCE is an entity (network node, component, or application) that is capable of obtaining network routing information, constructing a network topology graph, and applying computational constraints to compute paths or routes through the network based on the network graph. A Path Computation Client (PCC) is any client application that sends requests to a PCE for path computation to be performed. The PCE and PCC architectures are described in **[RFC4655]**. An external PCE may collect routing information from PCCs (e.g., routers) and compute paths based on a number of criteria, and then instructs routers to use the computed paths **[RFC4655]**.

The PCEP is the protocol that enables communications and interactions between a PCC and a PCE, and handles interactions such as path computation requests and path computation replies, plus notifications of specific states related to the use of a PCE for traffic engineering purposes **[JUNIPCEPCON20] [RFC5440]**. Segment Routing **[RFC8402]** is a form of source routing where a data packet source selects a path in a network, and encodes that path in the packet header as an ordered list of network segments to serve as forwarding instructions for the packet. Each segment in the packet header is identified by a unique Segment ID (SID).

PCEP runs over TCP and defines messages and objects that are used to manage PCEP sessions, and allows path requests and replies to be sent for multidomain traffic engineered MPLS LSPs (TE LSPs). PCEP provides a mechanism for a PCC to request a PCE to perform path computation for external MPLS LSPs. The interactions using PCEP include the PCC sending LSP status reports to the PCE, and the PCE sending updates for the external LSPs.

3.22.3 EXAMPLE LINK-STATE DISTRIBUTION ARCHITECTURE

In JUNOS (Juniper Network Operating System), IGPs (like OSPF and IS-IS) install the aggregated topology information of a network in a TED **[JUNILINKST20]** as illustrated in Figure 3.60. The process of distributing link-state information using BGP involves advertising the TED into BGP (the Import Process), and installing entries from BGP into the TED (the Export Process).

Before advertising the TED into BGP, the link and node entries in the TED are converted into routes. The TED then installs these converted routes into a Link-State Distribution Table (a user-visible Routing Table) on behalf of the corresponding IGP, subject to some routing policies. The TED installs IGP topology information in addition to traffic engineering topology information in the Link-State Distribution Table. The process of leaking entries from the TED into Link-State Distribution Table is called TED Import as illustrated in Figure 3.60. The routing polices govern the TED Import Process. To implement inter-domain traffic engineering, the routes in the Link-State Distribution Table are leaked into the TED via a routing policy (the TED Export Process as illustrated in Figure 3.60).

Using BGP-LS, the BGP router reads IGP entries from Link-State Distribution Table and advertises these to its BGP peers. The BGP router can be configured to import or export routes from the Link-State Distribution Table, subject to routing policy. The BGP propagates these routes to peers like any other standard BGP NLRI. BGP peers that receive these traffic engineering NLRIs store them in the form of routes in the local Link-State Distribution Table, which is the same table that stores locally originated traffic engineering routes.

The BGP-installed routes in Link-State Distribution Table are then propagated to other BGP peers like any other route. Thus, the standard BGP route selection process applies to the traffic engineering NLRIs received from multiple BGP routers. Once

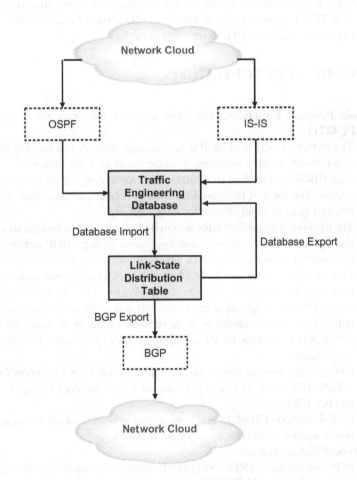

FIGURE 3.60　Link-State Distribution using BGP in Junos

the entries are installed in the TED, the BGP-learned link-state and traffic engineering information is made available for CSPF path computation.

When multiple routing protocols (e.g., OSPF and IS-IS) provide information about the same link for the TED, some of the traffic engineering attributes can differ. Therefore, to determine which information source should be given preference, JUNOS assigns *Credibility Values* to each information source [JUNILINKST20]. However, Cisco IOS sets the preferences in the TED to align with the Administrative Distances of the routing information source. The higher the Administrative Distance of a protocol, the less it is preferred for both the TED and the Routing Table.

In JUNOS, the TED uses an information source preference scheme based on Credibility Values. A protocol that has a higher Credibility Value is preferred over one that has a lower Credibility Value. A BGP router with traffic engineering capability can advertise information learned from multiple routing protocols at the same time, and so, in addition to the IGP entries that are installed in the TED, there can be BGP-installed traffic engineering entries that correspond to more than one routing protocol. The TED Export component creates a TED Protocol and Credibility Level for each routing protocol that BGP router supports.

3.23 SUMMARY OF BGP FEATURES

- **Basic Protocol Functions**: The basic protocol functions are described in [RFC4271].
  - o The primary function of BGP is to exchange network reachability information between routing domains or autonomous systems. Internal or interior BGP (iBGP) is a form of BGP that is used for routing within an autonomous system. The form of BGP used for routing between autonomous systems is referred to as external or exterior BGP (eBGP).
  - o The network reachability information includes a list of autonomous systems that a piece of routing information has passed through. BGP routers use this to construct a loop-free graph of autonomous system connectivity and how network destinations can be reached, while enforcing some routing policy decisions. BGP supports mechanisms for routing loop detection and avoidance when routes are advertised between autonomous systems, advertised within a BGP Confederation, or advertised within an autonomous system with Route Reflectors. BGP can import and/or export routes subject to routing policies.
  - o Unlike other routing protocols like RIP that use User Datagram Protocol (UDP), BGP uses TCP as its Transport Layer protocol (using TCP port number 179).
  - o BGP-4 supports CIDR, in addition, to mechanisms that allow the aggregation of routes, as well as, aggregation of AS-Paths.
- **Protocol Enhancements**:
  - o BGP Communities [RFC1997] and Extended Communities [RFC4360].
  - o Route Reflection [RFC4456].

     o BGP Confederations [**RFC5065**].

     o Route Refresh [**RFC2918**].

     o Route Flap Damping [**RFC2439**].

     o Capabilities Advertisement [**RFC5492**].

     o BGP Route Server [**RFC7947**].

     o BGP-4 Multi-Protocol Extensions for IPv6 Inter-Domain Routing [**RFC2545**].

     o Multi-Protocol Extensions for BGP-4 [**RFC4760**].

     o Subcodes for the BGP Cease Notification message [**RFC4486**].

     o Outbound route filtering capability for BGP-4 [**RFC5291**].

- **Authentication**: TCP MD5 Signature [**RFC5925**] obsoleted by the TCP-AO [**RFC5925**].
- **VPN Support**:
  - o BGP/MPLS VPNs [**RFC4364**]
  - o BGP-MPLS IP VPN Extension for IPv6 VPN [**RFC4659**].

REVIEW QUESTIONS

1. What is the difference between interior BGP (iBGP) and exterior BGP (eBGP)?
2. What is the difference between an internal BGP peer and an exterior BGP peer?
3. Explain the differences between a BGP Stub Autonomous System, Transit Autonomous System, and a Multihomed Autonomous System.
4. What is a multihop eBGP peering?
5. What Transport Layer Protocol and port number does BGP use?
6. Explain briefly the purposes of the BGP OPEN, UPDATE, KEEPALIVE, and NOTIFICATION messages.
7. Explain briefly the purpose of the BGP ROUTE-REFRESH message.
8. Does BGP send periodic routing updates like in RIP?
9. What is the purpose of the BGP Hold Time? Which BGP message type carries the Hold Time?
10. What is the purpose of the BGP Keepalive Time?
11. What is a BGP Withdrawn Route? Which BGP message type carries Withdrawn Routes?
12. What is a BGP path attribute? Which BGP message type carries path attributes?
13. Explain the differences between a well-known mandatory BGP attribute, a well-known discretionary BGP attribute, an optional transitive BGP attribute, and an optional non-transitive BGP attribute.
14. Explain the purpose of the BGP Origin Attribute (ORIGIN).
15. Explain the purpose of the BGP AS-Path Attribute (AS_PATH).
16. Explain the purpose of the BGP Next-Hop Attribute (NEXT_HOP).
17. Explain the difference between the BGP MED Attribute (MULTI_EXIT_DISC) and the BGP Local Preference Attribute (LOCAL_PREF).
18. What is a BGP Community? Explain the purpose of the BGP Communities attribute.

19. Explain the difference between the BGP Atomic Aggregate Attribute (ATOMIC_AGGREGATE) and the BGP Aggregator Attribute (AGGREGATOR).
20. What is an Autonomous System Number (ASN)? What is a Private ASN?
21. Which BGP path attributes carry ASNs?
22. Why do iBGP routers in a standard or normal autonomous system require full mesh BGP peering sessions?
23. What is a BGP Confederation? What are the benefits of using BGP Confederations?
24. What is BGP Route Refection? What are the benefits of using BGP Route Reflection?
25. What is a BGP Route Server? What are the benefits of using BGP Route Servers?
26. What is the importance of authentication in BGP?
27. Why is it recommended to use TCP-AO **[RFC5925]** instead of TCP MD5 Signature Option **[RFC2385]** for BGP authentication?
28. What is the purpose of the Sequence Number Extensions (SNEs) in TCP-AO?
29. Explain the purposes of the BGP Adj-RIBs-In, Loc-RIB, and Adj-RIBs-Out databases.
30. Why is BGP Route Flap Dampening **[RFC2439]** no longer considered good BGP practice?
31. What is the difference between a hard reset and soft reset of a BGP session?
32. What is a BGP peer group? What are the benefits of configuring BGP peer groups?

REFERENCES

[AWEYA1BK18]. James Aweya, *Switch/Router Architectures: Shared-Bus and Shared-Memory Based Systems*, Wiley-IEEE Press, ISBN 9781119486152, 2018.

[AWEYA2BK19]. James Aweya, *Switch/Router Architectures: Systems with Crossbar Switch Fabrics*, CRC Press, Taylor & Francis Group, ISBN 9780367407858, 2019.

[CISCBGPCOMD19]. Cisco Systems, *Cisco IOS IP Routing: BGP Command Reference*, December 1, 2019.

[CISCBGPLINKS19]. Cisco Systems, *IP Routing: BGP Configuration Guide*, Chapter "Border Gateway Protocol Link-State", September 12, 2019.

[CISCID13753]. Cisco Systems, "BGP Best Path Selection Algorithm", Document ID:13753, September 12, 2016.

[CISCCoPPIMPL]. Cisco Systems, "Control Plane Policing Implementation Best Practices", White Paper.

[CISCCoPPFG06]. Cisco Systems, "Control Plane Policing", Feature Guide, December 5, 2006.

[CISCDONSTEW10]. Denise Donohue and Brent Stewart, *CCNP Routing and Switching Quick Reference (642-902, 642-813, 642-832)*, Chapter "CCNP Routing and Switching Quick Reference: BGP and Internet Connectivity", Cisco Press, January 25, 2010.

[CISCFUNNETD]. *Cisco Fundamentals Network Design*, CCIE Manual.

[CISCHALABS00]. Sam Halabi, *Internet Routing Architectures*, 2nd Edition, Cisco Press, August 23, 2000.

[CISCJAINEDGE16]. Vinit Jain and Bradley Edgeworth, *Troubleshooting BGP: A Practical Guide to Understanding and Troubleshooting BGP*, Chapter "BGP Fundamentals", Cisco Press, December 20, 2016.

[CISCMASCOMD14]. Cisco Systems, *Cisco IOS Master Command List, All Releases*, January 27, 2014.

[CISCMPLSVPN18]. Cisco Systems, *Cisco ASR 9000 Series Aggregation Services Router MPLS Layer 3 VPN Configuration Guide, Release 5.3.x*, Chapter "Implementing MPLS Layer 3 VPNs", October 4, 2018.

[CISCNETACAD17]. Cisco Networking Academy, *Connecting Networks v6 Companion Guide*, Chapter "Branch Connections", Cisco Press, September 11, 2017.

[CISCPICONGUI19]. Cisco Systems, *IP Routing: Protocol-Independent Configuration Guide, Cisco IOS XE Gibraltar 16.12.x*, Chapter "TCP Authentication Option", October 31, 2019.

[CISCROUTESERV]. Cisco Systems, *IP Routing: BGP Configuration Guide*, Cisco IOS Release 15M&T, Chapter "Configuring a BGP Route Server", September 12, 2019.

[CISCUSINGBGP]. Cisco Systems, *Using the Border Gateway Protocol for Interdomain Routing*, 1996.

[CISCWHMCDA04]. Russ White, Danny McPherson, and Srihari Sangli, *Practical BGP*, Chapter "Introduction to the Border Gateway Protocol", Addison-Wesley Professional, July 6, 2004.

[JUNBGPGUIDE20]. Juniper Networks, *BGP User Guide*, March 26, 2020.

[JUNIL2VPNGU18]. Juniper Networks, *Layer 2 VPNs Feature Guide for EX9200 Switches*, Chapter 5 "Configuring Path Selection for Layer 2 VPNs and VPLS", October 13, 2018.

[JUNIL3VPNGU19]. Juniper Networks, *Layer 3 VPNs User Guide for Routing Devices*, December 10, 2019.

[JUNILINKST20]. Juniper Networks, *MPLS Applications User Guide*, Section "Link-State Distribution Using BGP Overview", January 8, 2020.

[JUNIPCEPCON20]. Juniper Networks, *MPLS Applications User Guide*, Section "PCEP Configuration", March 27, 2020.

[RFC1930]. J. Hawkinson and T. Bates, "Guidelines for creation, selection, and registration of an Autonomous System (AS)", IETF RFC 1930, March 1996.

[RFC1997]. R. Chandra, P. Traina, and T. Li, "BGP Communities Attribute", IETF RFC 1997, August 1996.

[RFC1998]. E. Chen and T. Bates, "An Application of the BGP Community Attribute in Multi-home Routing", IETF RFC 1998, August 1996.

[RFC2385]. A. Heffernan, "Protection of BGP Sessions via the TCP MD5 Signature Option", IETF RFC 2385, August 1998.

[RFC2439]. C. Villamizar, R. Chandra, and R. Govindan, "BGP Route Flap Damping", IETF RFC 2439, November 1998.

[RFC2545]. P. Marques and F. Dupont, "Use of BGP-4 Multiprotocol Extensions for IPv6 Inter-Domain Routing", IETF RFC 2545, March 1999.

[RFC2918]. E. Chen, "Route Refresh Capability for BGP-4", IETF RFC 2918, September 2000.

[RFC3345].   D. McPherson, V. Gill, D. Walton, and A. Retana, "Border Gateway Protocol (BGP) Persistent Route Oscillation Condition", IETF RFC 3345, August 2002.

[RFC3630].   D. Katz, K. Kompella, and D. Yeung, "Traffic Engineering (TE) Extensions to OSPF Version 2", IETF RFC 3630, September 2003.

[RFC4098].   H. Berkowitz, E. Davies, Ed., S. Hares, P. Krishnaswamy, and M. Lepp, "Terminology for Benchmarking BGP Device Convergence in the Control Plane", IETF RFC 4098, June 2005.

[RFC4271].   Y. Rekhter, T. Li, and S. Hares, Eds., "A Border Gateway Protocol 4 (BGP-4)", IETF RFC 4271, January 2006.

[RFC4360].   S. Sangli, D. Tappan, and Y. Rekhter, "BGP Extended Communities Attribute", IETF RFC 4360, February 2006.

[RFC4364].   E. Rosen and Y. Rekhter, "BGP/MPLS IP Virtual Private Networks (VPNs)", IETF RFC 4364, February 2006.

[RFC4451].   D. McPherson and V. Gill, "BGP MULTI_EXIT_DISC (MED) Considerations", IETF RFC 4451, March 2006.

[RFC4456].   T. Bates, E. Chen, and R. Chandra, "BGP Route Reflection: An Alternative to Full Mesh Internal BGP (IBGP)", IETF RFC 4456, April 2006.

[RFC4486].   E. Chen and V. Gillet, "Subcodes for BGP Cease Notification Message", IETF RFC 4486, April 2006.

[RFC4655].   A. Farrel, J.-P. Vasseur, and J. Ash, "A Path Computation Element (PCE)-Based Architecture", IETF RFC 4655, August 2006.

[RFC4659].   J. De Clercq, D. Ooms, M. Carugi, and F. Le Faucheur, "BGP-MPLS IP Virtual Private Network (VPN) Extension for IPv6 VPN", IETF RFC 4659, September 2006.

[RFC4724].   S. Sangli, E. Chen, R. Fernando, J. Scudder, and Y. Rekhter, "Graceful Restart Mechanism for BGP", IETF RFC 4724, January 2007.

[RFC4760].   T. Bates, R. Chandra, D. Katz, and Y. Rekhter, "Multiprotocol Extensions for BGP-4", IETF RFC 4760, January 2007.

[RFC5065].   P. Traina, D. McPherson, and J. Scudder, "Autonomous System Confederations for BGP", IETF RFC 5065, August 2007.

[RFC5291].   E. Chen and Y. Rekhter, "Outbound Route Filtering Capability for BGP-4", IETF RFC 5291, August 2008.

[RFC5305].   T. Li and H. Smit, "IS-IS Extensions for Traffic Engineering", IETF RFC 5305, October 2008.

[RFC5440].   J.P. Vasseur and J.L. Le Roux, Eds., "Path Computation Element (PCE) Communication Protocol (PCEP)", IETF RFC 5440, March 2009.

[RFC5492].   J. Scudder and R. Chandra, "Capabilities Advertisement with BGP-4", IETF RFC 5492, February 2009.

[RFC5693].   J. Seedorf and E. Burger, "Application-Layer Traffic Optimization (ALTO) Problem Statement", IETF RFC 5693, October 2009.

[RFC5925].   J. Touch, A. Mankin, and R. Bonica, "The TCP Authentication Option", IETF RFC 5925, June 2010.

[RFC6793].   Q. Vohra and E. Chen, "BGP Support for Four-Octet Autonomous System (AS) Number Space", IETF RFC 6793, December 2012.

[RFC6996].   J. Mitchell, "Autonomous System (AS) Reservation for Private Use", IETF RFC 6996, July 2013.

[RFC7300].   J. Haas and J. Mitchell, "Reservation of Last Autonomous System (AS) Numbers", IETF RFC 7300, July 2014.

[RFC7471].   S. Giacalone, D. Ward, J. Drake, A. Atlas, and S. Previdi, "OSPF Traffic Engineering (TE) Metric Extensions", IETF RFC 7471, March 2015.

[RFC7752].   H. Gredler, Ed., J. Medved, S. Previdi, A. Farrel, and S. Ray, "North-Bound Distribution of Link-State and Traffic Engineering (TE) Information Using BGP", IETF RFC 7752, March 2016.

[RFC7947].   E. Jasinska, N. Hilliard, R. Raszuk, and N. Bakker, "Internet Exchange BGP Route Server", IETF RFC 7947, September 2016.

[RFC8402].   C. Filsfils, Ed., S. Previdi, Ed., L. Ginsberg, B. Decraene, S. Litkowski, and R. Shakir, "Segment Routing Architecture", IETF RFC 8402, July 2018.

[RFC8570].   L. Ginsberg, Ed., S. Previdi, Ed., S. Giacalone, D. Ward, J. Drake, and Q. Wu, "IS-IS Traffic Engineering (TE) Metric Extensions", IETF RFC 8570, March 2019.

[RFC8571].   L. Ginsberg, Ed., S. Previdi, Q. Wu, J. Tantsura, and C. Filsfils, "BGP - Link State (BGP-LS) Advertisement of IGP Traffic Engineering Performance Metric Extensions", IETF RFC 8571, March 2019.

[RIPEBRFD06].   RIPE Network Coordination Centre, "RIPE Routing Working Group Recommendations On Route-flap Damping", May 10, 2006.

[RIPEBRFD13].   RIPE Network Coordination Centre, "RIPE Routing Working Group Recommendations On Route-flap Damping", January 7, 2013.

[ZHANICDCS05].   Beichuan Zhang, Pei Dan, Daniel Massey, and Lixia Zhang, "Timer Interaction in Route Flap Damping", IEEE 25th International Conference on Distributed Computing Systems. June 2005.

[ZHOUMSIG02].   Morley Mao Zhuoqing, Ramesh Govindan, George Varghese and Randy H. Katz, *Route Flap Damping Exacerbates Internet Routing Convergence*", *SIGCOMM '02, Proceedings of the 2002 Conference on Applications, Technologies, Architectures, and Protocols for Computer Communications*, August 2002, pp. 221–233.

Index

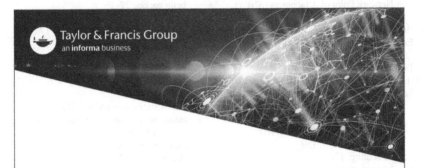